Research Notes in Mathematics

Submission of proposals for consideration
Suggestions for publication, in the form of outlines and representative samples, are invited by the editorial board for assessment. Intending authors should contact either the main editor or another member of the editorial board, citing the relevant AMS subject classifications. Refereeing is by members of the board and other mathematical authorities in the topic concerned, located throughout the world.

Preparation of accepted manuscripts
On acceptance of a proposal, the publisher will supply full instructions for the preparation of manuscripts in a form suitable for direct photo-lithographic reproduction. Specially printed grid sheets are provided and a contribution is offered by the publisher towards the cost of typing.

Illustrations should be prepared by the authors, ready for direct reproduction without further improvement. The use of hand-drawn symbols should be avoided wherever possible, in order to maintain maximum clarity of the text.

The publisher will be pleased to give any guidance necessary during the preparation of a typescript, and will be happy to answer any queries.

Important note
In order to avoid later retyping, intending authors are strongly urged not to begin final preparation of a typescript before receiving the publisher's guidelines and special paper. In this way it is hoped to preserve the uniform appearance of the series.

Titles in this series

1. Improperly posed boundary value problems
   A Carasso and A P Stone
2. Lie algebras generated by finite dimensional ideals
   I N Stewart
3. Bifurcation problems in nonlinear elasticity
   R W Dickey
4. Partial differential equations in the complex domain
   D L Colton
5. Quasilinear hyperbolic systems and waves
   A Jeffrey
6. Solution of boundary value problems by the method of integral operators
   D L Colton
7. Taylor expansions and catastrophes
   T Poston and I N Stewart
8. Function theoretic methods in differential equations
   R P Gilbert and R J Weinacht
9. Differential topology with a view to applications
   D R J Chillingworth
10. Characteristic classes of foliations
    H V Pittie
11. Stochastic integration and generalized martingales
    A U Kussmaul
12. Zeta-functions: An introduction to algebraic geometry
    A D Thomas
13. Explicit *a priori* inequalities with applications to boundary value problems
    V G Sigillito
14. Nonlinear diffusion
    W E Fitzgibbon III and H F Walker
15. Unsolved problems concerning lattice points
    J Hammer
16. Edge-colourings of graphs
    S Fiorini and R J Wilson
17. Nonlinear analysis and mechanics: Heriot-Watt Symposium Volume I
    R J Knops
18. Actions of fine abelian groups
    C Kosniowski
19. Closed graph theorems and webbed spaces
    M De Wilde
20. Singular perturbation techniques applied to integro-differential equations
    H Grabmüller
21. Retarded functional differential equations: A global point of view
    S E A Mohammed
22. Multiparameter spectral theory in Hilbert space
    B D Sleeman
24. Mathematical modelling techniques
    R Aris
25. Singular points of smooth mappings
    C G Gibson
26. Nonlinear evolution equations solvable by the spectral transform
    F Calogero
27. Nonlinear analysis and mechanics: Heriot-Watt Symposium Volume II
    R J Knops
28. Constructive functional analysis
    D S Bridges
29. Elongational flows: Aspects of the behaviour of model elasticoviscous fluids
    C J S Petrie
30. Nonlinear analysis and mechanics: Heriot-Watt Symposium Volume III
    R J Knops
31. Fractional calculus and integral transforms of generalized functions
    A C McBride
32. Complex manifold techniques in theoretical physics
    D E Lerner and P D Sommers
33. Hilbert's third problem: scissors congruence
    C-H Sah
34. Graph theory and combinatorics
    R J Wilson
35. The Tricomi equation with applications to the theory of plane transonic flow
    A R Manwell
36. Abstract differential equations
    S D Zaidman
37. Advances in twistor theory
    L P Hughston and R S Ward
38. Operator theory and functional analysis
    I Erdelyi
39. Nonlinear analysis and mechanics: Heriot-Watt Symposium Volume IV
    R J Knops
40. Singular systems of differential equations
    S L Campbell
41. N-dimensional crystallography
    R L E Schwarzenberger
42. Nonlinear partial differential equations in physical problems
    D Graffi
43. Shifts and periodicity for right invertible operators
    D Przeworska-Rolewicz
44. Rings with chain conditions
    A W Chatters and C R Hajarnavis
45. Moduli, deformations and classifications of compact complex manifolds
    D Sundararaman
46. Nonlinear problems of analysis in geometry and mechanics
    M Atteia, D Bancel and I Gumowski
47. Algorithmic methods in optimal control
    W A Gruver and E Sachs
48. Abstract Cauchy problems and functional differential equations
    F Kappel and W Schappacher
49. Sequence spaces
    W H Ruckle
50. Recent contributions to nonlinear partial differential equations
    H Berestycki and H Brezis
51. Subnormal operators
    J B Conway
52. Wave propagation in viscoelastic media
    F Mainardi
53. Nonlinear partial differential equations and their applications: Collège de France Seminar. Volume I
    H Brezis and J L Lions
54. Geometry of Coxeter groups
    H Hiller
55. Cusps of Gauss mappings
    T Banchoff, T Gaffney and C McCrory

56 An approach to algebraic K-theory
A J Berrick

57 Convex analysis and optimization
J-P Aubin and R B Vintner

58 Convex analysis with applications in the differentiation of convex functions
J R Giles

59 Weak and variational methods for moving boundary problems
C M Elliott and J R Ockendon

60 Nonlinear partial differential equations and their applications: Collège de France Seminar. Volume II
H Brezis and J L Lions

61 Singular systems of differential equations II
S L Campbell

62 Rates of convergence in the central limit theorem
Peter Hall

63 Solution of differential equations by means of one-parameter groups
J M Hill

64 Hankel operators on Hilbert space
S C Power

65 Schrödinger-type operators with continuous spectra
M S P Eastham and H Kalf

66 Recent applications of generalized inverses
S L Campbell

67 Riesz and Fredholm theory in Banach algebra
B A Barnes, G J Murphy, M R F Smyth and T T West

68 Evolution equations and their applications
F Kappel and W Schappacher

69 Generalized solutions of Hamilton-Jacobi equations
P L Lions

70 Nonlinear partial differential equations and their applications: Collège de France Seminar. Volume III
H Brezis and J L Lions

71 Spectral theory and wave operators for the Schrödinger equation
A M Berthier

72 Approximation of Hilbert space operators I
D A Herrero

73 Vector valued Nevanlinna Theory
H J W Ziegler

74 Instability, nonexistence and weighted energy methods in fluid dynamics and related theories
B Straughan

75 Local bifurcation and symmetry
A Vanderbauwhede

76 Clifford analysis
F Brackx, R Delanghe and F Sommen

77 Nonlinear equivalence, reduction of PDEs to ODEs and fast convergent numerical methods
E E Rosinger

78 Free boundary problems, theory and applications. Volume I
A Fasano and M Primicerio

79 Free boundary problems, theory and applications. Volume II
A Fasano and M Primicerio

80 Symplectic geometry
A Crumeyrolle and J Grifone

81 An algorithmic analysis of a communication model with retransmission of flawed messages
D M Lucantoni

82 Geometric games and their applications
W H Ruckle

83 Additive groups of rings
S Feigelstock

84 Nonlinear partial differential equations and their applications: Collège de France Seminar. Volume IV
H Brezis and J L Lions

85 Multiplicative functionals on topological algebras
T Husain

86 Hamilton-Jacobi equations in Hilbert spaces
V Barbu and G Da Prato

87 Harmonic maps with symmetry, harmonic morphisms and deformations of metrics
P Baird

88 Similarity solutions of nonlinear partial differential equations
L Dresner

89 Contributions to nonlinear partial differential equations
C Bardos, A Damlamian, J I Díaz and J Hernández

90 Banach and Hilbert spaces of vector-valued functions
J Burbea and P Masani

91 Control and observation of neutral systems
D Salamon

92 Banach bundles, Banach modules and automorphisms of C*-algebras
M J Dupré and R M Gillette

93 Nonlinear partial differential equations and their applications: Collège de France Seminar. Volume V
H Brezis and J L Lions

Nonlinear partial differential equations and their applications
Collège de France Seminar
VOLUME V

H Brezis & J L Lions (Editors)
D Cioranescu (Coordinator)
Université Pierre et Marie Curie (Paris VI)

Nonlinear partial differential equations and their applications Collège de France Seminar VOLUME V

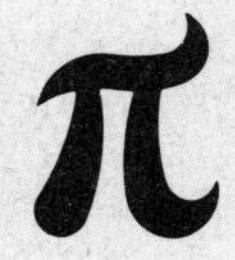

Pitman Advanced Publishing Program
BOSTON · LONDON · MELBOURNE

Library of Congress Cataloging in Publication Data
(Revised for vol. 5)
Main entry under title:

Nonlinear partial differential equations and their applications.

(Research notes in mathematics ; 60, 70,)
Lectures presented at the weekly Seminar on Applied Mathematics, College de France, Paris.
English and French.
1. Differential equation, Partial – Congresses.
2. Differential equations, Nonlinear – Congresses.
I. Brezis, H. (Haim) II. Lions, Jacques Louis.
III. Cioranescu, D. (Doïna) IV. Seminar on Applied Mathematics. V. Series.
QA377.N67 1982 515.3′53 81-4350
ISBN 0-273-08491-7 (v. 1)
ISBN 0-273-08541-7 (v. 2)

PITMAN BOOKS LIMITED
128 Long Acre, London WC2E 9AN

PITMAN PUBLISHING INC
1020 Plain Street, Marshfield, Massachusetts 02050

Associated Companies
Pitman Publishing Pty Ltd, Melbourne
Pitman Publishing New Zealand Ltd, Wellington
Copp Clark Pitman, Toronto

First published 1983

AMS Subject Classifications: 35-XX, 34-XX, 46-XX

Library of Congress Cataloging in Publication Data

—COPY TO BE SUPPLIED—

British Library Cataloguing in Publication Data

Nonlinear partial differential equations and their
applications.—(Research notes in mathematics; 93)
Vol. 5
1. Differential equations, Partial—Congresses
2. Differential equations. Nonlinear—Congresses
I. Brézis, H. II. Lions, J. L.
III. Cioranescu, D. IV. Collège de France
V. Series
515.3′53 QA377

ISBN 0-273-08620-0

Reproduced and printed by photolithography
in Great Britain by Biddles Ltd, Guildford

Preface

The present volume consists of written versions of lectures held during the year 1981-1982 at the weekly Seminar on Applied Mathematics at the College de France. They mostly deal with various aspects of the theory of non linear partial differential equations.

We thank :

- the speakers who kindly accepted to write up their lectures.
- Mrs Doïna Cioranescu who has coordinated the activities of the Seminar and prepared the material for publication; without her patience and determination this volume would never have appeared.
- Mrs Arpin, Mrs Force, Mrs Negic for their competent typing of the manuscripts.

Paris, Frebruary 1983

Haïm BREZIS Jacques Louis LIONS

P.S. The Seminar is partially supported by a Grant from the C.N.R.S.

Préface

Ce volume regroupe les textes des conférences données en 1981-1982 au Séminaire de Mathématiques appliquées qui se réunit chaque semaine au Collège de France. Elles concernent principalement l'étude d'équations aux dérivées partielles non linéaires sous des éclairages variés.

Nous remercions vivement :

- les conférenciers qui ont bien voulu accepter de rédiger leurs exposés,
- Mme Doïna Cioranescu qui s'est chargée de coordonner les activités du Séminaire et de la préparation matérielle de cet ouvrage; sans sa patience et sa persévérance cette publication n'aurait pas vu le jour,
- Mme Arpin, Mme Force, Mme Négic qui ont tapé avec compétence les manuscripts.

Paris, Février 1983

Haïm BREZIS Jacques Louis LIONS

P.S. Le séminaire était subventionné en partie par les crédits d'une RCP du C.N.R.S.

Contents

I J BAKELMAN

Variational problem connected with the Monge-Ampère equation

1. INTRODUCTION.

This paper is devoted to proving an existence Theorem for functionals whose Eulerian equations are given by the Monge-Ampere equation

$$\det \| u_{x_i x_j} \| = f(x_1, x_2, \ldots, x_n) \tag{1.1}$$

Such a functional is given by Courant and Hilbert [1] for the case of functions of two variables :

$$E(u) = \iint_G \{[u_x^2 u_{yy} - 2u_x u_y u_{xy} + u_y^2 u_{xx}] + 6fu\} dxdy \tag{1.2}$$

where G is an open domain in the x,y-plane. Unfortunately the functional (1.2) doesn't have a good geometric or analytic interpretation and is not convenient for investigations; the integrand in (1.2) doesn't give any idea concerning the functional spaces in which this variational problem belongs, also it doesn't suggest ideas concerning the generalization for functions of n-variables.

We found in [2] , [3] another functional

$$I(u) = -\iint_G \{u(u_{xx}u_{yy} - u_{xy}^2) - 3fu\} dxdy \tag{1.3}$$

whose Eulerian equation is

$$u_{xx}u_{yy} - u_{xy}^2 = f(x,y) \tag{1.4}$$

If ∂G is a C^2-curve, and if $u(x,y) \in C^2(\bar{G})$ satisfies the condition $u|_{\partial G} = 0$, then

$$E(u) = 2I(u)$$

The functional (1.3) is closely connected with the Monge-Ampere operator $u_{xx}u_{yy} - u_{xy}^2$ and admits a simple geometric interpretation in terms of convex functions. This functional also admits a simple natural generalization to functions of n variables.

In the papers [2], [3] we studied the two-dimensional variational problem for the functional (1.3) and proved that the absolute minimum for this problem is a generalized solution of the equation (1.4) with the boundary condition

$$u|_{\partial G} = 0 \qquad (1.5)$$

But some points used in [2], [3] don't generalize to the n-dimensional case and new ideas and techniques are therefore required. In the present paper we study the n-dimensional variational problem connected with n-dimensional Monge-Ampere equation (1.1) and prove the existence and uniqueness of an absolute minimum for this problem. This minimum is a generalized solution of the equation (1.1). The proof of this result is based on deeper ideas and geometric constructions than those used previously in [2], [3] ; these ideas and constructions are based on new estimates for the main functional, and on the concept and properties of dual convex hypersurfaces in the special space. They uncover the fundamental geometric contents of this variational problem.

If the function f is sufficiently smooth and strictly positive and if the boundary of the domain G is also sufficiently smooth and strictly convex (all principal normal curvatures of G are positive), then it follows from our results and those of Pogorelov [10] and Cheng-Yau [11], [12] that the generalized solution of our variational problem is also smooth.

2. THE MAIN VARIATIONAL PROBLEM.

2.1. Preliminary considerations. Let G be an open bounded domain in n-dimensional Euclidean space E^n. The two-dimensional functional I(u) (see Introduction, formula (1.3)) admits the following n-dimensional generalization

$$I_n(u) = -\int_G [u \det \| u_{x_i x_j} \| - (n+1)f(x)u]dx. \tag{2.1}$$

where $x_1, x_2, \ldots, x_n$ are Cartesian coordinates in E^n.

The formal Eulerian equation for $I_n(u)$ is

$$\det \| u_{x_i x_j} \| = f(x) \tag{2.2}$$

It is clear that all elliptic solutions of (2.2) are convex or concave functions. We shall consider only the convex solutions. The variational problem for the functional $I_n(u)$ is degenerate since the Eulerian equation (2.2) has the same order as the functional $I_n(u)$. Therefore we should consider only one boundary condition

$$u|_{\partial G} = 0 \tag{2.3}$$

instead of two. Thus the convexity of functions realizing the extremum of the functional $I_n(u)$, replaces to some extent one of the eliminated boundary conditions.

The functional $I_n(u)$ can be conveniently investigated if we extend it to the set of all non-positive continuous functions. Such an extension is based on the concepts of normal mapping and the R-curvature of convex functions. We give an exposition of these concepts and their properties in the next section. For a detailed exposition see [3], [4], [5], [6], [7], [8], [9].

2.2. Normal mapping and R-curvature of Convex Functions.

A). *Normal mapping of Convex Functions.*

Let $x_1, x_2, \ldots, x_n, x_{n+1}$ be Cartesian coordinates in (n+1)-dimensional Euclidian space E^{n+1}. We let E^n denote the hyperplane $x_{n+1} = 0$ in E^{n+1}, and let G be an open convex bounded domain in E^n. We introduce the notations : $x_{n+1} = z$; $x = (x_1, x_2, \ldots, x_n)$ is a point of E^n, $(x,z) = (x_1, x_2, \ldots, x_n, z)$ is a point of E^{n+1} ; $z(x)$ is a function $z : G \to \mathbb{R}$ with graph S_z ; $W^+(G)$ is the set of all convex functions on G ; $W^-(G)$ is the set of all concave functions on G. If $z(x) \in W^{\pm}(G)$, then s_z is called a *convex (concave) hypersurface*.

Pick some arbitrary convex functions $z(x) \in W^+(G)$ (*), and let α be a supporting hyperplane to S_z, with equation

$$Z - z^o = \sum_{i=1}^{n} p_i^o(X_i - x_i^o)$$

where $(x_1^o, \ldots, x_n^o; z^o) \in S_z \cap \alpha$, and (X,Z) is an arbitrary point of α.

(*) Analogous for concave functions .

The point $\chi_z(\alpha) = p_o = (p_1^o, p_2^o, \ldots, p_n^o) \in \mathbb{R}^n$ is called the *normal image of the supporting hyperplane* α.

We construct the set

$$\chi_z(x_o) = \bigcup_{\alpha} \chi_z(\alpha),$$

where α runs through all supporting hyperplanes to S_z at the point $(x_o, z(x_o)) \in S_z$. The set $\chi_z(x_o)$ is called the *normal image of the point* x_o (relative to the function $z(x)$). It is clearly a closed convex subset of $\mathbb{R}^n$. Finally we put for any subset $e \subset G$:

$$\chi_z(e) = \bigcup_{x_o \in e} \chi_z(x_o)$$

and call it *the normal image of the subset* $e \subset G$ (with respect to the function $z(x)$).

The main properties of the normal mapping.

These are as follows :

(a) For each closed subset e of the domain G the set $\chi_z(e)$ is a closed subset of $\tilde{R}^n$; for each Borel subset $e \subset G$ the set $\chi_z(e)$ is Borel subset of $\tilde{R}^n$.

(b) Let $z_1(x)$ and $z_2(x)$ be convex functions, for which :

$$z_1|_{\partial G} = z_2|_{\partial G},$$

and

$z_1(x) \leq z_2(x)$ for $x \in G$.

Then

$$\chi_{z_2}(G) \subset \chi_{z_1}(G). \tag{2.5}$$

(c) If $z(x) \in W(G) \cap C^2(G)$, then the normal mapping can be considered as a mapping of points, namely, the tangential mapping

$$\chi_z(x) = (\frac{\partial z}{\partial x_1}, \ldots, \frac{\partial z}{\partial x_n}).$$

B) R-*curvature*.

Let $R(p) > 0$ be a locally summable function on R^n. The function of sets

$$\omega(R,z,e) = \int_{\chi_z(e)} R(p)dp, \quad e \subset G \tag{2.6}$$

is non-negative and completely additive on the ring of Borel subsets of the convex domain G for any $z(x) \in W^+(G)$. This function is called *the* R-*curvature of the convex function* $z(x) \in W^+(G)$.

If $R(p) \equiv 1$, then the 1-curvature of $z(x) \in W^+(G)$ is called the measure (or area) of the normal mapping and simply denoted by $\omega(z,e)$. We set

$$A(R) = \int_{\tilde{R}^n} R(p)dp. \tag{2.7}$$

It is clear that $A(R) > 0$; note the case $A(R) = +\infty$ is not excluded.

The main properties of R-curvature :

(a) The inequality

$$\omega(R,z,G) \leq A(R) \tag{2.8}$$

holds for any convex function $z(x) \in W^{+}(G)$. □

(b) If $z(x) \in W^{+}(G) \cap C^{2}(G)$, then

$$\omega(R,z,e) = \int_{e} \det \|z_{ij}\| . R(Dz)dx \quad □ \tag{2.9}$$

(c) Weak convergence of R-curvatures.

If the sequence of convex functions $z_n(x) \in W^{+}(G)$ converges to the convex function $z(x) \in W^{+}(G)$ in any point $x \in G$, then

$$\lim_{n \to \infty} \int_{G} \phi(x)\omega(R,z_n,de) = \int_{G} \phi(x)\omega(R,z,de) \tag{2.10}$$

where $\phi(x)$ is any continuous function on some set M such that $\bar{M} \subset G$ and is identically zero on $G \setminus \bar{M}$.

Note $\bar{M}$ is the closure of M, G is an open convex bounded domain. Therefore

$$\operatorname{dist}(\bar{M},\partial G) > 0. \quad □ \tag{2.11}$$

(d) Estimates of convex functions.

Let the following conditions be fulfilled : If $V(\omega_o) = \{z(x)\}$ is the set of all convex and concave functions belonging to $W^{\pm}(G)$ and satisfying the conditions

(d.1) $-\infty < m \leq z|_{\partial G} \leq M < +\infty,$ (2.12)

(d.2) $\omega(R,z,G) \leq \omega_o < A(R)$ (2.13)

Then the following inequalities hold in G :

$$m - T_R(\omega_o)d(G) \leq z(x) \leq M \text{ for } z(x) \in W^+(G) \; ; \tag{2.14}$$

$$m \leq z(x) \leq M + T_R(\omega_o)d(G), \text{ for } z(x) \in W^-(G). \tag{2.15}$$

Here $T_R : [0,A(R)) \to R$ denote the inverse for the function

$$g_R(r) = \int_{|p| \leq r} R(p)dp. \quad \square \tag{2.16}$$

and d(G) is the diameterof G.

Example. If $R(p) \equiv 1$, then

$$g_R(r) = \int_{|p| \leq r} dp = \mu_n r^n,$$

where μ_n is the volume of the unit n-ball. Therefore

$$T_R(g) = T_1(g) = \left(\frac{g}{\mu_n}\right)^{\frac{1}{n}} \tag{2.17}$$

(e) Compactness and uniformly convergence of convex functions in $C(\bar{G})$.

Suppose the following conditions are fulfilled :

$$(e.1) : \quad R(p) \geq C_o(1+|p|^2)^{-k} \tag{2.18}$$

everywhere in R^n and $C_o = \text{const} > 0$, $k = \text{const} \geq 0$;

(e.2) : For any $x_o \in \partial G$ there exists a n-ball U_{x_o} such that $x_o \in \partial U_{x_o}$, $\bar{G} \subset \bar{U}_{x_o}$ and

$$r_{x_o} \leq r_o = \text{const} < +\infty .$$

(e.3) : Let $z_n(x) \in W^+(G)$ be a sequence of convex functions such that :

a) $z_n(x)$ converge to $z(x) \in W^+(G)$ in any point $x \in G$;

$$\text{b)} \quad z_n(x)|_{\partial G} = 0 \text{ and } z(x)|_{\partial G} = 0 \tag{2.19}$$

c) There exists the neighborhood S_{x_o} for every point $x_o \in \partial G$ such that the inequality

$$\varlimsup_{n \to \infty} \omega(R,z_n,e) \leq a\{\sup_e\{\text{dist}(x,\partial G)\}\}^{\lambda}\text{mes } e \tag{2.20}$$

holds for any Borel subset $e \subset S_{x_o} \cap G$, where $\lambda \geq 0$ and $a > 0$ are the common constants for all $x_o \in \partial G$. Then we have

<u>The main result</u> : If all conditions (e.1), (e.2) and (e.3) are fulfilled and

$$k \leq \frac{n+1+\lambda}{2} \tag{2.21}$$

then the sequence $z_n(x)$ converges to $z(x)$ uniformly.

Remark. If $R(p) \equiv 1$, then $k = 0$ and the inequality (2.21) takes the form

$$0 \leq \frac{n+1+\lambda}{2} .$$

Therefore we can take $\lambda = 0$, so that (2.20) takes the form

$$\overline{\lim_{n \to \infty}} \; \omega(z_n, e) \leq a.\text{mes } e. \tag{2.20a}$$

2.3. The operator F_H and its properties.

Let G be an open convex bounded domain, and let H denote an open convex subdomain of G such that

$$\text{dist}(\bar{H}, \partial G) = h_H > 0, \tag{2.22}$$

where $\bar{G}$ and $\bar{H}$ are closures of G and H. Further, let $C_o^-(\bar{G})$ be the set of all continuous non-positive functions in the space $C(\bar{G})$, and let $v(x)$ be the product $u(x).\phi_{\bar{H}}(x)$, where $u(x)$ is any function from $C_o^-(\bar{G})$ and $\phi_{\bar{H}}(x)$ is the characteristic function of the set $\bar{H}$:

$$v(x) = \begin{cases} u(x) & \text{if } x \in \bar{H} , \\ 0 & \text{if } x \in \bar{G} \setminus \bar{H}. \end{cases} \tag{2.23}$$

If S_v is the graph of $v(x)$, then the boundary of $\overline{Co}\{S_v\}$ consists of two parts $\bar{G}$ and the graph S_w of some convex function $w(x)$; clearly

$$w(x) \in W^+(\bar{G}) \cap C_o^-(\bar{G}). \tag{2.24}$$

If β is any supporting hyperplane to S_v passing through the point $(x_o, w(x_o))$, where $x_o \in G \setminus \bar{H}$, then the set $\beta \cap S_v$ contains at least some line segment AB such that $A \in \partial G$ and $B \in \bar{H}$. Therefore β is a singular supporting hyperplane to S_v. It is well known [see [15], Section 4] that the spherical image of all singular supporting hyperplanes to every convex hypersurface has zero measure. From this fact it follows that

$$\omega(w, G \setminus \bar{H}) = 0. \tag{2.25}$$

Thus any open convex domain $H \subset G$ satisfying the condition (2.22) generates the set $W_H^+(\bar{G})$ of convex functions $w(x)$ by means of the construction considered above. From (2.24) it follows that

$$W_H^+(\bar{G}) \subset W^+(\bar{G}) \cap C_o^-(\bar{G}). \tag{2.26}$$

We call the function $w(x) \in W_H^+(\bar{G})$ corresponding to a function $u(x) \in C_o^-(\bar{G})$ the convex function spanned on $u(x)$ from below on the set $\bar{H}$.

Thus the operator

$$F_H : C_o^-(\bar{G}) \to W_H^+(\bar{G}) \tag{2.27}$$

maps any $u(x) \in C_o^-(\bar{G})$ in the convex function $w(x)$ spanned on $u(x)$ from below on the set $\bar{H}$.

Theorem 1 : *The operator* F_H *has the following properties*

a) $$F_H(C_o^-(\bar{G})) = W^+(\bar{G}) ; \tag{2.28}$$

b) $$F_H(w(x)) = w(x) \tag{2.29}$$

for every $w(x) \in W_H^+(\bar{G})$;

c) *the set* $W_H^+(\bar{G})$ *is closed in the space* $C_o^-(\bar{G})$.

Proof. The observations a) and b) are obvious.

Let $w_1(x), w_2(x), \ldots, w_n(x), \ldots$ be any sequence of $W_H^+(\bar{G})$ which is fundamental in $C(\bar{G})$. Let $w(x) \in C_o^-(\bar{G})$ be the limit of $w_n(x)$ in the space $C(\bar{G})$. Since

$$w_n(x) = F_H(w_n(x)),$$

then $w(x)$ is some convex function in $C_o^-(\bar{G})$. The assertion c) will be proved, if we establish the equality

$$F_H(w(x)) = w(x). \tag{2.30}$$

We denote by $v(x) = \phi_{\bar{H}}(x).w(x)$, then (2.30) is equivalent to equality

$$\overline{Co}\{S_w\} = \overline{Co}\{S_v\} \tag{2.31}$$

where S_w and S_v are correspondingly the graphs of the functions w(x) and v(x). From the properties of the functions w(x) and v(x) it follows that

$$1) \qquad \overline{Co}\{S_v\} \subset \overline{Co}\{S_w\} \; ; \tag{2.32}$$

$$2) \qquad \overline{Co}\{S_w\} = Q_o \cap \bigcap_\alpha Q_\alpha \bigcap_\beta Q'_\beta \; ; \tag{2.33}$$

$$3) \qquad \overline{Co}\{S_v\} = Q_o \cap \bigcap_\alpha Q_\alpha \bigcap_\gamma Q''_\gamma \tag{2.34}$$

where a) Q_o is the closed halfspace $z \leq 0$;

b) Q_α is any closed halfspace such that

$$Q_\alpha \supset \overline{Co}\{S_w\}$$

and the hyperplane $P_\alpha = \partial Q_\alpha$ is a supporting hyperplane to S_w having at least one common point with $S_{w,H}$, where $S_{w,H}$ is the graph of the restriction of the function w(x) to the set $\bar{H}$;

c) Q' is any closed halfspace such that

$$Q' \supset \overline{Co}\{S_w\}$$

and the hyperplane $P'_\beta = \partial Q'$ is a supporting hyperplane to $S_w \setminus S_{w,H}$;

d) Q" is any closed halfspace such that

$$Q''_\gamma \supset \overline{Co}\{S_v\}$$

and the hyperplane $P''_\gamma = \partial Q''_\gamma$ is a supporting hyperplane to $\overline{Co}\{S_v\}$ and passing through at least one point of ∂G and one point of $S_{w,H}$.

From the properties of the functions $w_k(x) \in W_H^+(\bar{G})$ and the function $w(x)$ it follows that there exists some halfspace Q" such that

$$Q'_\beta = Q''_\gamma$$

for every halfspace Q'_β considered in (2.33). Then from (2.33) and (2.34) we obtain

$$\overline{Co}\{S_w\} \subset \overline{Co}\{S_v\} \tag{2.35}$$

Now from (2.32) and (2.35) it follows the equality (2.31).

Theorem 1 is proved. □

Theorem 2 : *Let* $w(x) = F_H(u(x))$, *where* $u(x)$ *is any function from* $C_o^-(\bar{G})$. *Then*

a) $\chi_w(G) = \chi_w(\bar{H})$, (2.36)

b) $\omega(w,G) \leq \mu_n \dfrac{\|u\|^n}{h_H^n}$; (2.37)

c) *The operator* $F_H : C_o^-(\bar{G}) \to W_H^+(G)$ *is continuous.*

Remarks. 1) $\|u\|$ is the norm of $u(x)$ in the space $C(\bar{G})$; it is clear that $\|w(x)\| = \|u(x)\|$;

2) χ_w is the normal mapping respectively of the function $w(x)$, $\omega(w,G)$ is the measure of the normal image for the domain G ;

3) $h_H = \text{dist}(\bar{H},\partial G) > 0$, μ_n is the volume of the n-dimensional unit ball.

Proof : Since every supporting hyperplane to the graph of the function $w(x) = F_H(u(x))$ has at least one common point with the set $S_{w,H}$ (see the proof of Theorem 1), then

$$\chi_w(G) = \chi_w(\bar{H}).$$

The assertion a) is proved.

Now let α be any supporting hyperplane to the graph of $w(x)$. Denote by $(x_o,w(x_o))$ the common point of α and $S_{w,H}$. Then

$$x_o \in \bar{H} \text{ and } \mathrm{dist}(x_o,\partial G) \geq h_H.$$

Let K_{x_o} be the convex cone with the vertex $(x_o,u(x_o))$ and the base $U(x_o,h_H)$, where $U(x_o,h_H)$ is the closed n-ball with the center x_o and the radius h_H. Let $k_{x_o}(x)$ be the convex function defining K_{x_o}. Then

$$\chi_w(\alpha) \in \chi_{k_o(x)}(U(x_o,h_H)).$$

This pertains to every supporting plane to the graph of $w(x)$. The cone K_{x_o} can only move as rigid body, but the set $\chi_{k_o(x)}(U(x_o,h_H))$ is independent of the point $x_o \in \bar{H}$.

Thus

$$\chi_w(G) \subset \chi_{k_o(x)}(U(x_o,h_H))$$

and

$$\omega(u,G) \le \mathrm{mes}(\chi_{k_o(x)}(U(x_o,h_H)).$$

The set $\chi_{k_o(x)}(U(x_o,h_H))$ is the ball with the center $O(0,0,\dots,0)$ and the radius $\rho = \dfrac{|w(x_o)|}{h_H}$ in the space $\tilde{R}^n$.

Therefore

$$(w,G) \le \mu_n \frac{|w(x_o)|^n}{h_H^n} \le \mu_n \frac{\|u(x)\|^n}{h_H^n}.$$

This proves assertion b).

Now let the sequence of the functions $u_n(x) \in W^+(\bar{G}) \cap C_o^-(G)$ converge uniformly to the function $u(x) \in W^+(\bar{G}) \cap C_o^-(\bar{G})$. Let

$$w_n(x) = F_H(u_n(x))$$

and

$$w(x) = F_H(u(x)).$$

Since

$$\|u_n(x)\| = \|w_n(x)\|$$

and

$$\|u(x)\| = \|w(x)\|,$$

then $\|w_n(x)\| \to \|w(x)\|$ and we can suppose that

$$\|w_n(x)\| \le \|w\| + 1 = \|u\| + 1$$

and the sets $\chi_{w_n}(G)$ and $\chi(G)$ are contained in one and the same ball with the center $0(0,\ldots, 0)$ and the radius $\rho = \frac{\|u(x)\|+1}{h_H}$ in the space $\tilde{R}^n$.

Therefore the functions $w_n(x)$, $w(x)$ have uniformly bounded C-norms in $\bar{G}$ and satisfy Lipschitz condition with degree 1 and common constant $(\|u(x)\|+1)h_H^{-1}$. Therefore we can take the subsequence $\{w_{n_k}(x)\}$ from $\{w_n(x)\}_{n=1}^{\infty}$ which converges uniformly to some convex function $g(x) \in W_H^+(\bar{G})$ (see theorem 1).

If $w(x) = g(x)$ for all $x \in \bar{H}$, then the assertion c) is proved.

Let x_o be the point of $\bar{H}$ such that $w(x_o) \neq g_o(x_o)$. We have two possibilities :

a) $w(x_o) > g(x_o)$

or

b) $w(x_o) < g(x_o)$.

If $w(x_o) > g(x_o)$, then

$$w(x_o) - g(x_o) = \delta > 0.$$

Let α be the supporting plane to the graph S_w of the function $w(x)$ passing through the point $(x_o, w(x_o))$. But

$$w(x) = F_H(u(x)).$$

Therefore there exists at least one point $x^* \in \bar{H}$ such that

$$(x^*, u(x^*)) \in \alpha ;$$

i.e. α is also a supporting plane to the graph of the function $u(x)$ if $x \in \bar{H}$.

The hyperplane α cuts off some cap L from the graph S_g of the function g(x), because

$$w(x_o) - g(x_o) = \delta > 0.$$

There exists the supporting hyperplane α' to S_g which has at least one contact point $(x', g(x'))$ with L, where $x' \in \bar{H}$.

It is evident that

$$\delta_1 = \text{dist}(\alpha, \alpha') > 0 \quad ^{(*)}.$$

From the definition of the function g(x) there exists the natural integer K such that

$$|w_{n_k}(x') - g(x')| < \frac{\delta_1}{10}$$

if k > K. Since $w_{n_k}(x)$ is the convex function spanned on the function $u_{n_k}(x)$ from below, then there exist points $\tilde{x} \in \bar{H}$ such that

$$\text{dist}((\tilde{x}, u_{n_k}(\tilde{x})), \alpha') < \frac{\delta_1}{5}$$

if k > K. But the points $(\tilde{x}, u_{n_k}(\tilde{x}))$ lie under α and

$$\text{dist}((\tilde{x}, u_{n_k}(\tilde{x})), \alpha) \geq \frac{4\delta_1}{5}$$

(*) δ_1 depends only on δ and the common Lipschitz constant $(\|u(x)\|+1)h_H^{-1}$ for all the functions $w_n(x)$ and w(x).

if $k > K$. Since points $(x,u(x))$ lie over the hyperplane α for all $x \in G$, the function $u(x)$ cannot be limit for the functions $u_n(x)$. This contradicts the condition $u(x) = \lim_{n\to\infty} u_n(x)$.

Therefore the inequality $w(x_o) > g(x_o)$ is impossible. The inequality $w(x_o) < g(x_o)$ is also impossible. We do not give the proof of the last assertion because it is based practically on the same idea as in the first case.

This proves Theorem 2. □

2.4. The functional $I_H(u)$.

Let $u(x) \in C_o^-(\bar{G})$ and let H be an open convex subdomain of G such that

$$\mathrm{dist}(\bar{H},G) = h_H > 0.$$

Let

$$w(x) = F_H(u)$$

be the convex function spanned on $u(x)$ from below on the set $\bar{H}$, and define the functional

$$\Phi_H(u) = -\int_G u\omega(w,de). \tag{2.38}$$

on the set $C_o^-(\bar{G})$.

Letting $\psi(e)$ be a non-negative completely additive set function on the subsets of G, we define the new function of sets

$$\Psi_H(e) = \psi(e \cap H). \tag{2.39}$$

Clearly $\Psi_H(e)$ is a non-negative completely additive set function on the subsets of G and

$$\Psi_H(G \setminus H) = 0 \tag{2.40}$$

We now introduce the functionals

$$\tau_H(u) = \int_G u\Psi_H(de) \tag{2.41}$$

and

$$I_H(u) = \Phi_H(u) + (n+1)\tau_H(u) \tag{2.42}$$

on the set $C_o^-(\bar{G})$.

Theorem 3 : *Let* $u(x)$ *be any function belonging to* $C_o^-(\bar{G})$ *and* $w(x) = F_H(u(x))$ *be the convex function spanned on* $u(x)$ *from below on the set* $\bar{H}$. *Then*

a) $$\Phi_H(w) = \Phi_H(u) , \tag{2.43}$$

b) $$\tau_H(w) \le \tau_H(u) , \tag{2.44}$$

c) $$I_H(w) \le I_H(u) . \tag{2.45}$$

Remark. Let $\psi(e) \ge C_o \text{ mes}(e)$ for every Borel subset $e \subset G$, where $C_o = \text{const} > 0$.

Then equality can hold in (2.44 - 2.45) if and only if the restriction $u(x)$ on the n-dimensional convex body $\bar{H}$ is some convex function, that is

$$u(x)|_{\bar{H}} = w(x)|_{\bar{H}} \tag{2.46}$$

where

$$w(x) = F_H(u(x)).$$

Proof : From the definitions of the functionals $\phi_H(u)$ and $\tau_H(e)$ it follows that

$$\phi_H(u) = - \int_{\bar{H}} u(x)\omega(w(x),de) \tag{2.47}$$

and

$$\tau_H(u) = \int_H u(x)\Psi_H(de) \tag{2.48}$$

because $\omega(w(x),G \setminus \bar{H}) = 0$ (see (2.25)) and $\psi_H(G \setminus H) = 0$ (see (2.40)).

Since

$$u(x) \geq w(x)$$

for any $x \in \bar{H}$, then from (2.48) we obtain

$$\tau_H(u) \geq \int_H w(x)\psi_H(de) = \int_G w(x)\Psi_H(de) = \tau_H(w).$$

It is clear that

$$\tau_H(u(x)) = \tau_H(w(x)) \tag{2.49}$$

if $u(x)|_{\bar{H}} = w(x)|_{\bar{H}}$. The condition

$$\psi(e) \geq C_o \operatorname{mes}(e)$$

mentioned in the above Remark and the equality (2.49) yield

$$u(x)|_{\bar{H}} = w(x)|_{\bar{H}} .$$

We denote by H_u the set of points $x \in \bar{H}$ where

$$u(x) = w(x)$$

and by S_{H_u} the part of the graph of $u(x)$ for $x \in H_u$.

Every supporting plane α of the graph of the function $w(x)$ has at least one common point with the set S_{H_u}.

Therefore $\chi_w(w, G \setminus H_u)$ consists only of the images of singular supporting hyperplanes to the graph of $w(x)$.

Therefore (see (2.25)) we obtain

$$\omega(w, G \setminus H_u) = 0 .$$

Thus

$$\phi_H(u) = \int_G u\omega(w,de) = \int_{H_u} u\omega(w,de) = \int_{H_u} w\omega(w,de) = \int_G w\omega(w,de) = \phi_H(w).$$

The inequality (2.45) now follows directly from (2.43) and (2.44).

Theorem 3 is proved. □

Theorem 4 : *The functionals* $\phi_H(u)$, $\tau_H(u)$ *and* $I_n(u)$ *are continuous on the set* $C_o^-(\bar{G})$.

Proof : Let $u_1(x), u_2(x), \ldots, u_n(x), \ldots$ be the sequence of functions from $C_o^-(\bar{G})$ uniformly converging to the function $u(x) \in C_o^-(\bar{G})$.

Then from well known theorems of functional analysis

$$\lim_{n \to \infty} \tau_H(u_n) = \lim_{n \to \infty} \int_G u_n(x)\Psi_H(de) = \int_G u(x)\Psi_H(de) = \tau_H(u).$$

From Theorem 2 it follows that the functions $w_n(x) = F_H(u_n(x))$ uniformly converge to the functions $w(x) = F(u(x))$. Now we can write (see Theorem 3) that

$$\phi_H(u_n) = \phi(w_n) = \int_G w_n\omega(w_n, de)$$

and

$$\phi_H(u) = \phi(w) = \int_G w\omega(w, de).$$

Since

$$\text{dist}\{\bar{H}, \partial G\} = h_H > 0$$

we can find a n-dimensional open domain Q such that

$$\bar{H} \subset Q \subset \bar{Q} \subset G$$

and

$$\text{dist}\{\bar{H}, \partial Q\} = \frac{1}{4} h_H, \ \text{dist}\{\bar{Q}, \partial G\} = \frac{1}{4} h_H.$$

We denote by $\phi(x)$ the following continuous function

$$\phi(x) = \begin{cases} 1 & \text{if } x \in \bar{H} \\ \text{takes values between} & \\ 0 \quad \text{and} \quad 1 & \text{if } x \in Q \setminus \bar{H} \\ 0 & \text{if } x \in \bar{G} \setminus Q . \end{cases}$$

Then

$$\int_G w\omega(w_n, de) = \int_{\bar{H}} w\omega(w_n, de) = \int_{\bar{H}} \phi(x) w\omega(w_n, de) = \int_G \phi(x) w(x) \omega(w_n, de)$$

and

$$\int_G w\omega(w, de) = \int_{\bar{H}} w\omega(w, de) = \int_{\bar{H}} \phi w\omega(w, de) = \int_G \phi(x) w(x) \omega(w, de).$$

The continuous function $\phi(x)u(x)$ takes zero values if $x \in \bar{G} \setminus \bar{Q}$. Therefore from Section (2.2) (formula (2.10)) it follows

$$\lim_{n \to \infty} \int_G w\omega(w_k, de) = \int_G w\omega(w, de). \tag{2.50}$$

Now

$$|I_H(u_n) - I_H(u)| \le \|w_n - w\| . \omega(u_n, G) + \left| \int_G w\omega(w_n, de) - \int_G w\omega(w, de) \right| \tag{2.51}$$

Since $\lim_{n \to \infty} \|u_n - u\| = 0$ then we can suppose that

$$\|u_n\| \le \|u\| + 1$$

for $k \ge k_o$, where $k_o > 0$ is a sufficient large integer. Then from Theorem 1 we obtain

$$\omega(u_n,G) \le \mu_n \frac{[\|u\|+1]^n}{h_H^n} . \tag{2.52}$$

Thus from (2.51), (2.50) and (2.52) we obtain

$$\lim_n I_H(u) = I(u). \qquad \square$$

Theorem 4 is proved.

2.5. Generalized solutions of elliptic Monge-Ampere equations and the main variational problem.

First of all from Theorem 3 and 4 it follows that we can seek the functions realizing the absolute minimum of the continuous functional $I_H(u)$ only in the set of convex functions $W_H^+(\bar{G})$.

The first principal result is contained in the Theorem A.

Theorem A : *There is only one function* $w_H(x) \in W_H^+(\bar{G})$ *such that*

$$I_H(w_H(x)) = \inf_{W_G^+(\bar{G})} I_H(w(x)). \tag{2.53}$$

The function $w_H(x)$ *is the solution of the following Dirichlet problem*

$$\begin{cases} \omega(w_H,e) = \Psi_H(e) & (2.54) \\ w_H|_{\partial G} = 0 \qquad \square & (2.55) \end{cases}$$

The proof of Theorem A will be obtained on the base of the results of

Sections 3 and 4.

The equation (2.54) is the extension of the Monge-Ampere equation (2.2) by means of non-negative completely additive functions of sets. Therefore the solutions of the equation (2.54) can be considered as generalized solutions of the equation (2.2).

Now I give briefly the exposition of some important facts of the theory of elliptic Monge-Ampere equations. This text can be considered as the continuation of Section 2.2. The details and proofs can be seen in [3], [4], [5], [7], [8]. All C^2-elliptic solutions of the n-dimensional Monge-Ampere equation

$$\det \|u_{x_i x_j}\| = f(x,u,\text{grad } u) \quad (2.56)$$

are convex or concave functions. We consider only convex solutions of (2.56). Therefore the function f must be positive in $G \times \mathbb{R} \times \mathbb{R}^n$. The concept of generalized solutions of (2.56) is constructed in the class of all convex functions $W^+(G)$. For simplicity we consider only the case

$$f(x,u,\text{grad } u) = \frac{\phi(x)}{R(\text{grad } u)} \quad (2.57)$$

where $\phi(x) \geq 0$ and $\phi(x) \in L(G)$ and $R(p) > 0$ is a locally summable function in $\tilde{R}^n$. We can write the equation (2.56) by assumption (2.57) in the form

$$R(\text{grad } u)\det \|u_{x_i x_j}\| = \phi(x) \quad (2.58)$$

and replace it by the equation

$$\omega(R,u,e) = \mu(e) \quad (2.59)$$

if we want to seek the generalized solutions of (2.58) in the class of all convex functions $W^+(G)$. In the right side of (2.59) $\mu(e)$ is a non-negative completely additive function of subsets e of the domain G and $\mu(G) < +\infty$. If $\mu(e)$ is absolutely continuous, then $\phi(x)$ is the density of this function and

$$\mu(e) = \int_e \phi(x)dx \qquad (2.60)$$

We consider the general case of the function R(p) in order to show deeply the character of the following existence and uniqueness theorem.

Theorem B : *The Dirichlet problem*

$$\begin{cases} \omega(R,u,e) = \mu(e) & (2.61) \\ u|_{\partial G} = 0 & (2.62) \end{cases}$$

has the unique generalized solution $u(x) \in W^+(G)$ *if :*

1) *the domain* G *satisfies the condition* (e.2) *of Section* 2.2 ;

2) *the function* R(p) *is locally summable and satisfies the condition* (e.1) *of Section* 2.2.

3) *the function* $\mu(e)$ *is non-negative and completely additive and satisfies the condition : there exists the neighborhood* S_{x_0} *for every point* $x_0 \in \partial G$ *such that the inequality*

$$\mu(e) \le a\{\sup_e[\text{dist}(x,\partial G)]\}^\lambda \text{ mes } e \qquad (2.63)$$

holds for any Borel subset $e \subset S_{x_0} \cap G$, *where* $\lambda \geq 0$ *and* $a > 0$ *are the common constants for all* $x_0 \in \partial G$.

4) $$k \leq \frac{n+1+\lambda}{2} . \tag{2.64}$$

5) $$\mu(G) < A(R) = \int_{\tilde{R}^n} R(p)dp. \quad \square \tag{2.65}$$

Remarks : A) The condition (2.64) is sharp ; the condition (2.65) is sufficient and interlocks with the necessary condition $\mu(G) = \omega(R,z,G) \leq A(R)$ (see Section 2.2).

B) If $R(p) \equiv 1$, then it is possible to take $\lambda = 0$ and not consider the inequality (2.64). The conditions (2.65) and (2.63) take the form

$$\mu(G) < + \infty \tag{2.65a}$$

and

$$\mu(e) \leq a.\text{mes } e \tag{2.63a}$$

for $e \subset S_{x_0} \cap G$, where x_0 is any point of ∂G.

Supplement to Theorem B : *If* $u(x)$ *and* $v(x) \in W^+(G)$ *are generalized solutions for the following Dirichlet problems*

$$\begin{cases} \omega(R,u,e) = \mu_1(e) , \\ u|_{\partial G} = 0 \end{cases} \quad \textit{and} \quad \begin{cases} \omega(R,v,e) = \mu_2(e) , \\ u|_{\partial G} = 0 , \end{cases}$$

and $$\mu_1(e) \leq \mu_2(e)$$

for any Borel subsets e *of the domain* G *; then*

$$u(x) \geq v(x)$$

for all $x \in G$.

(Note that the Supplement is also correct if the domain G is only convex and bounded and $\mu_1(e)$ and $\mu_2(e)$ are any non-negative completely additive functions).

Now we apply the results of elliptic Monge-Ampere equations to our variational problem. Let H_1 and H_2 be two convex subdomains of a convex bounded domain G such that

$$H_1 \subset H_2 \subset \bar{H}_2 \subset G$$

and

$$\text{dist}(\bar{H}_2, \partial G) = h_{H_2} > 0 \ .$$

Then we say that $H_1 < H_2$. If $H_1 < H_2$, then

$$W^+_{H_1}(\bar{G}) \subset W^+_{H_2}(\bar{G}) \tag{2.66}$$

and

$$\psi_{H_1}(e) \leq \psi_{H_2}(e) \tag{2.67}$$

for any Borel subset e of G. Therefore

$$\inf_{C_o^-(\bar{G})} I_{H_1}(u) = \inf_{W_{H_1}^+(\bar{G})} I_{H_1}(w) \geq \inf_{W_{H_2}^+(\bar{G})} I_{H_2}(w) = \inf_{C_o^-(\bar{G})} I_{H_2}(u) \qquad (2.68)$$

These relations follow from (2.39), (2.48) and Theorem 3. Let $w_1(x)$ and $w_2(x)$ be the functions realizing correspondingly

$$\inf_{W_{H_1}^+(\bar{G})} I_{H_1}(w) \text{ and } \inf_{W_{H_2}^+(\bar{G})} I_{H_2}(w)$$

Then from Theorem A and the Supplement to Theorem B it follows that

$$w_1(x) \geq w_2(x) \qquad (2.69)$$

for any $x \in G$.

Now if H_1 and H_2 are convex subdomains of a convex bounded domain G such that

$$\mathrm{dist}(\bar{H}_1, \partial G) = h_{H_1} > 0$$

and

$$\mathrm{dist}(\bar{H}_2, \partial G) = h_{H_2} > 0$$

then we can find some convex subdomain H of G such that

$$H_1 \subset H \subset \bar{H} \subset G$$

$$H_2 \subset H \subset \bar{H} \subset G$$

and

$$\mathrm{dist}(\bar{H},\partial G) = h_H > 0 \ ;$$

that is $H_1 < H$ and $H_2 < H$.

Thus the set of all convex subdomains of G is a semi-ordered set with respect to the relation $<$. We denote this set by Ξ.

Now we suppose that the domain G satisfies the condition (e.2) of Section 2.2 and the function $\psi(e)$ satisfies the condition (2.63a).

Then from Theorem A and B, Supplement to Theorem B and (2.66-69), it follows that :

a) $$\lim_{H\in\Xi} w_H(x) = w(x) \in W^+(\bar{G}) \cap C_o^-(\bar{G})$$

where $w_H(x)$ is the solution of the Dirichlet problem (2.54-55) realizing $\inf\limits_{C_o^-(\bar{G})} I_H(u)$.

b) $w(x)$ is the solution of the Dirichlet problem

$$\omega(w,e) = \psi(e) \tag{2.70}$$

$$w|_{\partial G} = 0 \tag{2.71}$$

c) $$\inf_{H\in\Xi} \{\inf_{C_o^-(\bar{G})} I_H(\bar{G})\} = \lim_{H\in\Xi} I_H(w_H)$$

d) $$\lim_{H\in\Xi} I_H(w(x)) = \inf_{H\in\Xi} \{\inf_{C_o^-(\bar{G})} I_H(\bar{G})\}$$

Thus if we understand $\inf\{\inf_{\Xi\ C_0^-(\bar{G})} I_H(\bar{G})\}$ as the absolute minimum of our functional, then the generalized solution of the Dirichlet problem (2.70-71) realizes this minimum.

This is the principal result in our main variational problem.

3. VARIATIONAL PROBLEM FOR THE FUNCTIONAL $I_H(u)$.

Let G be a convex bounded domain and H a convex subdomain of G such that dist $\bar{H},\partial G) = h_H > 0$.

Lemma 1 : *The inequality*

$$|w(x)| \geq \frac{h_H}{\text{diam } G} \|w(x)\| \tag{3.1}$$

holds for every convex function $u(x) \in W_H^+(\bar{G})$ *and every* $x \in \bar{H}$.

Proof : The inequality (3.1) holds trivially for the function $w(x) = 0$ for $x \in G$. Therefore we assume that

$$\|w(x)\| > 0 \tag{3.2}$$

Since any supporting hyperplane to the graph of $w(x) \in W_H^+(\bar{G})$ has at least one common point with the convex hypersurface $S_{w,H}$ (see Theorem 1), then there exists the point $x_o \in \bar{H}$ such that

$$\|w(x)\| = |w(x_o)| \tag{3.3}$$

Now we consider the convex cone K with the vertex $(x_o,w(x_o))$ and the base

∂G. Let K be the graph of the convex function $k(x)$. Then

$$w(x) \leq k(x) \leq 0 \tag{3.4}$$

for any $x \in \bar{G}$ and

$$w(x)|_{\partial G} = k(x)|_{\partial G} = 0 \tag{3.5}$$

The equality

$$k(x) = \frac{|xx'|}{|x_o x'|} |k(x_o)| \tag{3.6}$$

holds for any point $x \in \bar{H}$, where x' is the point of intersection of the ray $x_o x$ (with origin x_o) and ∂G, and $|xx'| = \mathrm{dist}(x,x')$, $|x_o x'| = \mathrm{dist}(x_o,x')$. Since

$$|xx'| \geq h_H \tag{3.7}$$

and

$$|x_o x'| \leq \mathrm{diam}\ G \tag{3.8}$$

then from (3.6), (3.7) and (3.8) we obtain

$$|k(x)| \geq \frac{h_H}{\mathrm{diam}\ G} |k(x_o)|. \tag{3.9}$$

But

$$|k(x_o)| = |w(x_o)| = \|w\| . \tag{3.10}$$

since the cone K has the vertex in the point $(x_o, w(x_o))$ and

$$\|w(x)\| = |w(x_o)|.$$

Now from (3.4), (3.9) and (3.10) we obtain the inequality (3.1).

Lemma 1 is proved. □

<u>Lemma 2</u> : *The inequality*

$$\|w(x)\| \leq \left[\frac{\omega(w,G)}{\mu_n}\right]^{1/n} . \text{ diam } G \tag{3.11}$$

holds for every convex function $w(x) \in W^+(\bar{G}) \cap C_o^-(\bar{G})$, *where* μ_n *is the volume of the* n*-unit ball.*

This lemma is the special case of Lemma 2 of the paper [8] (see also [3], [9]).

<u>Theorem 5</u> : *The inequality*

$$I_H(w) \geq \frac{\mu_n h_H}{(\text{diam } G)^{n+1}} \|w(x)\|^{n+1} - \psi_H(G)(n+1)\|w(x)\| \tag{3.12}$$

holds for any $w(x) \in W_H^+(\bar{G})$.

<u>Proof</u> : From Lemma 1 we obtain

$$\int_G [-w(x)]\omega(w,de) \geq \frac{h_H}{\text{diam } G} \|w(x)\| \omega(w,\bar{H}). \tag{3.13}$$

But from Theorem 2 (see Section 2.3) it follows that

$$\omega(w,\bar{H}) = \omega(w,G).$$

Now from Lemma 2 we obtain

$$\omega(w,G) \geq \frac{\mu_n}{(\text{diam } G)^n} \|w(x)\|. \tag{3.14}$$

Thus the inequalities (3.13) and (3.14) lead to the inequality

$$\int_G [-w(x)]\omega(w,de) \geq \frac{\mu_n h_H}{(\text{diam } G)^{n+1}} \|w(x)\|^{n+1}. \tag{3.15}$$

From (3.15) we obtain finally the inequality (3.12) for

$$I_H(w) = - \int_G w(x)\omega(w,de) + (n+1)\int_G w(x)\psi_H(de).$$

Theorem 5 is proved. □

<u>Theorem 6</u> : *The inequality*

$$I_H(w) \leq \frac{\mu_n}{h_H} \|w\|^{n+1} - \frac{(n+1)h_H}{\text{diam } G} \psi_H(G) \|w\| \tag{3.16}$$

holds for every convex function $w(x) \in W_H^+(\bar{G})$.

<u>Proof</u> : First we estimate from above the integral

$$\int_G \{-w(x)\}\omega(w,de) .$$

We have

$$0 \le \int_G \{-w(x)\}\omega(w,de) \le \|w\|\omega(w,G)$$

and from Theorem 2 we obtain

$$0 \le \int_G \{-w(x)\}\omega(w,de) \le \frac{\mu_n}{h_H^n}\|w\|^{n+1} . \tag{3.17}$$

Now we estimate from below $\int_G |w(x)|\psi_H(de)$. Since

$$\psi_H(G \setminus \bar{H}) = 0$$

then

$$\int_G |w(x)|\psi_H(de) = \int_{\bar{H}} |w(x)|\psi_H(de).$$

Now from Lemma 1 it follows that

$$\int_G |w(x)|\psi_E(de) \ge \frac{h_H}{\operatorname{diam} G}\|w(x)\|\psi_H(G). \tag{3.18}$$

Thus from (3.17) and (3.18) we finally obtain

$$I_H(w) = -\int_G w\omega(w,de) + (n+1)\int_G w\psi_H(de) \le \frac{\mu_n}{h_H^n}\|w\|^{n+1} - \frac{(n+1)h_H}{\operatorname{diam} G}\psi_H(G)\|w\|$$

because $w(x) \le 0$ in G.

Theorem 6 is proved. □

Let $U(H,m,M)$ denote the subset of function $w(x) \in W_H^+(\bar{G})$ satisfying the condition

$$m \leq \|w(x)\| \leq M \tag{3.19}$$

where $0 \leq m < M < +\infty$ are constants. If $m = 0$, the $U(H,0,M)$ consists of functions $w(x) \in W_H^+(\bar{G})$ satisfying the inequality

$$\|w(x)\| \leq M . \tag{3.20}$$

<u>Theorem 7</u> : *Every set* $U(H,m,M)$ *is compact in* $C(\bar{G})$.

<u>Proof</u> : The set $U(H,m,M)$ is bounded and closed in $C(\bar{G})$ and any function $w(x) \in U(H,m,M)$ satisfies the Lipschitz condition of the degree one and constant $M\mu_n^{1/n}(h_H)^{-1}$. Thus $U(H,m,M)$ is compact in $C(\bar{G})$.

Theorem 7 is proved. □

<u>Theorem 8</u> : (<u>Main theorem about the absolute minimum of the functional</u> $I_H(u)$). *The function* $I_H(u)$ *has at least one absolute minimum and the function* $w_o(x)$ *belonging to* $W_H^+(\bar{G})$ *and realizing this minimum satisfies the inequalities*

$$m_o \leq \|w_o(x)\| \leq M_o ,$$

where

$$m_o = \frac{1}{2}\left(\frac{h_H^{n+1}\,\psi_H(G)}{\mu_n(\text{diam } G)}\right)^{\frac{1}{n}} ,$$

$$M_o = \max \{1, \left(\frac{(n+1)\psi_H(G)+1}{\mu_n h_H} (\text{diam } G)^{n+1}\right)^{\frac{1}{n}} \}.$$

Proof : From Theorem 5 it follows that

$$\lim_{k\to\infty} I_H(w_k) = +\infty$$

if $w_k(x) \in W_H^+(\bar{G})$ and $\|w_k(x)\| \to +\infty$.

Therefore we can find a positive number M_o such that

$$I_H(w) > 1 \text{ if } \|w(x)\| > M_o .$$

For example we can take M_o to be the number M_o mentioned in Theorem 8. Now from the expression of $I_H(u)$ and Theorem 6 we can see that $I_H(0) = 0$ and $I_H(w) < 0$ if $w \in W_H^+(\bar{G})$, $\|w\| > 0$ and $\|w\|$ is sufficiently small.

Therefore the functional $I_H(u)$ is bounded from below and $I_H(u)$ takes negative values.

Now we consider the function

$$\phi(t) = \frac{\mu_n}{h_H} t^{n+1} - \frac{(n+1)h_H}{\text{diam } G} \psi_H(G)t$$

for $t \in [0,+\infty)$. This function has only two roots 0 and some positive number t_o and takes negative values only inside the interval $(0,t_o)$. Let t^* be the point such that

$$\phi(t^*) = \inf_{[0,t_o]} \phi(t).$$

Then $\phi'(t^*) = 0$ and $t^* = \left(\frac{h_H^{n+1} \psi_H(G)}{\mu_n (\text{diam } G)} \right)^{\frac{1}{n}}$.

Let $m_o = \frac{t^*}{2}$. Then

$$\underset{C_o^-(\bar{G})}{\text{Inf}} I_H(u) = \underset{W_H^+(\bar{G})}{\text{Inf}} I_H(w) \le \phi(t^*) < \phi(m) < 0.$$

Recall that $\underset{W_H^+(\bar{G})}{\text{Inf}} I_H(w)$ is a finite negative number. It is clear that

$$\underset{W_H^+(\bar{G})}{\text{Inf}} I_H(w) = \underset{U(H,m_o,M_o)}{\text{Inf}} I_H(w)$$

where m_o and M_o were defined above.

From Theorem 7 it follows that there exists at least one function $w_o(x) \in U(H,m,M)$ such that

$$I_H(w_o(x)) = \underset{U(H,m_o,M_o)}{\text{Inf}} I_H(w) = \underset{C_o^-(\bar{G})}{\text{Inf}} I_H(u).$$

This proves Theorem 8. □

In the next Section we prove that $w_o(x)$ is the generalized solution of the following Dirichlet problem

$$\begin{cases} \omega(w,e) = \psi_H(e) \\ w|_{\partial G} = 0 . \end{cases}$$

4. DUAL CONVEX HYPERSURFACES. EULERIAN EQUATION FOR THE FUNCTIONAL $I_H(u)$.

4.1. Special map on the hemisphere S^n_-.

Let G be an open convex bounded domain in $E^n = \{x = (x_1,\dots, x_n)\}$. Let $\tilde{R}^{n+1} = \{p = (p_1,\dots, p_{n+1})\}$ be an (n+1)-dimensional Euclidean space and S^n_- be the unit n-hemisphere :

$$p_{n+1} < 0, \quad p_1^2 + p_2^2 + \dots + p_{n+1}^2 = 1 \tag{4.1}$$

in $\tilde{R}^{n+1}$. We consider the map

$$\gamma : S^n_- \to E^n$$

defined by

$$x_1 = \frac{p_1}{|p_{n+1}|}, \quad x_2 = \frac{p_2}{|p_{n+1}|}, \dots, x_n = \frac{p_n}{|p_{n+1}|}, \tag{4.2}$$

where

$$x = (x_1,\dots, x_n) \in E^n,$$

$$p = (p_1, p_2,\dots, p_{n+1}) \in S^n_-$$

and

$$x = \gamma(p).$$

We can also consider γ as a diffeomorphism between the smooth manifolds S^n_- and E^n with natural differential structures. Then the diffeomorphism $\gamma^{-1} : E^n \to S^n_-$ maps any point $x = (x_1, x_2,\dots, x_n) \in E^n$ in the point

$$p = \gamma^{-1}(x) = (\frac{x_1}{q}, \frac{x_2}{q}, \ldots, \frac{x_n}{q}, -\frac{1}{q}) \qquad (4.3)$$

where $q = (1+x_1^2+x_2^2+\ldots+x_n^2)^{1/2}$. We denote γ^{-1} by γ_1.

The set $\bar{G}^* = \gamma_1(\bar{G})$ is a closed convex domain in S_-^n, where $\bar{G}$ is the closure of G and

$$\text{dist}(\partial S_-^n, \bar{G}^*) = \delta_o > 0 \qquad (4.4)$$

in the intrinsic spherical meaning.

4.2. Dual convex hypersurfaces.

Let u(x) be any continuous non-positive function in $\bar{G}$ satisfying the condition

$$u|_{\partial G} = 0 .$$

The function u(x) defines the new function $u^*(p)$ in $\bar{G}^*$ by the formula

$$u^*(p) = (1-p_1^2-\ldots-p_n^2)^{1/2}u(\gamma(p)) \qquad (4.5)$$

for $p = (p_1, \ldots, p_n, p_{n+1}) \in \bar{G}^* \subset S_-^n$, where $x = \gamma(p)$ (see (4.2)). Conversely if we define

$$\tilde{u}^*(x) = u^*(\gamma_1(x)) \qquad (4.6)$$

where $p = \gamma_1(x)$, then

$$\tilde{u}^*(x) = \frac{1}{(1+x_1^2+\ldots+x_n^2)^{1/2}}\, u(x) \ . \tag{4.7}$$

We denote by H and $\bar{H}$ an open convex subdomain of G and its closure and assume that

$$\mathrm{dist}(\bar{H}, G) = h_H > 0.$$

Then $H^* = \gamma_1(H)$ and its closure $\bar{H}^* = \gamma_1(\bar{H})$ are respectively open and closed spherical convex domains and the intrinsic distance h_{H^*} between $\bar{H}^*$ and ∂G^* is positive. Clearly h_{H^*} depends only on h_H.

The inequality

$$(p,z) \le u^*(p) \tag{4.8}$$

defines the closed half-space $U_p \subset R^{n+1}$ for each fixed vector $p \in \bar{G}^*$ and any vector $z \in \tilde{R}^{n+1}$, satisfying the inequality (4.8). The set

$$Q_H(u) = \bigcap_{p\in\bar{H}^*} U_p \tag{4.9}$$

is a closed infinite convex body in $\tilde{R}^{n+1}$. The sets

$$K(\partial G^*) = \bigcap_{q\in\partial G^*} V_q \tag{4.10}$$

and

$$K(\bar{G}^*) = \bigcap_{q\in\bar{G}^*} V_q$$

are one and the same solid convex cone in $\tilde{R}^{n+1}$ with vertex $\tilde{O}(0,0,\ldots,0)$, where V_q is the closed half-space

$$(q,z) \leq 0 \qquad (4.11)$$

for any fixed $q \in \partial G^*$ (or $\bar{G}^*$) and any vector $z \in \tilde{R}^{n+1}$.

Now the sets

$$P_H(u) = \partial Q_H(u) \qquad (4.12)$$

and

$$L(\partial G^*) = \partial K(\partial G^*) \qquad (4.13)$$

are complete infinite convex hypersurfaces in R^{n+1} and the latter is a convex n-dimensional cone with the vertex $\tilde{O}(0,0,\ldots,0)$.

Theorem 9 : *Let* $w(x)$ *be the convex function spanned by* $u(x) \in C_o^-(\bar{G})$ *from below on the set* $\bar{H}$. *Then*

$$Q_H(u) = Q_H(w) \qquad (4.14)$$

and

$$P_H(u) = P_H(w).$$

Moreover the convex body $Q_H(u)$ *and the convex hypersurface* $P_H(u)$ *have one and the same supporting function* $w^*(p)$ *defined on* $\bar{H}^*$.

Proof : From definition of the function w(x) it follows that

$$w(x) \le u(x) \le 0$$

for any $x \in \bar{H}$. Therefore

$$w^*(p) \le u^*(p) \le 0$$

for any $p \in \bar{H}^*$. Thus

$$W_p \subset U_p$$

for any $p \in \bar{H}^*$, where W_p and U_p correspondingly are the closed half-spaces

$$(p,z) \le w^*(p) \text{ and } (p,z) \le u^*(p)$$

for every fixed vector $p \in \bar{H}^*$ and any vector $z \in \tilde{R}^{n+1}$. Therefore

$$Q_H(w) = \bigcap_{p \in \bar{H}^*} W_p \subset \bigcap_{p \in \bar{H}^*} U_p = Q_H(u). \tag{4.15}$$

From the theory of convex bodies it is well known that if M is an infinite closed convex body and $v^*(p) < 0$, $p \in \bar{H}^* \subset S^n_-$ is the supporting function of M, then the function

$$v(x) = (1 + \sum_{i=1}^{n} x_i^2)^{1/2} v^*(\gamma(x))$$

is a negative convex function for $x \in \bar{H}$, where

$$\gamma(x) = (\frac{x_1}{q}, \frac{x_2}{q}, \ldots, \frac{x_n}{q}, -\frac{1}{q}) \in \bar{H}^*$$

and $q = (1 + \sum_{i=1}^{n} x_i^2)^{1/2}$. Now we apply this fact to the case $M = Q_H(u)$. Let $v^*(p)$ be the supporting function of the convex body $Q_H(u)$, then clearly

$$0 \geq u^*(p) \geq v^*(p)$$

for any $p \in \bar{H}^*$. Therefore we obtain for negative convex function $v(x)$ the inequality

$$0 \geq u(x) \geq v(x)$$

for any $x \in \bar{H}$. From the definition of the convex function $w(x)$ spanned by $u(x)$ from below on $\bar{H}$ it follows that

$$u(x) \geq w(x) \geq v(x)$$

for any $x \in \bar{H}$. Repeating our reasoning for the functions $w(x)$ and $v(x)$ we obtain

$$Q_H(w) \supset Q_H(v) = Q_H(u) \tag{4.16}$$

From (4.15) and (4.16) it follows that

$$Q_H(u) = Q_H(v) \ ,$$

and hence

$$P_H(u) = P_H(v).$$

Theorem 9 is proved. □

Now we consider the new convex body

$$Q_G(w) = \bigcap_{p\in\bar{G}^*} W_p \qquad (4.17)$$

for every function $w(x) \in W_H^+(\bar{G})$, where the closed half-space W_p was defined above in this Section.

Theorem 10 :

$$Q_G(w) = Q_H(w) \cap K(\partial G^*). \qquad (4.18)$$

Proof : It follows from definitions of the sets $Q_G(w)$ and $Q_H(w)$ that

$$Q_G(w) = Q_H(w) \cap Q_{G\setminus H}(w) , \qquad (4.19)$$

where

$$Q_{G\setminus H}(w) = \bigcap_{p\in \bar{G}^*\setminus\bar{H}^*} W_p .$$

First of all we note that the asymptotic solid cone $K_H(w)$ to $Q_H(w)$ has the set $\bar{H}^*$ as a spherical image. If the vertex of $K_H(w)$ lies inside $Q_H(w)$, then the whole cone $K_H(w)$ lies inside $Q_H(w)$. Let $L_H(w)$ be the boundary of $K_H(w)$. We suppose that the vertex of $K_H(w)$ coincides with the nearest point of $P_H(w)$ to the origin $\tilde{O}$ of $\tilde{R}^{n+1}$.

Then the set

$$\lambda(w) = L_H(w) \cap L(\partial G^*)$$

is the (n-1)-dimensional hypersurface homeomorphic to (n-1)-sphere. Recall that $L(\partial G^*) = \partial K(\partial G^*)$, where $K(\partial G^*)$ is the convex solid cone.

Evidently $\sup_{Z \in \lambda(w)} \{dist\{0,Z\}\}$ can be estimated above by means of $\|w(x)\|$, $h_H = dist\,\{\bar{H},\partial G\}$ and $\delta_o = dist(\partial S^n_-,\bar{G}^*)$.

If

$$\nu(w) = P_H(w) \cap L(\partial G^*) \tag{4.20}$$

then $\nu(w)$ is also homeomorphic to (n-1)-sphere and $\nu(w)$ lies between $\lambda(w)$ and the origin of $\tilde{R}^{n+1}$. Thus

$$\sup_{Z \in \nu(w)} \{dist\{\tilde{0},Z\}\} \tag{4.21}$$

can also be estimated by means of $\|w(x)\|$, h_H and δ_o.

Now all supporting hyperplanes to the graph S_W of the convex function $w(x)$ of the points $(x,w(x))$ will be singular if x belongs to $\bar{G} \setminus \bar{H}$ (see the proof of Theorem 1, Section 2). Let α be such supporting hyperplane, then $\alpha \cap S_W$ is some closed bounded convex k-dimensional body, where $1 \leq k \leq n-1$. We denote by $\pi_\alpha \subset \bar{G}$ the closed k-dimensional convex body which is the projection of the set $\alpha \cap S_W$. Then π_α determines the singular point Y on $P_H(w)$ with k-dimensional set of supporting hyperplanes to $P_H(w)$, because $\pi_\alpha \cap \bar{H} \neq \emptyset$. The spherical image of this set of supporting hyperplanes coincides with $\gamma_1(\pi_\alpha) \subset S^n_-$. (For definition of the mapping γ_1 see in Section 4.1). Since α passes through some point $(x_o,0)$, where $x_o \in \partial G$, then Y_α belongs to the cone $L(\partial G^*)$ or more precisely $Y_\alpha \in \nu(w)$ (see (4.20)).

Clearly

$$\nu(w) = \bigcup_{\alpha} \gamma_{\alpha}$$

where α runs the set of all supporting hyperplanes to S_w having contact points $(x,w(x))$ with S_w, and $x \in \bar{G} \setminus \bar{H}$.

Therefore from (4.17), (4.18), (4.19) and the last considerations it follows that

$$Q_G(w) = Q_H(w) \cap K(\partial G^*).$$

Theorem 10 is proved. □

Thus the convex hypersurface

$$P_G(w) = \partial Q_G(w) \tag{4.22}$$

consists of two parts : the first one $S_H(w)$ lies inside the solid cone $K(\partial G^*)$ and the second one $T_{\partial G}(w)$ lies on the boundary $L(\partial G^*)$ of the cone $K(\partial G^*)$. Both hypersurfaces have one and the same boundary $\nu(w) \subset P_G(w)$. Let us agree to include $\nu(w)$ as a part of $S_H(w)$ and $T_{\partial G}(w)$ and consider both hypersurfaces asclosed hypersurfaces with boundary.

We call $S_H(w)$ the *dual convex hypersurface* (with respect to $\bar{H} \subset G$) of the convex function $w(x) \in W_H^+(\bar{G})$. The function

$$w^*(p) = (1-p_1^2 - \ldots - p_n^2)^{1/2} w(\gamma(p)) \tag{4.23}$$

is the supporting function for $S_H(w)$ for any $p \in \bar{G}^*$.

4.2. Expression of the functional $I_H(u)$ be means of dual convex hypersurfaces

Let $w(x)$ be any convex function belonging to $W_H^+(\bar{G})$ and $S_H(w)$ be its dual convex hypersurface. We denote by $\sigma(S_H(w),e')$ the surface function of $S_H(w)$ (see [14], [15]). The surface function $\sigma(S_H(w),e')$ is defined as the completely additive non-negative function on the ring of Borel's subsets e' of the domain $\bar{G}^* \subset S_-^n$ and the values of this function equal to the area of the sets $\tilde{e} \subset S_H(w)$ such that $\tilde{e}$ consist of all points of $S_H(w)$ having supporting hyperplanes with unit outside normals belonging to e'. From our considerations it follows that

$$\sigma(S_H(w),\bar{G}^* \setminus \bar{H}^*) = 0 . \tag{4.24}$$

It is well known (see [15]) that $\{\sigma(S_H(w_k),e')\}$ converges weakly to $\sigma(S_H(w_o),e')$ if

$$\lim_{k\to\infty} \|w_k - w_o\| = 0.$$

Let $V_H(w)$ be the volume of the part of the convex cone $K(\partial G^*)$ situated under the dual convex hypersurface $S_H(w)$.

Theorem 11 : *The equality*

$$V_H(w) = -\frac{1}{n+1}\int_{\bar{H}} w(x)\omega(w,de) \tag{4.25}$$

holds for every convex function $w(x) \in W_H^+(G)$.

Proof : If $w(x) \in W_H^+(\bar{G})$ then for the volume $V_H(w)$ there is the formula

$$V_H(w) = - \frac{1}{(n+1)} \int_{\bar{G}^*} w^*(p)\sigma(S_H(w),de') . \tag{4.26}$$

(see [14], [15]).

But the surface function $\sigma(S_H(w),e')$ has the following representation

$$\sigma(S_H(w),e') = \int_e (1+ \sum_{i=1}^{n} x_i^2)^{1/2}\omega(w,de) , \tag{4.27}$$

in the map γ (see Section 4.1). The formula (4.27) can first be proved for convex polyhedrons and extends for all class $W_H^+(\bar{G})$ of convex functions by approximation of polyhedrons. (*) From (4.26) and (4.27) it follows that

$$V_H(w) = - \frac{1}{(n+1)} \int_{\bar{G}^*} w^*(\gamma_1(x))(1+ \sum_{i=1}^{n} x_i^2)^{1/2}\omega(w,de) =$$

$$= - \frac{1}{(n+1)} \int_{\bar{G}} w(x)\omega(w,de) = - \frac{1}{(n+1)} \int_{\bar{H}} w(x)\omega(w,de) ,$$

because $\omega(w,\bar{G} \setminus \bar{H}) = 0$.

Theorem 11 is proved. □

Remark. Since any convex polyhedron can be approximated by C^2 convex hypersurfaces (function) with everywhere strictly principal normal curvatures, then it is sufficient to establish the formula (4.27) only for such class of hypersurfaces (functions).

(*) Of course we use the weak convergence of the surface functions.

From the Gauss theorem it follows that

$$\sigma(S_H(w),e') = \int_{e'} \frac{ds_p}{K(p)} \tag{4.28}$$

where ds_p is the element of area on S^n_- and $K(p)$ is the Gauss curvature of $S_H(w)$ in the point of $S_H(w)$ with the outside unit normal p. We find

$$\int_{e'} \frac{ds_p}{K(p)} = \int_e (1+\sum_{i=1}^n x_i^2)^{1/2} \det \|w_{x_i x_j}\| dx , \tag{4.29}$$

where $e' = \gamma^{-1}(e)$ (see Section 4.1 and also [15]). From (4.28) and (4.29) we obtain (4.27) for the C^2-convex functions (hypersurfaces) with strictly positive principal normal curvatures.

<u>Theorem 12</u> : *The functional* $I_H(u)$ *in the* $C^-_o(\bar{G})$ *has the following representation*

$$\begin{aligned} I_H(u) &= (n+1)[V_H(F_H(u)) + \int_{H^*} u^*(p)\Psi^*_H(de')] = \\ &= [-\int_{H^*_u} u^*(p)\sigma(S_H(w),de') + (n+1)\int_{H^*} u^*(p)\Psi^*_H(de')], \end{aligned} \tag{4.30}$$

where $w(x) = F_H(u(x))$ *is the convex function spanned by* $u(x)$ *from below on* $\bar{H} \subset G$ *;* $\Psi^*_H(e')$ *is the non-negative completely additive function of Borel subsets* e' *of* G^* *determined by the formula*

$$\Psi^*_H(e') = \int_{\gamma(e)} (1+\sum_{i=1}^n x_i^2)^{1/2}\Psi_H(de) ; \tag{4.31}$$

$H^*_u = \gamma_1(H_u)$ *and* H_u *is the closed of* $\bar{H}$ *where* $u(x) = w(x)$.

Proof : It follows from definition of $I_H(u)$ that

$$I_H(u) = - \int_G u\omega(w,de) + (n+1) \int_G u\Psi_H(de) \; . \qquad (4.32)$$

Now

$$\int_G u(x)\Psi_H(de) = \int_H u^*(x)(1+\sum_{i=1}^{n} x_i^2)^{1/2}\Psi_H(de)$$

since $\Psi_H(G \setminus H) = 0$. Therefore

$$\int_G u(x)\Psi_H(de) = \int_{H^*} u^*(p)\Psi_H^*(de') \qquad (4.33)$$

if we use (2.40) and (4.31).

From Theorem 10 and Theorem 3 we obtain

$$\begin{aligned}(n+1)V_H(F_H(u)) &= - \int_{\bar{H}} w(x)\omega(w,de) = - \int_{H_u} w(x)\omega(w,de) = \\ &= - \int_{H_u} u(x)\omega(w,de) = - \int_{H_u^*} u^*(p)\sigma(S_H(w),de')\end{aligned} \qquad (4.34)$$

From Theorem 3 we obtain

$$- \int_G u(x)\omega(w,de) = - \int_{H_u} u(x)\omega(de) \qquad (4.35)$$

Now it follows from (4.34) and (4.35) that

$$- \int_G u(x)\omega(w,de) = (n+1)V_H(F_H(u)) = \int_{H_u^*} u^*(p)\sigma(S_H(w),de') \; . \qquad (4.36)$$

Thus from (4.32), (4.33) and (4.36) we obtain (4.30).

Theorem 12 is proved.

4.3. Expression of the variation of $I_H(u)$.

First of all we study the variation of the functional

$$\Phi(u) = -\int_G u(x)\omega(w,de) \tag{4.37}$$

where $u(x) \in C_o^-(\bar{G})$ and $w(x)$ is the convex function spanned by $u(x)$ from below on the convex closed domain $\bar{H} \subset G$. From Theorem 12 it follows that

$$\begin{aligned}\Phi(u) = (n+1)V_H(w) &= -\int_{H_u^*} w^*(p)\sigma(S_H(w),de') = \\ &= -\int_{H_u^*} u^*(p)\sigma(S_H(w),de')\end{aligned} \tag{4.38}$$

where H_u^* is a closed subset of $\bar{H}^* = \gamma_1(\bar{H})$, where $w^*(p) = u^*(p)$ and $V_H(w)$ is the volume of the part of the convex cone $K(\partial G^*)$ situated under the dual convex hypersurface $S_H(w)$.

Now we want to complement $S_H(w)$ to the whole closed convex hypersurface. The boundary $\nu(w)$ of $S_H(w)$ lies on the conic convex hypersurface $L(\partial G^*) = \partial\{K(\partial G^*)\}$ and homeomorphic to (n-1)-sphere. We can evidently find two numbers m_1 and m_2 depending on $\|w\|$, $\text{dist}\{\bar{H},\partial G\} = h_H > 0$ and $\text{dist}\{\bar{G}^*,\partial S_-^n\}$ such that

$$0 < m_1 \leq \text{dist}\{\tilde{O},\nu(w)\} \leq m_2 < +\infty. \tag{4.39}$$

We denote by $S^n_+(r)$ the hemisphere

$$\begin{cases} p_1^2 + p_2^2 + \ldots + p_{n+1}^2 = r^2 , \\ p_{n+1} \geq 0 \end{cases} \tag{4.40}$$

and by $U^n_+(r)$ the set

$$\begin{cases} p_1^2 + p_2^2 + \ldots + p_{n+1}^2 \leq r^2 , \\ p_{n+1} \geq 0 \end{cases} \tag{4.41}$$

We only consider the functions $u(x) \in C_o^-(\bar{G})$ such that for the convex functions $w(x) = F_H(u(x))$ spanned by $u(x)$ from below on $\bar{H}$ the following inequalities

$$m_o \leq \|w(x)\| \leq M_o \tag{4.42}$$

hold (see Theorem 8). Then there exist the common numbers $0 < m_1 < m_2 < +\infty$ such that for all functions $u(x) \in C_o^-(\bar{G})$ the following inequalities

$$0 < m_1 \leq \text{dist}\{\tilde{O}, \nu(w)\} \leq m_2 < +\infty \tag{4.43}$$

hold, if (4.42) are fulfilled $w(x) = F_H(u(x))$.

Thus we will be able to construct all the bounded convex bodies $\Pi_H(w)$.

Now consider the supporting function of $\Pi_H(w)$. We denote this function by $h_H(p)$ where p runs the whole unit sphere S^n : $p_1^2 + p_2^2 + \ldots + p_n^2 + p_{n+1}^2 = 1$.

The closed convex hypersurfaces $\Lambda_H(w)$ has at least two ribs $\nu(w)$ and $\nu(m_1+2m_2)$ which are the boundaries for three domains $S_H(w)$, $Z(m_1+2m_2)$ and $T(w,U_+^n(m_1+2m_2)) \subset L(\partial G^*)$.

Therefore

$$h_H^*(p) = \begin{cases} w^*(p) < 0 \quad \text{if} & \text{if } p \in G^* \\ 0 & \text{if } p \in \partial G^* \\ \text{takes positive values} & \text{if } p \in S^n \setminus (\bar{G}^* \cup \{\frac{1}{m_1+2m_2} Z(m_1+2m_2)\}), \\ m_1 + 2m_2 & \text{if } p \in \frac{1}{m_1+2m_2} Z(m_1+2m_2)) \subset S_+^n. \end{cases}$$

Note that $\sigma(\Lambda_H(w), S^n \setminus (\bar{G}^* \cup \{\frac{1}{m_1+2m_2} Z(m_1+2m_2))\} = 0$.

Therefore the volume of $\Pi_H(w)$ can be found by the formula

$$V(\Pi_H(w)) = \frac{1}{n+1} \int_{S^n} h_H^*(p)\sigma(\Lambda_H(w),de') = \frac{\sigma(K(\partial G^*))}{n+1} (m_1+2m_2)^{n+1} +$$

$$+ \frac{1}{n+1} \int_{\bar{G}^*} w^*(p)\sigma(S_H(w),de) = \frac{\sigma(K(\partial G^*))}{n+1} - \frac{1}{n+1}\int_{H_u^*} u^*(p)\sigma(S_H(w),de),$$

where $\sigma(K(\partial G^*))$ is the solid angle of the convex cone $K(\partial G^*)$.

Thus

$$V(\Pi_H(w)) = \frac{\sigma(K(\partial G^*))}{n+1} + \frac{1}{n+1} \phi(u) \quad . \tag{4.45}$$

If we change the point of the reference of distances with the sign to supporting hyperplanes to any convex body, then the supporting function of this body changes its values. If such a point p_0 coincides with the inner point of $\Pi_H(w)$, then the supporting function takes only positive values and is some strictly positive function on S^n.

Minkowski, Alexandrov, Fenchel and Jessen investigated the variation of the volume in the class of bounded convex bodies and established the formulas for the weak differential (the first variation) by different conditions (see [13],[14],[15],[16],[17]). The main methods and techniques of these investigations were the theory of Minkowski mixed volumes and the Brunn-Minkowski inequality.

Alexandrov proved that if $h_o(p)$ is any strictly positive continuous function on the unit sphere S^n : $|p| = 1$ and H_o is the closed convex body defined by intersection of all the halfspaces

$$(p,z) \leq h_o(p) \ , \ p \in S^n \ ,$$

then

$$\lim_{t \to 0} \frac{V(H_t) - V(H)}{t} = \int_{S^n} \eta(p)\sigma(H_o, de') \ , \tag{4.47}$$

where $\eta(p)$ is any continuous function on S^n, t is real parameter converging to zero, $\sigma(H_o, e')$ is the surface function of H_o and H_t is the closed bounded convex body defined by intersection of all the halfspaces

$$(p,z) \leq h_o(p) + t\eta(p) \ .$$

Remark. Since $h_o(p)$ and $\eta(p)$ are continuous on S^n and $h_o(p)$ is strictly positive, then $h_o(t) + t\eta(p)$ is also positive for sufficiently small t and the bodies H_t will be constructable.

Since 1) all terms of (4.46) are independent on the point of reference to the supporting hyperplanes and 2) it is possible to take any function $\eta \neq 0$ only on any closed set $\bar{H}_1 \subset H$, then from (4.46), (4.45) and Theorem 12 it follows that

$$\lim_{t\to 0} \frac{I_H(u+t\eta) - I_H(u)}{t} = (n+1)\left\{\int_{H_u} \eta[-\omega(w,de)+\psi_H(de)]\right\}. \qquad (4.48)$$

where $w = F_H(u)$.

From (4.48) and Theorem 8 it follows that the function

$$w_o(x) \in U(H,m_o,M_o) \subset W_H^+(\bar{G})$$

realizing the absolute minimum of $I_H(u)$ in $C_o^-(G)$ is a generalized solution of the Dirichlet problem

$$\begin{cases} \omega(w,e) = \psi_H(e) \\ w|_{\partial G} = 0 \,. \end{cases}$$

Thus the Theorem A of Section 2.5. is proved. This completes the proofs of all Theorems of the present paper.

REFERENCES.

[1] R. Courant, D. Hilbert, Methoden der Mathematischen Physik, Vol. II, Springer Verlag (1931) ; Methods of Math. Physics. Vol. II. Interscience Publ. (1975).

[2] I. Bakelman, Variational problem connected with Monge-Ampere equations. Dokl. Akad. Nauk USSR, 141, 5 (1961).

[3] I. Bakelman, Geometric methods of solution of elliptic equations. Edition Nauka. Moscow (1965). (Monograph).

[4] I. Bakelman, Generalized solutions of the Monge-Ampere equations. Dokl. Akad. Nauk USSR, 111, 6 (1957).

[5] I. Bakelman, The Dirichlet problem for equations of Monge-Ampere type and their n-dimensional analogues. Dokl. Akad. Nauk USSR. 126, 5 (1959).

[6] I. Bakelman, Equations with the Monge-Ampere operator. Proc. Fourth All-Union Math. Congress 1961. Leningrad. Vol. II. (1964).

[7] I. Bakelman, Generalized solutions of the Dirichlet problem for the n-dimensional elliptic equations. Universität Bonn, SFB 40 (1980) (preprint).

[8] I. Bakelman, Applications of the Monge-Ampere operators to the Dirichlet problem for quasilinear equations. Annals of Math. Studies, 102, Seminar on differential geometry. Princeton (1982).

[9] I. Bakelman, R-curvature, estimates and stability of solutions of the Dirichlet problem for elliptic equations. Journ. of diff. equations. 43, 1 (1982).

[10] A. Pogorelov, The n-dimensional Minkowski problem. Edition Nauka. Moscow (1975). (Monograph).

[11] S.Y. Cheng, S.T. Yau, On the regularity of the solution of the n-dimensional Minkowski problem. Comm. Pure and Appl. Math., 29 (1976).

[12] S.Y. Cheng, S.T. Yau, On the regularity of the solution of the Monge-Ampere equation det $\|u_{ij}\| = F(x,u)$. Comm. Pure and Appl. Math. 30 (1977).

[13] H. Minkowski, Volumen und Oberfläche. Math. Ann. 57 (1903).

[14] T. Bonnesen and W. Fenchel, Theorie der convexen Körper. Berlin (1934). New York (1938). (Monograph).

[15] H. Busemann, Convex surfaces. Interscience Publishers. Inc. New York. (1958). (Monograph).

[16] A. Alexandrov, On the theory of mixed volumes of convex bodies. Math. Sbornik 2,5; 2,6 (1937). 3,1 ; 3,2 (1938).

[17] W. Fenchel, W. Jessen, Mengenfunktionen und konvexe Körper. Danske Vid. Selsk. Math. Fys. Medd. 16,3 (1938).

Ilya J. BAKELMAN

Texas A&M University
COLLEGE STATION
Texas 77843

USA

M BEN-ARTZI

Some new results in the spectral and scattering theory of Stark-like Hamiltonians

I. INTRODUCTION

The Stark effect in Quantum mechanics deals with the application of a uniform external electric field to a Hydrogen atom. Using standard perturbation techniques (where the electric field is regarded as a perturbation of the basic Coulomb potential), physicists calculated the shifted energy levels corresponding to perturbed bound states (see [7]).
It turns out, however, that once an electric field, weak as it may be, is added to a Coulomb potential, there can be no eigenvalues (in the L^2 sense) and the above "calculated eigenvalues" are in fact "pseudo-eigenvalues", i.e., values associated with spectral concentration (see [6]). As a matter of fact, from the mathematical point of view it is more appropriate to view the Coulomb potential as a perturbation imposed on the electric field.

The Stark Hamiltonian in R^n is given by $-\Delta - \frac{1}{|x|} - Ex_1$. We shall discuss here a family of Hamiltonians (i.e., potential perturbations of the Laplacian) which includes the Stark case. Thus, our basic operator is of the form $H_0 = -\Delta + V_0(x_1)$ and the perturbed operator is $H = H_0 + V(x)$, where V depends in general on all coordinates $x = (x_1,...,x_n)$.

Let $\{E_0(\lambda)\}$ be the spectral family of projections associated with H_0. Recall that the subspace of absolute continuity with respect to H_0 consists of the set of all elements $u \in L^2(R^n)$ for which $(E_0(\lambda)u,u)$ is absolutely continuous with respect to the Lebesgue measure $d\lambda$ on the line. It is

Lecture given at the French-Israeli conference, March 22-27 (1982)

a closed subspace [6] which reduces H_0 . The projection on this subspace is denoted by P_0 .

For the pair of operators (H,H_0) we define the wave-operators by :

$$W_{\pm}(H,H_0) = \text{Strong} - \lim_{t \to \pm\infty} \exp(itH)\exp(-itH_0)P_0$$

when these limits exist.

We shall be concerned with the following two questions :

Q1 : Do $W_{\pm}(H,H_0)$ exist ?

Q2 : Is H absolutely continuous apart from a sequence of eigenvalues, and how are these eigenvalues distributed ?

Obviously, the answers depend on our assumptions on V_0, V . We list now our assumptions on V_0 , which will enable us to study Q1. In order to study Q2, it will be necessary to impose further restrictions on V_0 (see (III) below).

(A) $\quad V_0(t) \in L^2(R)_{loc}$, $V_0(t) = V_L(t) + V_S(t)$

(B) (i) $\quad V_L(t) \in C^{2k}(R)$

(ii) $\quad (\frac{d}{dt})^j V_L(t).(1+|V_L(t)|)^{-1} = O(|t|^{-j\delta})$ as $|t| \to \infty$

for $j = 1,\dots,2k$, with $\delta > 0$ and $2k\delta > 1$.

(iii) $\quad V_L(t) \to \mp\infty$ as $t \to \pm\infty$.

(iv) $\quad |V_L(t)| = O(t^2)$ as $|t| \to \infty$

(C) $\quad |V_S(t)|.(1+|V_L(t)|)^{-1/2} = O(|t|^{-1-\varepsilon})$ $\varepsilon > 0$ $|t| \to \infty$

Assumptions B(iii) (iv) can be relaxed [3] .

In particular, the case $V_{0,\alpha}(t) = -(\operatorname{sgn} t).|t|^{\alpha}$, $0 < \alpha \leq 2$ is included. We denote $H_{0,\alpha} = H_0 + V_{0,\alpha}$. We will write down the results obtained by specializing the general theorems to this case. Of course, $\alpha = 1$ is again the Stark case.

For the pure Stark case, Q1 was treated by Avron-Herbst [2] , who also proved the non-existence of eigenvalues for a certain class of perturbations. The case Q2 was treated by Herbst [5] and Yajima [9]. . For a one-dimensional model, with V_0 satisfying assumptions similar to ours (with $k = 1$) Q2 was treated by Rejto-Sinha [8] .

We study Q1 in Section (II). Proofs may be found in [3] . Q2 is studied in Section (III) (see [4] for proofs).

Observe that our operators can be shown to be essentially self-adjoint in the sense that the closures of their restrictions to smooth compactly supported functions are self-adjoint operators. These "realizations" are denoted by H_0 , H, etc.

II. EXISTENCE OF WAVE-OPERATORS

We formulate here two theorems about the existence of $W_{\pm}(H,H_0)$. In the first we impose pointwise estimates on the perturbation potential V, whereas the second assumes only that $V \in L^2(R^n)$, but $V_0(x_1)$ is further restricted.

For simplicity we assume that $V_0(t) \in C^{\infty}(R)$ and condition (B)(ii) is satisfied for all $j \geq 1$.

We set :

$$w(y) = \int_0^y (1+|V_L(t)|)^{-1/2} dt \qquad y \geq 0$$

Theorem 2.1. : *Assume that for some* $r > \frac{1}{2}$, $s > 0$

$$(1+y)^{-s} (1+|V_L(y)|)^{-1/4} w(y)^r \in L^2(0,\infty).$$

Suppose that for some $t_0 > 0$

$$|V(x_1,x')| \leq C(1+x_1)^{-s} (1+|x'|)^{r_1} \qquad x_1 \geq t_0$$

where $x = (x_1,x')$ *and* $0 \leq r_1 < r - \frac{1}{2}$

Suppose also that for some integer N

$$(1+|x|)^{-N} V(x) \in L^2(R^n)$$

Then $W_{\pm}(H,H_0)$ *exist, where* $H = H_0 + V$.

Theorem 2.2. : *Assume that* $V_L(y)$ *satisfies the following conditions :*

$$w(y)^2 \{|V_S(y)| (1+|V_L(y)|)^{-1/2} + (1+|V_L(y)|)^{-k+1/2} (1+y)^{-2k\delta}\} \in L^1(0,\infty) \quad (1)$$

for some positive integer k *and* $2k\delta > 1$.

for $\varepsilon > 0$, *set* $g_\varepsilon(y) = \sup_{s > \varepsilon y} (1+|V_L(s)|)^{-1/4}$. (2)

and assume that for every $\varepsilon > 0$, $y^{-1} g_\varepsilon(y) \in L^1(1,\infty)$.

Then, if $V(x) \in L^2(R^n)$, *the wave-operators* $W_{\pm}(H,H_0)$ *exist.* □

We now specialize our results to the case that

$$V_0(t) = V_{0,\alpha}(t) + V_S(t) = (-\text{sgn } t)\ |t|^{\alpha} + V_S(t)$$

where $0 < \alpha \leq 2$ and $V_S(t)$ is compactly supported square-integrable. We denote $H_0 \equiv H_{0,\alpha}$ in this case.

Theorem 2.3. : *Suppose for some integer* N, $(1+|x|)^{-N} V(x) \in L^2(R^n)$ *and that* $V = V_1 + V_2$ *where :*

$$V_1(x) \in L^2(R^n) .$$

$$|V_2(x_1,x')| \leq C(1+|x'|)^r (1+x_1)^{-s} \qquad x_1 \geq 0$$

where $r \geq 0$, $s > (r+1)(1 - \frac{\alpha}{2})$

Then $W_{\pm}(H,H_{0,\alpha})$ *exist.* □

The case $\alpha = 1$ yields the result of Avron-Herbst [2] .

Detailed proofs of the theorems are given in [3]. The idea is to use Cook's lemma that gives the following sufficient condition for the existence of $W_{\pm}$: There exists a fundamental set S such that $\psi \in S \Rightarrow |V.\exp(-itH_0)P_0\ \psi| \in L^1(R;dt)$. (See [6]). Thus, an eigenfunction expansion is established for H_0 and the asymptotic behavior of $\exp(-itH_0)\psi$ is studied. The above norm is then estimated by stationary phase methods.

III. SPECTRAL STRUCTURE OF H.

With V_0 satisfying the assumption in the Introduction it can be shown easily that H_0 is spectrally absolutely continuous and its spectrum consists of the entire real axis [3] .

Let $R(z)$, $E(\lambda)$ denote the resolvent and spectral projection, respectively, corresponding to N . The following formula is well-known [6] :

$$(E(\lambda_1,\lambda_2)u,u) = \frac{1}{\pi} \lim_{\varepsilon \to 0+} \int_{\lambda_1}^{\lambda_2} \operatorname{Im}(R(\mu+i\varepsilon)d\mu$$

Thus, suppose that $Y \subseteq L^2(R^n) \subseteq Z \subseteq Y^*$ are spaces such that $R(z)$ can be extended as a continuous function valued in $B(Y,Z)$ "down" to the interval (λ_1,λ_2) . If Y is dense in $L^2(R^n)$ it follows from the above formula that H is absolutely continuous in (λ_1,λ_2) . We then say that H has the "limiting absorption property".

We formulate this property first for H_0 . In order to do that, we need a further restriction on $V_L(t)$, namely :

(D) $V_L(t)$ is monotone non-increasing and for some $\varepsilon > 0$, $c > 0$,

$$|V_L(t)| \geq c|t|^{\varepsilon}$$

$$- V_L'(t) \geq c(1+|V_L(t)|)t^{-1} \qquad \text{as } t \to +\infty .$$

Let $\sigma,\tau \in R$. For $u \in C_0^\infty(R^n)$ we define the following norms :

$$\|u\|^2_{\tau,\sigma} = \int_{-\infty}^{0} \int_{R^{n-1}} (1+|x_1|)^{2\tau} (1+|V_L(x_1)|)^{-1} |u(x_1,x')|^2 \, dx_1 dx'$$

$$+ \int_0^\infty \int_{R^{n-1}} (1+|x_1|)^{2\sigma}(1+|V_L(x_1)|)^{-1/2}|u(x_1,x')|^2 \, dx_1 dx'$$

$$|||u|||^2_{\tau,\sigma} = \|\omega(x_1)u\|^2_{\tau,\sigma} + \|H_0 u\|^2_{\tau,\sigma}$$

where :

$$\omega(x_1) = \begin{cases} 1 + |V_L(x_1)| & x_1 < 0 \\ (1 + |V_L(x_1)|)^{1/2} & x_1 > 0 \end{cases}$$

We denote by $\Psi_{\tau,\sigma}$, $\Phi_{\tau,\sigma}$ the completions of $C_0^\infty(R_n)$ under $\| \ \|_{\tau,\sigma}$, $||| \ |||_{\tau,\sigma}$ respectively.

For a compact interval $I \subset R$ we set

$$\Omega(I) = \{z | \text{Re } z \in I, \quad 0 \leq \text{Im } z \leq 1\} \ .$$

<u>Theorem 3.1.</u> : *Let* $\sigma > \frac{1}{2}$ *and* $\tau > 1$. *The map*

$$z \to R_0(z) = (H_0 - z)^{-1} \qquad \text{Im } z \neq 0$$

can be extended as a continuous map defined on $\Omega(I)$ *, with values in* $B(\Psi_{\tau,\sigma}, \Phi_{\tau',-\sigma})$ *, equipped with the uniform operator topology ,* $\forall \tau' < \tau$.

Passing to the operator H , we use the resolvent equation :

$$R(z) = (I + R_0(z)V)^{-1} R_0(z) \ .$$

If $V : \Phi_{\tau',-\sigma} \to \Psi_{\tau,\sigma}$ is compact, it follows from the Fredholm-Riesz theory that $R(z)$ can be extended continuously to a neighborhood of $\lambda \in R$ if (-1) is not an eigenvalue of $R_0(\lambda)V$. It can be shown that the set of λ' s for which $I + R_0(\lambda)V$ is non-inversible is discrete and the corresponding eigen-functions are rapidly decaying. In the following theorem, we incorporate an explicit condition for the compactness of V .

<u>Theorem 3.2.</u> : *Let* $V(x) = V_1(x) + V_2(x)$ *where* $V_1 : H^2(R^n) \to L^2(R^n)$ *is compactly supported and compact, and* $V_2(x)$ *is bounded and satisfies*

$$|V_2(x_1, x')| \leq \begin{cases} C(1+|x_1|)^{-\varepsilon} (1+|V_L(x_1)|)(1+|x'|)^{-\varepsilon} & x_1 \leq 0 \\ & \qquad \varepsilon > 0 \\ C(1+x_1)^{-1-\varepsilon} (1+|V_L(x_1)|)^{1/2} (1+|x'|)^{-\varepsilon} & x_1 > 0 \end{cases}$$

Let $\frac{1}{2} < \sigma < \frac{1}{2} + \frac{\varepsilon}{2}$, $\tau > 1$. *Then* $R(z)$ *can be extended as a continuous function to* $\Omega(I)$, *with values in* $B(\Psi_{\tau,\sigma}, \Phi_{\tau',-\sigma})$, *except possibly for a finite number of eigenvalues of* H *in* I. *The eigenfunctions corresponding to these eigenvalues are rapidly decaying in the sense that they belong to* $\Phi_{\eta,\theta}$ *for all* $\eta, \theta \in R$.

Corollary 3.3. : *Under the conditions of the previous theorem,* H *is spectrally absolutely continuous, except possibly for a discrete sequence of eigenvalues. Furthermore, the wave-operators* $W_{\pm}(H,H_0)$ *exist and are complete.*

It is easy to see how to implement those conditions for the potential $V_{0,\alpha}$. In particular, for the Stark case $\alpha = 1$, the perturbation potential is allowed to blow up as $x_1 \to -\infty$, while it is assumed to fall off like $(1+x_1)^{-1/2-\varepsilon}$ as $x_1 \to +\infty$.

Proofs of these results are given in [4]. Note that, as first observed by Agmon [1] for the short-range case, the discreteness of the set of eigenvalues is closely related to the rapid decay of the corresponding eigenfunctions. Indeed, let G be a subspace of $L^2(R^n)$, with norm $\| \ \|_G$, and suppose that the inclusion map $i : G \to L^2(R^n)$ is compact. Suppose further that if u is an eigenfunction of H, $(H-\lambda)u = 0$, $\lambda \in I$, then $u \in G$ and $\| u \|_G \leq C \| u \|_{L^2(R^n)}$, where C depends on I only. Then any sequence of such (normalized) eigenfunctions is precompact and hence finite (by the self-adjointness of H).

By Rellich's compactness theorem, such a space G is furnished by a subspace of $L^2(R^n)$ which is the product of $H^2(R^n)$ and a function that falls off at infinity.

We are thus reduced to establishing a "division" theorem, namely, proving that solutions of

$$R_0(\lambda)Vu = -u \qquad\qquad u \in \Phi_{\tau',-\sigma}$$

are rapidly decaying.

Incorporating the result of [2] about the non-existence of eigenvalues we obtain :

The spectrum of $-\Delta - \frac{1}{|x|} - Ex_1$ is $(-\infty,\infty)$ and is purely absolutely continuous.

REFERENCES

[1] S. Agmon, Spectral properties of Schrödinger operators and scattering theory, Ann. Scuola Norm. Sup. Pisa 2(1975), 151-218.

[2] J.E. Avron and I.W. Herbst, Spectral and scattering theory of Schrödinger operators related to the Stark effect, Comm. Math. Phys. 52 (1977), 239-254.

[3] M. Ben-Artzi, An application of asymptotic techniques to certain problems of spectral and scattering theory of Stark-Like Hamiltonians, Trans. A.M.S., to appear.

[4] M. Ben-Artzi, Unitary equivalence and scattering theory for Stark-Like Hamiltonians, preprint.

[5] I.W. Herbst, Unitary equivalence of Stark Hamiltonians, Math. Z. 155 (1977), 55-70.

[6] T. Kato, "Perturbation theory for linear operators", Berlin, Heidelberg, New-York; Springer 1976.

[7] E. Merzbacher, "Quantum Mechanics", New-York, Wiley 1961.

[8] P.A. Rejto and K. Sinha, Absolute continuity for a 1-dimensional model of the Stark-Hamiltonian, Helv. Phys. Acta 49 (1976), 389-413.

[9] K. Yajima, Spectral and scattering theory for Schrödinger operators with Stark effect. J. Fac. Sci. Univ. Tokyo 1A Math. 26 (1979), 377-389.

Matania BEN-ARTZI

Department of Mathematics
Technion - Israel

32000 - HAIFA

ISRAEL

P M FITZPATRICK

Global multidimensional existence results for *m*-parameter compact vector fields

In this lecture I will describe some recent joint work with I. Massabò and J. Pejsachowicz ([3], [4], [5], [6]).

1. INTRODUCTION

As is well known, the degree theory of Leray and Schauder has played a central role in the resolution of problems in nonlinear differential and integral equations which may be formulated as equations of the form

$$h(x) = 0, \ x \in U, \tag{1.1}$$

where U is an open subset of a Banach space, X, $h^{-1}(0) \cap U$ is compact, and $h(x) = x - H(x)$, for $x \in U$, H being a compact mapping.

Now there are a number of problems in nonlinear differential and integral equations which have a natural formulation as m-parameter family of problems of type (1.1). Specifically, they appear as

$$f(\lambda,x) = 0, \quad (\lambda,x) \in \mathcal{O},$$

where $\mathcal{O}$ is an open subset of $\mathbb{R}^m \times X$, and $f(\lambda,x) = x - F(\lambda,x)$, for $(\lambda,x) \in \mathcal{O}$, F being a compact mapping.

In the case when m = 1, for equations of type (1.2) let us recall two types of phenomena to whose investigation classical Leray-Schauder degree techniques have successfully been applied.

First, there is the problem of *continuation*, where one has information about (1.2) at some fixed value of the parameter, $\lambda = \lambda_0$; namely, that the degree of the section of f, f_{λ_0}, over the slice, $\mathcal{O}_{\lambda_0} \equiv \{x | (\lambda_0, x) \in \mathcal{O}\}$, is nonzero. From this, one concludes that there is a component of solutions of equation (1.2) which intersects $\mathcal{O}_{\lambda_0}$ and either becomes unbounded or approaches the boundary of $\mathcal{O}$, $\partial\mathcal{O}$. This is the Leray-Schauder continuation principle (see [2] and [8]).

Secondly, there is the problem of bifurcation. Here, one knows that $T \equiv \mathcal{O} \cap \{\mathbb{R} \times \{0\}\} \subseteq f^{-1}(0)$, and one wishes to describe the manner in which $f^{-1}(0)\backslash T$ branches away from T. If there are two parameter values, $\underline{\lambda}$ and $\overline{\lambda}$, such that $\{(\lambda,0) | \underline{\lambda} \leq \lambda \leq \overline{\lambda}\} \subseteq \mathcal{O}$, and such that neither $(\underline{\lambda},0)$ nor $(\overline{\lambda},0)$ lies in $\overline{f^{-1}(0)\backslash T}$, then the Leray-Schauder index, $\text{ind}(f_\lambda, 0)$, is well defined at $\lambda = \underline{\lambda}$ and $\lambda = \overline{\lambda}$. If these two indices differ, then there is a connected component of $f^{-1}(0)\backslash T$, C, with $\overline{C} \cap [\underline{\lambda},\overline{\lambda}] \neq \emptyset$, and either C is unbounded, $\overline{C} \cap \partial\mathcal{O} \neq \emptyset$, or $\overline{C} \cap \{(\lambda,0) | \lambda \notin [\underline{\lambda},\overline{\lambda}]\} \neq \emptyset$. This is the global bifurcation theorem of Rabinowitz ([10]).

Now then $m > 1$ there is an obvious way in which one can glean some information about (1.2) from the case just described : for each one-dimensional curve, Γ, in $\mathcal{O} \cap \{\mathbb{R}^m \times \{0\}\}$, let $\mathcal{O}_\Gamma = \{(\lambda,x) \in \mathcal{O} | \lambda \in \Gamma\}$ and $f_\Gamma = f|_{\mathcal{O}_\Gamma}$, and then use the above results to describe $(f_\Gamma)^{-1}(0)$. [In using these results over a general curve, Γ, a certain amount of care has to be taken since the conclusions are not purely topological : being unbounded is a property not necessarily preserved under C^1 changes of variable].

This procedure is somewhat unsatisfactory. In the case when f is continuously Fréchet differentiable, with $(\lambda_0,x_0) \in \mathcal{O}$ such that $f(\lambda_0,x_0) = 0$ and $\frac{\partial f}{\partial x}(\lambda_0,x_0)$ is invertible, it would be as if one concluded that on each line, Γ, in $\mathbb{R}^m \times \{0\}$, passing through (λ_0,x_0), $(f_\Gamma)^{-1}(0)$, in a neighborhood of (λ_0,x_0), is a C^1 curve. While this is certainly so, there is a much more precise conclusion : these curves fit together, and $f^{-1}(0)$, in a neighborhood of (λ_0,x_0), is an m-manifold.

Reasoning analogously, it is not unreasonable to expect that the one-dimensional slices of solutions, in the continuation problem and in the bifurcation problem fit together and give a global, "m-dimensional" conclusion. We will describe a theorem, whose conclusions give a precise description of how these slices mesh together. The theorem we shall describe has, as its only assumption, the existence of a map $g : \mathcal{O} \to \mathbb{R}^m$, which we call a *complement* for f, such that the Leray-Schauder degree of the map (g,f) on $\mathcal{O}$ is nonzero, where $(g,f)(\lambda,x) = (g(\lambda,x),f(\lambda,x))$, for $(\lambda,x) \in \mathcal{O}$.

In Section 2 we discuss the idea of complementing maps. In Section 3 we describe our main results. The final section briefly deals with the application of our continuation result to nonlinear problems which may be formulated as an equation for the zero of a compact perturbation of a linear mapping which is Fredholm, of positive index.

2. COMPLEMENTING MAPS

Observe that if we identify X with $\{0\} \times X \subseteq \mathbb{R}^m \times X$, then the mapping f, in equation (1.2) is a compact perturbation of the identity on $\mathbb{R}^m \times X$. However, its range is contained in a subspace of $\mathbb{R}^m \times X$ of codimension m, so that the Leray-Schauder degree, when it is defined, is necessarily zero.

In order to use degree theory on the whole equation we will fill in these missing dimensions.

<u>Definition 2.1</u> ([3]) : Let f, F, $\mathcal{O}$, m and X be as defined in the introduction. A continuous mapping $g : \mathcal{O} \to \mathbb{R}^m$, which maps bounded sets into bounded sets, will be called a complement for $f : \mathcal{O} \to X$ provided that $\{(\lambda,x) | f(\lambda,x) = 0,\ g(\lambda,x) = 0,\ (\lambda,x) \in \mathcal{O}\}$ is compact and

$$\deg((g,f),\mathcal{O},0) \neq 0 ,$$

where $(g,f)(\lambda,x) = (g(\lambda,x),f(\lambda,x))$, whenever $(\lambda,x) \in \mathcal{O}$.

Observe that in the above definition we make no assumption about $(g,f)^{-1}(0) \cap \partial\mathcal{O}$: in fact, (g,f) is not assumed to be defined on $\partial\mathcal{O}$.

The following simple consequence of the product formula for the degree shows how continuation problems fall into the context of complementing maps.

<u>Proposition 3.1</u> : ([3]). *Let* f, F, $\mathcal{O}$, X *and* m *be as in the introduction. For* $\lambda_0 \in \mathbb{R}^m$, *let* $\mathcal{O}_{\lambda_0} = \{x \in X | (\lambda_0,x) \in \mathcal{O}\}$ *and let* $f_{\lambda_0} : \mathcal{O}_{\lambda_0} \to X$ *be defined by* $f_{\lambda_0}(x) = f(\lambda_0,x)$. *If* $\deg(f_{\lambda_0},\mathcal{O}_{\lambda_0},0) \neq 0$, *then* $f : \mathcal{O} \to X$ *is complemented by* $g : \mathcal{O} \to \mathbb{R}^m$ *defined by* $g(\lambda,x) = \lambda - \lambda_0$.

In order to apply the notion of complementing mapping to bifurcation problems, one needs a less obvious complement (see the remarks following the bifurcation corollary).

As the conclusion of our main theorem will assert, if a map $f : \mathcal{O} \to X$ has a complement, $g : \mathcal{O} \to \mathbb{R}^m$, then $f^{-1}(0)$ must have a very definite structure. The converse question immediately comes to mind. What assumptions of $f^{-1}(0)$ imply that there exists a complement, g, for f ? The next proposition gives one possible answer.

It is easy to see that in the case that the f in equation (1.2) is linear then it can be complemented on a neighborhood of the origin if and only if it is onto. On the other hand, recall that for a mapping $\varphi : M \to \mathbb{R}^k$, where M is a C^1 manifold and φ is C^1, 0 is called a regular value of φ if $d\varphi(x) : T_M(x) \to \mathbb{R}^k$ is onto for each $x \in \varphi^{-1}(0) \cap M$. Consequently, the following result gives necessary and sufficient conditions for the existence of a complement provided that the linearization of f has a complement at each point of $f^{-1}(0)$.

<u>Proposition 2.2</u> : ([3]). *Let $\mathcal{O}$ be an open, bounded subset of $\mathbb{R}^{m+k}$ whose boundary is smooth. Let $f : \overline{\mathcal{O}} \to \mathbb{R}^k$ be C^1 and such that 0 is a regular value both of $f : \mathcal{O} \to \mathbb{R}^k$ and of $f : \partial\mathcal{O} \to \mathbb{R}^k$. Then there exists a complement $g : \mathcal{O} \to \mathbb{R}^m$ for $f : \mathcal{O} \to \mathbb{R}^k$ if and only if $f^{-1}(0) \cap \mathcal{O} \neq \emptyset$ and $f^{-1}(0) \cap \partial\mathcal{O} \neq \emptyset$.*

3. STATEMENT OF THE MAIN RESULTS.

For topological spaces there are a number of generalizations of the linear concept of dimension. Foremost among these are the inductive dimension, due to Poincaré, and the covering dimension, due to Lebesgue. When the topological space is a separable metric space, these concepts of dimension coincide. All of the sets whose dimension we consider here are such spaces, so our use of the word "dimension" is unambiguous.

When Z is a topological space and $z \in Z$, the space Z will be said to have dimension at least k at z if each neighborhood, in Z, of z has dimension at least k.

A classic reference on dimension theory is Hurewicz and Wallman ([7]).

For a pair of normal topological spaces, (A,B), with $A \subseteq B$, $\check{H}^m(A,B)$ will denote the m-th Čech cohomology group with integral coefficients. A map of pairs $h : (A,B) \to (\mathbb{R}^m, \mathbb{R}^m \setminus \{0\})$ will be called cohomologically nontrivial provided that the induced homomorphism

$$h^* : \check{H}^m(A,B) \to \check{H}^m(\mathbb{R}^m, \mathbb{R}^m \setminus \{0\})$$

is nontrivial.

Our basic result is the following

Theorem : *Let* X *be a Banach space,* m *be a positive integer and* $\mathcal{O} \subseteq \mathbb{R}^m \times X$ *be open. Let* $f : \mathcal{O} \to X$ *be an* m*-parameter compact vector field. Assume there exists a complement,* $g : \mathcal{O} \to \mathbb{R}^m$*, for* $f : \mathcal{O} \to X$*. Then there exists a closed connected subset,* C*, of* $f^{-1}(0)$*, which intersects* $g^{-1}(0)$*, which has dimension at least* m *at each point, and*

$$\left\{ \begin{array}{l} \textit{for each subset, } D \textit{ of } C\textit{, with } g^{-1}(0) \cap D = \emptyset \textit{ and } C \setminus D \\ \textit{compact, the mapping } g : (C,D) \to (\mathbb{R}^m, \mathbb{R}^m \setminus \{0\}) \textit{ is} \\ \textit{cohomologically nontrivial.} \end{array} \right. \tag{3.1}$$

In particular, $g^{-1}(0) \cap C \neq \emptyset$*, and either* C *is unbounded or* $C \cap \partial\mathcal{O} \neq \emptyset$*.*

Corollary : *(Global Continuation). Let* m, X, $\mathcal{O}$ *and* f *be as in the theorem. Suppose* $\lambda_0 \in \mathbb{R}^m$ *is such that* $\deg(f_{\lambda_0}, \mathcal{O}_{\lambda_0}, 0) \neq 0$. *Then there exists a closed connected subset,* C, *of* $f^{-1}(0)$, *whose dimension at each point is at least* m, $C \cap \mathcal{O}_{\lambda_0} \neq \emptyset$, *and either* C *is unbounded or* $\bar{C} \cap \partial\mathcal{O} \neq \emptyset$. *If, in addition, both* f *and* g *are defined on* $\bar{\mathcal{O}}$, *with* $f^{-1}(0) \cap g^{-1}(0) \cap \partial\mathcal{O} = \emptyset$, *then if* C *is bounded* $\bar{C} \cap \partial\mathcal{O}$ *has dimension at least* $m - 1$, *when* $m > 1$, *while* $\bar{C} \cap \partial\mathcal{O}$ *contains at least two points when* $m = 1$. *Finally, if* $\dim(\bar{C} \cap \partial\mathcal{O}) < m - 1$ *and the following a-priori bound holds,*

when $\{(\lambda_n, x_n)\} \subseteq f^{-1}(0)$ *and* $\{\lambda_n\}$ *is bounded, then* $\{x_n\}$ *is bounded,*

it follows that for each $\lambda \in \mathbb{R}^m$ *there exists* $x \in X$ *with* $(\lambda, x) \in C$.

Corollary : *(Global Bifurcation). Let* m, X, $\mathcal{O}$ *and* f *be as in the Theorem. Let* $T = \mathcal{O} \cap \{\mathbb{R}^m \times \{0\}\}$, *and let* $\Gamma \equiv \{(\lambda_1, \lambda_2, \ldots, \lambda_m) \in \mathbb{R}^m \mid \lambda_2 = \ldots = \lambda_m = 0\}$. *Suppose that* $\underline{\lambda}$ *and* $\bar{\lambda}$ *lie on* Γ, *and that* $\Gamma \subseteq T$. *Assume that neither* $\underline{\lambda}$ *nor* $\bar{\lambda}$ *are bifurcation points of* f *and that* $\operatorname{ind}(f_{\underline{\lambda}}, 0) \neq \operatorname{ind}(f_{\bar{\lambda}}, 0)$. *Then there exists a closed connected subset,* C, *of* $f^{-1}(0) \backslash T$, *whose dimension at each point is at least* m, $C \cap [\underline{\lambda}, \bar{\lambda}] \neq \emptyset$, *and at least one of the following three properties hold* :

$$\begin{aligned} &(i) && C \text{ is unbounded,} \\ &(ii) && \bar{C} \cap \partial\mathcal{O} \neq \emptyset, \\ &(iii) && \bar{C} \cap \{\Gamma \backslash [\underline{\lambda}, \bar{\lambda}]\} \neq \emptyset. \end{aligned} \tag{3.2}$$

The proof of the theorem is rather long : briefly, one translates the assumption that g complements f into a statement about the action of (g,f)

in homology ; this, in turn, yields information about the action of (g,f) in cohomology ; by means of the cup-product one then gets, by using the continuity properties of cohomology, the requisite conclusions about f in cohomology.

The first part of the continuation corollary follows immediately from Proposition 2.1 and the theorem. To obtain the more detailed conclusions available when f and g are defined on $\bar{\mathcal{O}}$ one has to carefully exploit property (3.1).

The proof of the bifurcation corollary is based on the following construction : For $r > 0$ let $\varphi : \mathbb{R} \to [0,r]$ be continuous, equal r^2 on $[\underline{\lambda} - \frac{r}{2}, \bar{\lambda} + \frac{r}{2}]$, and have $\varphi^{-1}(0) = [\underline{\lambda} - r, \bar{\lambda} + r]$. Now define $g : \mathbb{R}^m \times X \to \mathbb{R}^m$ by $g(\lambda,x) = (\|x\|^2 - \varphi(\lambda_1), \lambda_2, \ldots, \lambda_m)$. One shows that if r is small, g complements f on $U \equiv \mathcal{O} \backslash \{\mathbb{R}^m \times \{0\}\}$, and so one can apply the theorem, with U playing the role of $\mathcal{O}$. The global properties of C follow from (3.1).

The theorem is proven in [5] ; an earlier version, under the additional hypotheses that f and g are defined on $\bar{\mathcal{O}}$ and $f^{-1}(0) \cap g^{-1}(0) \cap \partial\mathcal{O} = \emptyset$, appears in [3].

When m = 1, the continuation corollary is a very slight refinement of the Leray-Schauder continuation principle (see [2] and [8]). In [3] and in [9] one finds m-dimensional versions of the continuation principle which are now encompassed by our corollary.

When m = 1, the bifurcation corollary is essentially the Rabinowitz global bifurcation theorem ([10]). When m > 1, the bifurcation corollary is a refinement of a recent result of Alexander and Antman ([1]) : our assumptions on $\mathcal{O}$ are weaker, and our description of the bifurcation set somewhat more analytically precise, than the corresponding result in [1].

4. CONCLUSION

The application of our results to bifurcation and nonlinear eigenvalue problems, when there are explicit parameters appearing in the equation, will not be discussed here. Rather, we briefly outline how the continuation results can be applied to existence problems of the following form : let U and V be Banach spaces, with $L \in \mathcal{L}(U,V)$ a mapping which is Fredholm of index $m > 0$; let $G : U \to V$ be a compact, not necessarily linear, mapping ; consider the equation

$$L(u) + G(u) = 0, \quad u \in U . \tag{4.1}$$

The above L could represent a 2k-th order elliptic differential operator, $k > 1$, which is neither symmetric nor satisfies Gärding's inequality ; or, L could be a singular integral operator ; or, $L : U \to W$ could be Fredholm of index 0, $V = R(L)$ having codimension m, and $G(U) \subseteq R(L)$. The last situation can occur when L represents an ordinary differential equation (see [4]).

It is easy to see that equation (4.1) may be reformulated as an equation of the type (1.2), where the parameter space is an m-dimensional subspace of N(L). Indeed, since L is Fredholm of index m, we may choose a closed subspace, U_1, of U, a closed subspace, V_2, of V, and subspaces, H_1 and H_2, of N(L) such that

$$\begin{cases} U = N(L) \oplus U_1 \\ N(L) = H_1 \oplus H_2, \ \dim(H_1) = m, \text{ and} \\ V = R(L) \oplus V_1 . \end{cases} \tag{4.2}$$

Now let $X = H_2 \oplus H_1$, and let $C : X \to V_1$ be the composition of the projection of X onto H_1, along U_1, together with a linear bijection from H_1 to V_1. Then $L + C : X \to V$ is a bijection and clearly

$$L(u) + G(u) = 0,\ u \in U,\ u = \lambda + x,\ \lambda \in H_1,\ x \in X \tag{4.3}$$

if and only if

$$x = F(\lambda,x),\quad \lambda \in H_1,\ x \in X\ , \tag{4.4}$$

where $F(\lambda,x) = [(L+C)|_X]^{-1}[G(\lambda+x) - C(x)]$.

The above mapping, F, is clearly compact.

Let us apply the bifurcation corollary with $\lambda_0 = 0$ and $\mathcal{O}$ an open subset of U. In order to do so it is necessary to calculate

$$\deg(I-g, \mathcal{O} \cap X, 0)\ , \tag{4.5}$$

where $g = F|_{\mathcal{O}\cap X}$.

But the calculation of (4.5) is exactly the type of calculation on which most Landesman-Lazer type nonlinearity theorems are based : one wants to calculate the degree of a mapping which is a nonlinear perturbation of a linear map which is Fredholm of index 0. In [3] there are asymptotic assumptions on the nonlinearity which allow one to compute (4.5), and hence apply the continuation results.

Finally, suppose that the L in (4.1) is surjective, that $G(0) = 0$ and that there is an $r > 0$ such that $L + G : \{u \in U_1 \mid \|u\| \le r\} \to V$ is one-to-one.

Clearly, if G is linear, then the set of solutions of equation (4.1) which lie on $\{u \in U, \|u\| = r\}$ has dimension m - 1. The continuation corollary implies that when G is nonlinear one can conclude that the dimension of the set of solutions of equation (4.1) that lie on $\{u \in U, \|u\| = r\}$ is at least m = 1. In [6] we outlined how this result could be applied to a nonlinear Riemann-Hilbert problem to get multiplicity results under very weak assumptions on the nonlinearities.

REFERENCES

[1] J.C. Alexander and S.S. Antman, Global and local behavior of bifurcating multidimensional continua of solutions of multiparameter eigenvalue problems. Archive for Rational Mechanics and Analysis, Vol. 76, N°4 (1981), 339-355.

[2] F.E. Browder, On continuity of fixed points under deformation of continuous mappings. Summa Brasil Math., 4 (1960), 183-190.

[3] P.M. Fitzpatrick, I. Massabò and J. Pejsachowicz, On the covering dimension of the set of solutions of some nonlinear equations. Trans. Amer. Math. Soc., to appear.

[4] P.M. Fitzpatrick, I. Massabò and J. Pejsachowicz, A global description of the periodic solutions of some ordinary differential equations, submitted.

[5] P.M. Fitzpatrick, I. Massabò and J. Pejsachowicz, Global several-parameter bifurcation and continuation theorems : a unified approach via complementing maps, to appear in J. of the London Math. Soc.

[6] P.M. Fitzpatrick, I. Massabo and J. Pejsachowicz, Complementing maps, continuation and bifurcation, to appear in Math. Ann.

[7] W. Hurewicz and H. Wallman, Dimension Theory. Princeton University Press, 1948.

[8] J. Leray and J. Schauder, Topologie et équations fonctionnelles. Ann. Sci. Ecole Norm. Sup., 51 (1934), 45-78.

[9] I. Massabò and J. Pejsachowicz, On the connectivity properties of the solution set of parametrized families of compact vector fields. J. Funct. Anal., to appear.

[10] P.H. Rabinowitz, Some global results for nonlinear eigenvalue problems. J. Funct. Anal., 7 (1971), 487-513.

P. M. FITZPATRICK

University of Maryland
Division of Math. & Phys. Sciences &
Engineering
COLLEGE PARK
MARYLAND 20742
U.S.A.

S KAMIN

Elliptic singular perturbation problems with turning points

Let Ω be an open domain in R^m with smooth boundary $\partial\Omega$. We consider the solutions of the Dirichlet problem

$$\begin{cases} \varepsilon\left(\sum_{i,j=1}^{m} a_{ij}(x)u_{x_i x_j} + \sum_{i=1}^{m} a_i(x)u_{x_i}\right) + \sum_{i=1}^{m} b_i(x)u_{x_i} + c(x)u = f, \\ \qquad \text{in } \Omega, \end{cases} \tag{1}$$

$$u_{|\partial\Omega} = \psi(x) \tag{2}$$

where $x = (x_1, x_2, \ldots, x_m) \in R^m$ and

$$\sum_{i,j=1}^{m} a_{ij}\, \xi_i\, \xi_j \geq \nu \sum_{i=1}^{m} \xi_i^2 , \quad (\nu > 0) .$$

We shall study the asymptotic behavior of the solutions $u_\varepsilon(x)$ of (1), (2) as $\varepsilon \to 0$. Problems of this kind arise when one studies the effect of small random perturbations on dynamical systems. Consider the deterministic flow

$$\frac{dx_i}{dt} = b_i(x) \tag{3}$$

where $x(t) = (x_1(t), \ldots, x_m(t))$ is a trajectory. Suppose that the system is perturbed by white noise. This gives rise to a motion $x_\varepsilon(t)$, which is a solution of the stochastic differential equation

$$dx_\varepsilon(t) = b(x_\varepsilon)dt + \varepsilon\sigma(x_\varepsilon)dw(t) , \tag{4}$$

This text corresponds to a lecture given at the French - Israeli conference held during the week of March 22-27.

where $\sigma(x)$ is a diffusion matrix and $w(t)$ an n-dimensional Wiener process (Brownian motion), ε being a small number. Our problem is to study the limiting behavior of a process of this kind. That is, we wish to determine to what extent a small diffusion may change the behavior pattern of the process, or, in other words, how stable is the motion under small random perturbations.

Putting

$$a_{ij} = \frac{1}{2}(\sigma\sigma^*)_{ij} , \qquad a_{ij}(x) = c(x) = f(x) = 0 ,$$

we can write the solution of (1), (2) in the form $u_\varepsilon(x) = M_x\psi(x_\varepsilon(\tau))$, where τ is the first exit time of the trajectory $x_\varepsilon(t)$ from Ω and M_x the expectation given $x_\varepsilon(0) = x$. One sees from this probabilistic representation of the solution that singular perturbation problems can be studied by both analytical and probabilistic methods. Both methods are in fact used. As will be shown in the sequel, some results are obtained by probabilistic methods, some by analytical methods, and still others by a combination of both. It would seem that the probabilistic approach is extremely useful in preliminary, intuitive work. For more delicate structural investigations, however, the analytical method is evidently more convenient.

It is well known that the behavior of the trajectories of system (3) is of decisive value in the problem at hand. Note that these trajectories are the characteristic curves of the degenerate equation

$$\sum b_i U_{x_i} + c(x)U = f . \tag{5}$$

We begin with the "regular case", studied in the earliest research on this topic (Levinson [1]). The term "regular" means here that the vector field

$\bar{b} = (b_1,\ldots,b_m)$ does not have singular points or limit cycles in Ω; every characteristic curve that enters the domain leaves it after a finite time. The typical behavior pattern of the trajectories is shown in Fig. 1.

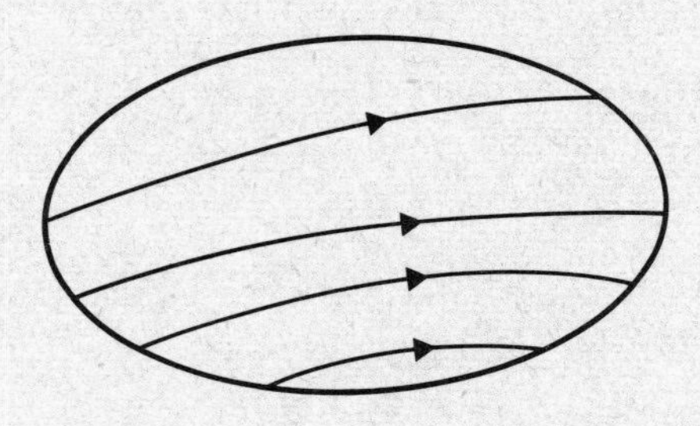

Fig. 1.

Suppose that a characteristic curve starting at a point $x \in \Omega$ leaves the domain at some point $P \in \partial\Omega$. A small diffusion may alter the motion of the particle, but only slightly. With probability close to unity the particle will leave the domain at P.

Levinson constructed an expansion of the solution for this case. Let $n = (n_1(x),\ldots,n_m(x))$ denote the outer normal to $\partial\Omega$ and $(b,n) = \sum b_i n_i$. We shall use the notation $\partial\Omega^+$ $(\partial\Omega^-)$ for the part of $\partial\Omega$ on which $(b,n) > 0$ $((b,n) < 0)$. Let $U(x)$ be a solution of equation (5) such that $U(x)|_{\partial\Omega^+} = \psi(x)$. Then, as proved in [1],

$$u_\varepsilon(x) = U(x) + h(x)e^{-g(x)/\varepsilon} + R_\varepsilon(x)\ , \tag{6}$$

where $R_\varepsilon(x) \to 0$ as $\varepsilon \to 0$,

$$g(x)|_{\partial\Omega^-} = 0\ , \tag{7}$$

$g(x) > 0$ inside Ω, and

$$h(x)\Big|_{\partial\Omega^-} = \psi(x) - U(x) . \tag{8}$$

The function

$$v_\varepsilon(x) = h(x)e^{-g(x)/\varepsilon} \tag{9}$$

appearing in (6) plays the role of a boundary layer near $\partial\Omega^-$. It describes the nonuniformity of the convergence of $u_\varepsilon(x)$ near $\partial\Omega^-$.

To find $g(x)$ and $h(x)$, we insert (9) into (1) and equate the coefficients of ε^{-1} and ε^0 to zero. This yields two equations, which will be written down here only for the special case

$$L_\varepsilon u = \varepsilon\Delta u + \sum b_i u_{x_i} = 0 . \tag{10}$$

In that case,

$$\sum g_{x_i}^2 = \sum b_i g_{x_i} , \tag{11}$$

$$\sum (b_i - 2g_{x_i})h_{x_i} - h\Delta g = 0 . \tag{12}$$

We first find $g(x)$ as a solution of the first-order nonlinear equation (11) satisfying the initial-value condition (7). We next solve equation (12) given (8). The functions $g(x)$ and $h(x)$ are thus defined in some neighborhood of $\partial\Omega^-$.

We now extend $h(x)$ to a smooth function vanishing outside some strip Ω_0 near $\partial\Omega^-$.

<u>Remark 1</u>. The following relationship may be derived from (6) - (9) , (11) :

$$\varepsilon \frac{\partial u}{\partial n}\Big|_{\partial\Omega^-} \approx - (\psi - U)(b,n)\Big|_{\partial\Omega^-} , \tag{13}$$

where $\partial/\partial n$ denotes the normal derivative and $\approx$ means that the difference between the right and left-hand sides of (13) tends to zero as $\varepsilon \to 0$.

Remark 2. Suppose that the field is not regular, but that a characteristic curve ℓ_0 through some point $x_0 \in \Omega$ leaves Ω at a point $P_0 \in \partial\Omega$. Then the field is regular near ℓ_0 and we can use Levinson's result in a suitable "tube" containing ℓ_0. This gives $u_\varepsilon(x_0) \to U(x_0)$, where $U(x)$ is the solution of (5) satisfying the condition $U(P_0) = \psi(P_0)$.

We now remove the restriction on the position of the trajectories of system (3). Under the additional assumption that the coefficient $c(x)$ in equation (1) is negative in $\overline{\Omega}$,

$$c(x) \leq - c_0 < 0$$

the problem of $\lim_{\varepsilon\to 0} u_\varepsilon(x)$ was solved in [2].

Without condition (14), however, the problem is much more difficult and, as I shall now show even for equation (10), is not solved yet. The degenerate equation in this case is

$$\sum b_i U_{x_i} = 0 . \tag{15}$$

Clearly, the number of different configurations of the field of characteristic curves for equation (15) is tremendous; each of these configurations evidently requires special treatment.

We now consider in greater detail the case in which the vector field $\overline{b} = (b_1,\ldots,b_m)$ has one singular point, usually termed a *turning point*. Let $x = 0$ be a turning point of attracting type (Fig. 2).

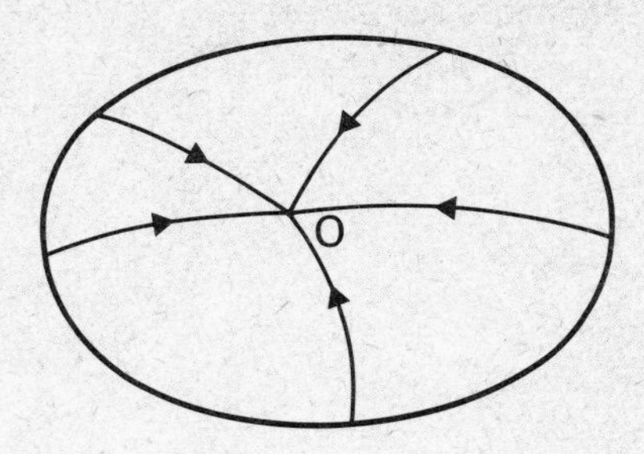

Fig. 2.

It is natural to assume that the leading term in the asymptotic expansion of u_ε as $\varepsilon \to 0$ should be a constant throughout the domain. Ventsel and Freidlin [3] proved by probabilistic methods that indeed $u_\varepsilon(x) \to$ const. in a certain special case. To state their result, we define a functional

$$I(\phi) = I_{T_1 T_2}(\phi) = \int_{T_1}^{T_2} \sum_{i,j} \left(\frac{d\phi_i}{dt} - b_i(\phi)\right)^2 dt \tag{16}$$

where $\phi(t) = (\phi_1(t),\ldots,\phi_m(t))$ is a smooth curve in R^m. Next we define

$$V(x) = \inf_{\phi \in H_x} I(\phi) \tag{17}$$

where H_x denotes the set of functions $\phi(t)$ defined on all possible intervals $[T_1,T_2]$ such that $\phi(T_1) = 0$, $\phi(T_2) = x$. The function $V(x)$ is continuous and satisfies a Lipschitz condition.

Let us assume that there is a unique point $x_0 \in \partial\Omega$ at which $V(x)$ achieves a minimum on $\partial\Omega$. Under this assumption, Ventsel and Freidlin prove that for $x \in \Omega$

$$\lim_{\varepsilon \to 0} U_\varepsilon(x) = \psi(x_0)$$

In probabilistic terms, this means that for small ε, with probability close to unity, the particle leaves the domain at x_0. It was proved in [3] that if the equation

$$\sum F_{x_i}^2 = \sum b_i F_{x_i} \tag{18}$$

has a smooth solution with $F(0) = 0, F(x) < 0$ for $|x| > 0$, then $V(x) = -4F(x)$. Note that this equation is identical to (11).

We now turn to analytical methods for the determination of $\lim u_\varepsilon(x)$, considering the same problem with one attracting point : Is it true that $u_\varepsilon(x) \to \text{const} = C_0$ and, if so, what is the value of C_0 ?.
Grasman and Matkowsky [4] and Matkowsky and Schuss [5], [6] recently proposed a formal asymptotic method for determing the leading term in the asymptotic expansion of the solution. They assume that $u_\varepsilon(x) \approx C_0$ inside Ω and

$$\varepsilon \frac{du_\varepsilon}{\partial n} \approx -(\psi - C_0)(b,n)$$

on $\partial\Omega$; then, using Green's identity, they calculate C_0. The approach adopted here is somewhat different from those of [4] - [6].

Our first step is to check the truth of the relationship $\lim u_\varepsilon(x) = \text{const}$. As an attempt to do this, it was proved in [7] (Theorem 2) that, under certain natural assumptions,

$$u_\varepsilon(x) = u_\varepsilon(0) + v_\varepsilon(x) - u_\varepsilon(0)v_\varepsilon^{(1)}(x) + o(1) \qquad (\varepsilon \to 0) \tag{19}$$

where $v_\varepsilon(x)$ and $v_\varepsilon^{(1)}(x)$ are boundary layer functions, as defined in (9), with $v_\varepsilon\big|_{\partial\Omega} = \psi$ and $v_\varepsilon^{(1)}\big|_{\partial\Omega} = 1$. The assumptions under which this is proved concern conditions for the attracting point and the regularity of the coefficients. By the maximum principle, $|u_\varepsilon(0)| \leq \max|\psi|$ and so it follows from (19) that there exists a subsequence $\{u_{\varepsilon_i}\}$ such that $u_{\varepsilon_i}(x) \to C_o$ for $x \in \Omega$.

The next step is to calculate C_o . A theorem of Ventsel and Freidlin yields the value of C_o under the assumption that $V(x)$ achieves its minimum on $\partial\Omega$ at a unique point. In order to determine C_o by analytical methods, we shall adopt another assumption, to be specified below.

We shall look for a solution in the entire domain, in the form

$$u_\varepsilon(x) - u_\varepsilon(0) = H_\varepsilon(x)e^{-G(x)/\varepsilon} \tag{20}$$

where $G(x)$ is independent of ε . Inserting (20) into (10), we obtain

$$\begin{aligned} L_\varepsilon(H_\varepsilon e^{-G/\varepsilon}) = e^{-G/\varepsilon}[\frac{1}{\varepsilon} H(\sum G_{x_i}^2 - \sum b_i G_{x_i}) + \varepsilon\Delta H + \\ + \sum (b_i - 2G_{x_i})H_{x_i} - H\Delta G] = 0 \end{aligned} \tag{21}$$

Condition (2) becomes

$$H_\varepsilon e^{-G/\varepsilon}|_{\partial\Omega} = \psi - u_\varepsilon(0)|_{\partial\Omega} \tag{22}$$

We shall assume that the equation

$$\sum G_{x_i}^2 = \sum b_i G_{x_i} \tag{23}$$

has a non-constant solution in Ω . Then $H_\varepsilon(x)$ must satisfy the equation

$$\tilde{L}_\varepsilon H = \varepsilon\Delta H + \sum(b_i - 2G_{x_i})H_{x_i} - H\Delta G = 0 \tag{24}$$

and also condition (22). Moreover, it follows from (19) that

$$\varepsilon\frac{\partial H}{\partial n} e^{-G/\varepsilon} \approx [\psi - u_\varepsilon(0)][\frac{\partial G}{\partial n} - (b,n)] \tag{25}$$

We must now use condition (25) to determine $\lim u_\varepsilon(0)$. We first consider the so-called potential case; i.e. we assume the existence of a function $G(x)$ such that

$$b_i = \frac{\partial G}{\partial x_i}, \qquad \forall\, i. \tag{26}$$

Then

$$\tilde{L}_\varepsilon H = \varepsilon \Delta H - \sum \frac{\partial}{\partial x_i}(b_i H) = 0 \tag{26'}$$

Integrating this equality over Ω and using (22) and (25), we obtain

$$\int_{\partial\Omega} [\psi - u_\varepsilon(0)]\,[\frac{\partial G}{\partial n} - 2(b,n)]e^{G/\varepsilon}dS = o(1)\int_{\partial\Omega} e^{G/\varepsilon}dS \tag{27}$$

The main contributions to the integrals in (27) come from the points on $\partial\Omega$ at which $G(x)$ achieves its maximum. At these points $\partial G/\partial n = (b,n)$, and therefore

$$\lim_{\varepsilon\to 0} u_\varepsilon(0) = \lim_{\varepsilon\to 0} \frac{\int_{\partial\Omega}\psi(b,n)e^{G/\varepsilon}dS}{\int_{\partial\Omega}(b,n)e^{G/\varepsilon}dS} \tag{28}$$

This formula yields C_o for the case of a potential field (i.e., (26)).

Next suppose there is a solution $G(x)$ of equation (23) such that $G(x) \neq$ const.

Remark 3. It should be noted that equation (23) is the same as (11) and (18); hence, if $G(x)$ exists, then $V(x) = -4G(x)$. Moreover, the characteristic curves of equation (12) are the projections of those of equation (11) onto x-space, and also extremals of the functional $I(\phi)$.

Let w be a solution of the equation adjoint to (24) ;

$$\tilde{L}^*_\varepsilon w = \varepsilon\Delta w + \sum (2G_{x_i} - b_i)w_{x_i} + (\Delta G - \text{div } b)w = 0\ , \tag{29}$$

and let $W(x)$ be a smooth solution of

$$\sum (2G_{x_i} - b_i) W_{x_i} + (\Delta G - \text{div } b) W = 0 .$$

As proved in [7], if $w_\varepsilon(x)$ is a solution of equation (29) such that

$$w_\varepsilon(x) - W(x)|_{\partial\Omega} = 0 , \qquad (30)$$

then

$$\varepsilon \frac{\partial w_\varepsilon}{\partial n} \Big|_{\partial\Omega} \to 0 , \qquad \text{as } \varepsilon \to 0 . \qquad (31)$$

Integrating the identity

$$\int\int_\Omega (w\tilde{L}H - H\tilde{L}^* w) dx = 0$$

by parts and using (30), (31), (25), (20), we get

$$\lim_{\varepsilon\to 0} u_\varepsilon(0) = \lim_{\varepsilon\to 0} \frac{\int_{\partial\Omega} e^{G/\varepsilon} W(b,n)\psi dS}{\int_{\partial\Omega} e^{G/\varepsilon} W(b,n) dS} \qquad (32)$$

This formula yields C_0 for the case in which there exists a smooth solution $G(x)$ in Ω. This condition does not appear in the Ventsel-Freidlin theorem. On the other hand, formula (32) is valid without any restrictions being made on the set of minimum points of $G(x)$ on $\partial\Omega$. The following question remains open : what is the limit of $u_\varepsilon(x)$ (if it exists !) in case $V(x)$ is not smooth but the set of its minimum points on $\partial\Omega$ contains more than one point?

We conclude with a few remarks about other configurations. If the domain contains a single turning point of saddle type, then the limit for every $x \notin \ell$ (see Fig. 3.) is known from the case treated by Levinson. In fact, for $x \in \ell$ we have

$$u_\varepsilon(x) \to \frac{1}{2} [\psi(P_1) + \psi(P_2)] . \qquad (33)$$

This formula was proved by Kifer [8] , using probabilistic methods, and by Kamin [9] , using analytical methods.

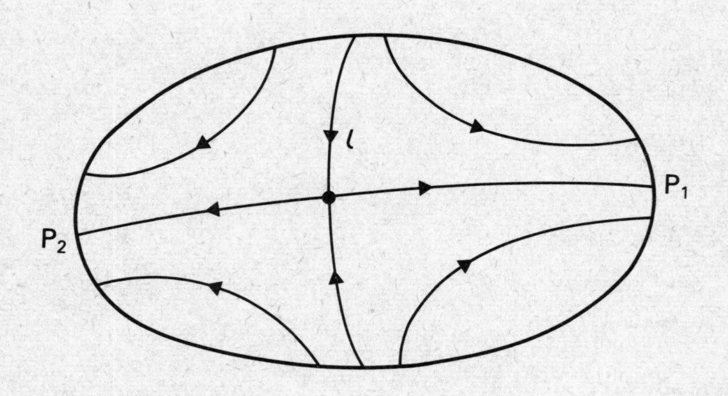

Fig. 3.

The case in which the domain contains three turning points was studied in [6] and [9] (see Fig. 4.).

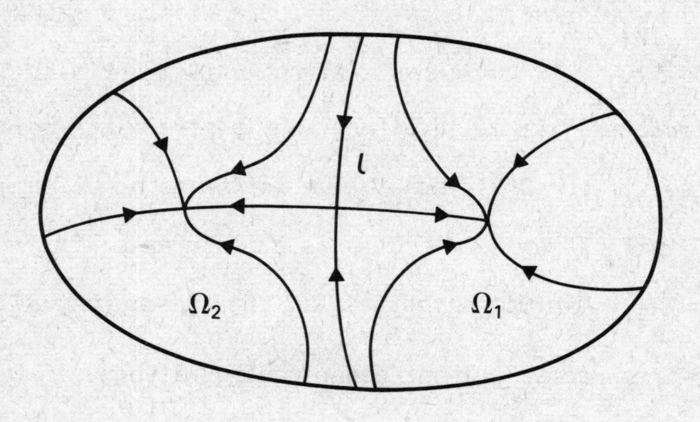

Fig. 4.

It is proved in [9] that

$$\lim u_\varepsilon(x) = \begin{cases} C_1 & \text{if} \quad x \in \Omega_1 , \\ C_2 & \text{if} \quad x \in \Omega_2 , \end{cases}$$

For this case one must construct, in addition to a boundary layer near $\partial\Omega$, an interior layer near ℓ . The existence of $G(x)$ is also essential in [9].

If the domain contains one turning point, of repelling type (Fig. 5.), then

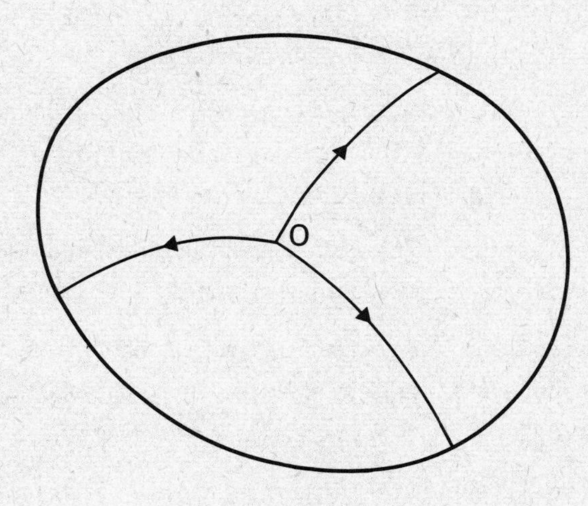

Fig. 5.

$\lim u_\varepsilon(x)$, $x \neq 0$, follows from (6), and the only open question is the existence of $\lim u_\varepsilon(0)$. As shown recently by a student of Kifer, this limit need not exist; in other words, there exist subsequences ε_i which yield different limits of $u_{\varepsilon_i}(0)$.

REFERENCES

[1] N. Levinson, The first boundary value problem for $\varepsilon\Delta u + A(x,y)u_x + B(x,y)u_y + C(x,y)u = D(x,y)$ for small ε. Ann. of Math. (2) 51 (1950), 428-445.

[2] S. Kamenomostskaya, The first boundary value problem for elliptic equations containing a small parameter. Izv. Akad. Nauk. U.R.S.R. ser. Mat. 19 (1955), 345-360.

[3] A.S. Ventcel, M.I. Freidlin, On small perturbations of dynamical systems, Uspehi Mat. Nauk 25 (1970), n°1 (151),3-55;Russian Math. Surveys 25 (1970) n°1, 1-56.

[4] J. Grasman, B. Matkowsky, A variational approach to boundary value problems and resonance for ordinary and partial differential equations, Siam, J. Appl. Math, 32 (1977), 588-597.

[5] B.J. Matkowsky, Z. Schuss, The exit problem for randomly perturbed dynamical systems, Siam, J. Appl. Math., 35, 3, 1977, 365-382.

[6] Z. Schuss, B. Matkowsky, The exit problem : A new approach to diffusion across potential barriers, Siam J. Appl. Math., 35, 3, (1979), 604-623.

[7] S. Kamin, Elliptic perturbation of a first order operator with a singular point of attracting type, Ind. Univ. Math. J. 27, 6 (1978). 935-952.

[8] K. Kifer, The exit problem for small random perturbations of dynamical systems with a hyperbolic fixed point, Isr. J. Math. 40 (1981), 1, 74-96.

[9] S. Kamin, On singular perturbation problems with several turning points; Ind. Univ. Math. J.31, 6 (1982), 819-841.

S. KAMIN

School of Mathematical Sciences
Tel Aviv University
TEL AVIV
ISRAEL

P L LIONS

Optimal control of diffusion processes and Hamilton–Jacobi Bellman equations. III—Regularity of the optimal cost function

INTRODUCTION

This paper deals with the same topic as Part 1 (P.L. Lions [36]) and Part 2 (P.L. Lions [37]) namely the *optimal control of diffusion processes* and the associated *Hamilton-Jacobi-Bellman equations*. We will therefore keep the notations (and main assumptions) of Parts 1-2 that we recall very briefly now.

An admissible system $\mathcal{A}$ is the collection of : i) a probability space (Ω,F,F_t,P) with the usual properties, ii) a Brownian motion B_t in $\mathbb{R}^m$ adapted to F_t , iii) a progressively measurable process α_t - *the control* - with values in a compact set of a given separable metric space A. The state of the system is given by the solution of the following stochastic differential equation :

$$dX_t = \sigma(X_t,\alpha_t)\, dB_t + b(X_t,\alpha_t)dt \ , \qquad X_o = x \tag{1}$$

where $\sigma(x,\alpha)$, $b(x,\alpha)$ are given vector-fields satisfying smoothness and boundedness assumptions that are detailed below (section I-1). We then introduce for each admissible system $\mathcal{A}$ a cost function $J(x,\mathcal{A})$ defined by :

$$J(x;\mathcal{A}) = E \int_0^\tau f(X_t,\alpha_t)\exp(- \int_0^t c(X_s,\alpha_s)ds) \tag{2}$$

where $f(x,\alpha)$, $c(x,\alpha)$ are real-valued functionssatisfying assumptions detailed in section I-1 and τ is the first exit time from $\bar{\mathcal{O}}$ of the process X_t . Finally $\mathcal{O}$ is a given smooth open set in $\mathbb{R}^N$. The optimal cost function to be determined is then :

$$u(x) = \inf_{\mathcal{A}} J(x,\mathcal{A}) \tag{3}$$

where the infimum is taken over all possible admissible systems.

From the heuristic dynamic progamming principle due to R. Bellman [3] , it is to be expected that u satisfies the following Hamilton-Jacobi-Bellman equation (HJB in short) :

$$\sup_{\alpha\in A} \{A_\alpha u(x) - f(x,\alpha)\} = 0 \quad \text{in } \mathcal{O} \tag{4}$$

where A_α is the second-order elliptic (possibly degenerate) operator given by :

$$A_\alpha = - a_{ij}(x,\alpha)\, \partial_{ij} - b_i(x,\alpha)\partial_i + c(x,\alpha)$$

and $a = \frac{1}{2}\sigma\sigma^T$. Here and below we use the usual convention of implicit summation for repeated indices. In addition u should vanish at least on some part Γ_0 of $\Gamma = \partial\mathcal{O}$. As we explained in [36], this derivation is rigorous if we know that $u \in C^2(\mathcal{O})$ (see W.H. Fleming and R. Rishel[17], N.V. Krylov[24]) Unfortunately this regularity is false in general.

In Parts 1-2, we developed various notions that make (4) meaningful : in [36] we proved that u is the maximum subsolution of (4) and in [37] we showed that, if $u \in C(\bar{\mathcal{O}})$, u is the unique solution in viscosity sense of (4) (with, say, zero boundary values); the notion of viscosity solutions of fully nonlinear elliptic equations being an extension of the one introduced by M.G. Crandall and P.L. Lions [14], [12] (see also M.G. Crandall, L.C. Evans and P.L. Lions [10], P.L. Lions [58]) for first-order Hamilton-Jacobi equations. Furthermore let us point out that in [36], we showed also general continuity results for u .

Another approach to the derivation of (4) consists in proving first regularity results for u which make it possible to check (4). This kind of approach was initiated by N.V. Krylov [25],[26],[24] in the case when $\mathcal{O} = \mathbb{R}^N$ (see also so M. Nisio [64], A. Bensoussan and J.L. Lions [4] for slightly simplified versions of Krylov's proofs) and in this special case ($\mathcal{O} = \mathbb{R}^N$) the most general results were obtained independently by P.L. Lions [38],[39] and N.V. Krylov [27],[28] . However in the general case Krylov approach is made more difficult because of the presence in the cost functions of τ which depends on x.

We complete here this approach and thus solve this difficulty. Let us give one example of our results (Theorem I-1 in section I-2, proved in section I-3) : assume that we have

$$\exists \nu > 0 \ , \ \forall (x,\alpha) \in \Gamma \times A \qquad a_{ij}(x,\alpha) n_i(x) n_j(x) \geq \nu \tag{5}$$

where $n(x)$ is the unit outward normal to $\Gamma = \partial\mathcal{O}$ at the point x . Then, there exists λ_1 which can be estimated as follows :

$$\lambda_1 = C_N \sup_{(x,\alpha)\in\mathbb{R}^N\times A} \{\sup_{i,j,k} \left| \frac{\partial\sigma_{ij}}{\partial x_k}(x,\alpha)\right|^2 + \sup_{i,k} \left|\frac{\partial b_i}{\partial x_k}(x,\alpha)\right|\}, \tag{6}$$

such that, if we have : $\lambda = \inf_{(x,\alpha)\in\mathbb{R}^N\times A} c(x,\alpha) > \lambda_1$, the following holds :

$$\exists C > 0 \ ; \ \forall \xi \in \mathbb{R}^N \ , \ |\xi| = 1 \ ; \quad \partial_\xi^2 u \leq C \quad \text{in} \quad \mathcal{D}'(\mathcal{O}) \tag{7}$$

we will say that u is *semi-concave* in $\mathcal{O}$; and in addition : $u \in W^{1,\infty}(\mathcal{O})$. This estimate is in fact a regularity result : indeed we will see that (7) , combined with the fact that u is a subsolution of (4), immediately yields

$$\sup_{\alpha\in A} ||A_\alpha u||_{L^\infty(\mathcal{O})} < +\infty. \tag{8}$$

Moreover if locally some nondegeneracy assumption - called in P.L. Lions [39] complementary nondegeneracy condition - holds, one obtains a local $W^{2,\infty}$ regularity from (7)-(8).
In turn, since u is a viscosity solution of (4), the results proved in Part 2 [37] imply that (4) then holds (a.e. in $\mathcal{O}$).

We show in section I.3 that the crucial estimate (7) may be obtained under much more general assumptions on σ, b, Γ and also for control problems with a pay-off when the state X_t exists from $\bar{\mathcal{O}}$. In section I.3, we explain the role of the assumption that λ is larger than λ_1 and consider in particular the case of uniformly nondegenerate matrices a.

Section II is devoted to related results : the local regularity of u (section II.1); regularity results for optimal stopping and time dependent problems (section II.2); different boundary conditions (section II.3); the case of unbounded coefficients in α (section II.4); the relations with Monge-Ampère equations (section II.5) or fully nonlinear degenerate elliptic equations with a convex nonlinearity (section II.6).

In section III, we consider uniqueness questions : a very particular case of ourresults yields : let $\tilde{u} \in W^{1,\infty}(\mathcal{O})$ satisfy (8), (4) (a.e. in $\mathcal{O}$) and :

$$\exists C > 0 , \quad \Delta\tilde{u} \le C \quad \text{in } \mathcal{D}'(\mathcal{O}) \tag{7'}$$

such a function is called *semi-super-harmonic*, then $\tilde{u}$ is the optimal cost function for a control problem similar to (1)-(3) (with an additional pay-off when the state exists from $\mathcal{O}$). In particular there is at most one such $\tilde{u}$ with prescribed boundary conditions. In section III.3 we give extensions of this typical result, while section III.4 is devoted to the uniqueness question for time-dependent problems.

Let us finally mention that the results stated and proved here were announced in P.L. Lions [40], [41].[42],[43]. The regularity results extend those obtained by N.V. Krylov (in the case $\mathcal{O} = \mathbb{R}^N$) [24]-[28]; P.L. Lions [38], [39]; M.V. Safonov [68],[69] (in the case $\mathcal{O} \subset \mathbb{R}^2$); P.L. Lions and J.L. Menaldi [59],[60]. The case when all the matrices are uniformly nondegenerate has been considered by many authors and was finally settled by P.L. Lions [55], [56], L.C. Evans and P.L. Lions[16]- cf. for more references section I.5 below.

The regularity results considered in sections I - II are proved by combinations of probabilistic and p.d.e. arguments : we already explained that the Krylov approach valid in $\mathbb{R}^N$ fails in the above general case because of the boundary (and the presence of τ). To solve this difficulty, after appropriate smoothing of the problem, we first obtain by p.d.e. methods an estimate on $\|D^2u\|_{L^\infty(\Gamma)}$; this estimate enables us to show (7) by an extension of an argument due to M.I. Freidlin [18] based upon the dynamic progamming principle - this argument was also used in P.L. Lions and J.L. Menaldi [59],[60]; M.V. Safonov [68],[69] and was first announced in P.L. Lions [40].
Finally the uniqueness results proved in section III are obtained with stochastic arguments very similar to the ones used in P.L. Lions [39] (for similar purposes) - once a localization procedure already used in Parts 1-2 [36], [37] has been applied.

Let us finally conclude this introduction by pointing out that we will not consider here ergodic control problems or stochastic control problems with state constraints, even if for those problems regularity and uniqueness results very similar to those proved here can be obtained. The topics will be studied in future publications - see also J.M. Lasry and P.L. Lions [34], P.L. Lions [57].

CONTENTS

I - REGULARITY RESULTS

I.1. Notations and assumptions

We will assume that the coefficients σ, b, c, f satisfy :

$$\begin{cases} \sup_{\alpha\in A} \|\phi(.,\alpha)\|_{W^{2,\infty}(\mathbb{R}^N)} < \infty, \quad \phi(x,.) \in C(A) \ , \ \forall\, x \in \mathbb{R}^N \\ \\ \text{for all } \ \phi = \sigma_{ij}\ (1\leq i\leq n,\ 1<j\leq m)\ ,\ = b_k\ (1\leq k\leq N)\ ,\ f,c\ ; \end{cases} \tag{9}$$

$$\inf\,\{c(x,\alpha)/\, x \in \mathbb{R}^N,\ \alpha \in A\} \ = \ \lambda > 0\ . \tag{10}$$

Regularity (9) insures that (1) has a unique solution and (10) insures that (2) has a meaning even if $\tau = +\infty$: recall that $\tau = \inf(t \geq 0,\ X_t \notin \bar{\mathcal{O}})$ and we set $\tau = +\infty$ if $X_t \in \bar{\mathcal{O}}$ for all $t \geq 0$.

From simple considerations on stochastic differential equations, one deduces that if $\phi(.,x)$ belongs to $C_b^2(\mathbb{R}^N)$ (for $\phi = \sigma_{ij}, b_i$) then the solution X_t of (1) depends smoothly on the initial point x and, more precisely, one can differentiate X_t twice with respect to x (in L^p sense for example - cf. Gihman and Skorohod for example [21]) and furthermore there exists C_N (depending only on N) such that setting

$$\lambda_1 = C_N \sup_{x\in\mathbb{R}^N,\,\alpha\in A} \{ \sup_{i,j,k} \left|\frac{\partial\sigma_{ij}}{\partial x_k}\right|^2 + \sup_{i,k} \left|\frac{\partial b_i}{\partial x_k}\right| \} \tag{6}$$

then we have :

$$
\begin{cases}
E\,[|\frac{\partial X_t}{\partial x_i}|^4] \le C^{2\lambda_1 t} & \forall i,\ \forall\ ,\ \forall t \ge 0 \\
E\,[|\frac{\partial^2 X_t}{\partial x_i \partial x_j}|^2] \le K(e^{2\lambda_1 t} - 1) & \forall i,\ \forall\ ,\ \forall t \ge 0 \\
E\,\{e^{-2\lambda_1 \theta}\{|\frac{\partial^2 X_\theta}{\partial x_i \partial x_j}|^2 + |\frac{\partial X_\theta}{\partial x_i}|^4\}\} \le K & \forall i,j,\ \forall \mathcal{A},\ \forall \theta \text{ bounded stopping time}
\end{cases} \quad (11)
$$

where K is some positive constant depending only on a bound on

$$\sup_{\alpha \in A} \|\phi(\cdot,\alpha)\|_{W^{2,\infty}(\mathbb{R}^N)} \quad \text{for } \phi = \sigma_{ij}, b_i .$$

From (11) and an easy approximation argument, one deduces that if (9) is satisfied and if $\lambda > \lambda_1$ then for all stopping time θ

$$v(x) = E\,\{\int_0^\theta f(X_t,\alpha_t)\exp(-\int_0^t c(X_s,\alpha_s)ds)dt\}$$

and

$$w(x) = E\{z(X_\theta)\exp(-\int_0^\theta c(X_t,\alpha_t)dt)\}$$

are functions which belong to $W^{2,\infty}(\mathbb{R}^N)$ as soon as $z \in W^{2,\infty}(\mathbb{R}^N)$.

This fact will be crucial in the following.

Let us finally point out that we can take (looking closely at the proofs below) for example :

$$\lambda_1 = \sup_{\substack{\alpha,x \\ |\xi| = 1}} \{2|\partial_i \sigma^T \xi_i|^2 + \mathrm{Tr}(\partial_i \sigma\ \partial_i \sigma T) + 2(\partial_i b^j)\xi^i \xi^j\}$$

I.2. Main result

In sections I.2 - I.3 - I.4 , $\mathcal{O}$ will always be a bounded smooth open set in $\mathbb{R}^N$.

Theorem I.1 : *Under assumptions* (9) *and*

$$\exists \nu > 0 , \quad \forall (x,\alpha) \in \Gamma \times A , \quad a_{ij}(x,\alpha) n_i(x)\, n_j(x) \geq \nu > 0 \tag{5}$$

and if $\lambda > \lambda_1$ *then* u *satisfies :* $u \in W^{1,\infty}(\mathcal{O})$, $u = 0$ *on* Γ and

$$\exists C > 0 , \ \forall \xi \in \mathbb{R}^N : |\xi| = 1 , \quad \partial_\xi^2 u \leq C \ \textit{in} \ \mathcal{D}'(\mathcal{O}) \tag{7}$$

Remark I.1 : The fact that $u \in W^{1,\infty}(\mathcal{O})$, $u = 0$ on Γ is a consequence of Part 1 [36] . We will also sketch briefly the proof of this fact in section I.3. The main new estimate is (7).

Remark I.2 : We will see in section I.4 similar results where (5) is relaxed, and the necessity of $\lambda > \lambda_1$ is discussed in sections I.3-4.

Applying Corollary I.1 of Part 2 [37] , we immediately deduce the following result:

Theorem I.2 : *Under the assumptions of Theorem* I.1.; $u \in W^{1,\infty}(\mathcal{O})$, $u = 0$ *on* Γ, u *satisfies* (7) *and :*

$$\sup_{\alpha \in A} \|A_\alpha u\|_{L^\infty(\mathcal{O})} < +\infty; \tag{8}$$

$$\sup_{\alpha \in A} (A_\alpha u(x) - f(x,\alpha)) = 0 \quad \textit{a.e. in } \mathcal{O} . \tag{4}$$

Let us recall that the proof of (4), given in [37] follows easily from the regularity implied by (7) and from the notion of viscosity solutions. And (8) also follows from (7) : let us explain how (8) is deduced from (7) and the following inequalities always satisfied by the optimal cost function :

$$A_\alpha u \leq f(.,\alpha) \quad \text{in} \quad \mathcal{D}'(\mathcal{O}) .$$

Since we have to argue pointwise, we first need to regularize u : we skip this technicality detailed in P.L. Lions [39] . Then let $x_o \in \mathcal{O}$, $\alpha \in A$: rotating the axis we may assume that $D^2u(x_o)$ is diagonal, then the above inequality yields :

$$A_\alpha u(x_o) = - \sum_i a_{ii}(x_o,\alpha)\, \partial_i^2 u(x_o) - \sum_i b_i(x_o,\alpha)\, \partial_i u(x_o) +$$

$$+ c(x_o,\alpha)u(x_o) \leq f(x_o,\alpha) \leq C ;$$

on the other hand $u \in W^{1,\infty}(\mathcal{O})$ and u satisfies (7), therefore :

$$A_\alpha u(x_o) \geq - C \operatorname{Tr} a(x_o,\alpha) - C \geq - C$$

and we obtain (8) . □

But as it is seen in the following result, (7) implies more information than merely (8) :

Corollary I.1 : *Under the assumptions of Theorem* I.1 *and if we assume that there exists an open set* $\omega \subset \mathcal{O}$, *an integer* $p \subset \{1,\dots,N\}$ *and a constant* $\nu > 0$ *such that :*

$$\left\{ \begin{array}{l} \forall\, x \in \omega ,\ \exists\, n \geq 1;\ \exists\, \alpha_1,\dots,\alpha_n \in A ,\ \exists\, \theta_1,\dots,\theta_n \in]0,1[\textit{ with } \sum_i \theta_i = 1 \\ \textit{such that} \\ \forall\, \xi \in \mathbb{R}^N,\ \sum_{i=1}^n \theta_i\, a_{k\ell}(x,\alpha_i)\xi_k\xi_\ell \geq \nu \sum_{j=1}^p \xi_j^2 \end{array} \right. \qquad (12)$$

then $\partial_{ij} u \in L^\infty(\omega)$ *for* $1 \le i, j \le p$.

This condition (12) in [38] is called complementary nondegeneracy condition (in the case $p = N$) . To prove the result we go back to the above proof and we first remark that we have :

$$- \sum_{i,j=1}^{p} a_{ij}(x_o,\alpha)\partial_{ij}u(x_o) \le C + \sum_{i \text{ or } j \ge p+1} a_{ij}(x_o,\alpha)\partial_{ij}u(x_o)$$

and since u satisfies (7), we finally obtain, rotating the axis as above :

$$- \sum_{i=1}^{p} a_{ii}(x_o,\alpha)\partial_i^2 u(x_o) \le C \quad .$$

Now, because of (12), we have : $0 < \nu \le a_{ii}(x_o,\alpha) \le C_o$ and thus we deduce for each $i \in \{1,\ldots,p\}$:

$$- \partial_i^2 u(x_o) \le \{C + \sum_{\substack{j=1 \\ j \ne i}}^{p} a_{jj}(x_o,\alpha)\partial_j^2 u(x_o)\} \, \nu^{-1} \le C$$

and we conclude using again (7) : $|\partial_i^2 u(x_o)| \le C$, $\forall\ 1 \le i \le p$ recall that $\partial_{ij} u(x_o) = 0$ if $1 \le i,j \le p$, $i \ne j$.

Remark I.3 : More generally if there exist an open set $\omega \subset \mathcal{O}$, a vector-field ξ Lipschitz and satisfying : $|\xi(x)| = 1$ in ω and a constant $\nu > 0$ such that :

$$\forall\ x \in \omega\ ; \quad \exists n > 1\ , \quad \exists\ \alpha_1,\ldots,\alpha_n \in A; \ \exists \theta_1,\ldots,\theta_n \in]0,1[\ \text{ with } \sum_i \theta_i = 1$$

$$\text{such that : } \sum_{i=1}^{n} \theta_i \, a_{k\ell}(x,\alpha_i)\xi_k(x)\xi_\ell(x) \ge \nu$$

then $\partial_\xi^2 u \in L^\infty(\omega)$.

We may now begin the proof of Theorem I.1. : we give in this section a few preliminary remarks to the proof given in section I.3. We first remark that, since all bounds below will depend only on ν, and on a bound on $\sup_{\alpha\in A} \|\phi(.,\alpha)\|_{W^{2,\infty}(\mathbb{R}^N)}$ for $\phi = \sigma_{ij}, b_i$, we may assume that without loss of generality σ,b,c,f are smooth, say $C^4(\mathbb{R}^N)$ uniformly in $\alpha\in A$. Indeed no difficulty arises in an approximation argument since (5) insures good passages to the limit for τ. Next, to simplify notations we will always take $c \equiv \lambda$. Finally let $\varepsilon > 0$, we introduce the $2N\times m$ matrix

$$\sigma^\varepsilon(x,\alpha) = (\sigma(x,\alpha),\varepsilon I_N)$$

and denote by X_t^ε, $J^\varepsilon(x,\mathcal{A})$, $u^\varepsilon(x)$ the quantities corresponding to σ^ε,b,f,c as X_t,J,u correspond to σ,b,f,c. Of course all assumptions satisfied by σ still hold for σ^ε : in particular (5) holds uniformly and the matrices $a^\varepsilon(x,\alpha) = \frac{1}{2}\sigma^\varepsilon(x,\alpha)\,\sigma^\varepsilon(x,\alpha)^T = a(x,\alpha) + \frac{\varepsilon^2}{2} I_N$ are uniformly definite positive. We claim that Theorem I.1 is proved if we prove uniform bounds for u^ε (for $\varepsilon \in]0,1[$) : indeed if we let ε go to 0, since we have : $E[\sup_{[0,T]} |X_t^\varepsilon - X_t|] \underset{\varepsilon\to 0_+}{\to} 0$, (for all $T<\infty$), (5) then implies easily : $\sup_{\mathcal{A},x} E[|\tau_\varepsilon\wedge T - \tau\wedge T|] \underset{\varepsilon\to 0_+}{\to} 0$ (for all $T<\infty$) and thus : u^ε converges uniformly towards u.

Therefore in the next section, we will assume :

$$\exists\, \nu > 0\ ,\ \forall (x,\alpha) \in \bar{\mathcal{O}}\times A\ ,\ a(x,\alpha) \geq \nu I_N$$

(in the sense of symmetric matrices) and we will show that the bounds on u only depend on ν in (5) (and on $\sup_{\alpha\in A} \|\phi(.,\alpha)\|_{W^{2,\infty}(\mathbb{R}^N)}$ for $\phi = \sigma_{ij},b_i$) .

Since A is separable there exists a dense sequence $(\alpha_j)_{j\geq 1}$ in A and if we denote by $A^n = \{\alpha_j / 1 \leq j \leq n\}$ and by u^n the associated optimal cost function, it is easily deduced from N.V. Krylov [24] and the results and methods of Part 1 [36] that u^n converges uniformly towards u as n goes to ∞. Therefore without loss of generality we may assume that $A = \{\alpha_1,\ldots,\alpha_n\}$ and we will denote indifferently A_i or A_{α_i}, f^i or $f(\cdot,\alpha_i)\ldots$

Finally to prove Theorem I.1, we will need to make some manipulation which performed upon u would require the apriori knowledge of some regularity of u. To do a rigorous proof, we consider the solution $(u^1,\ldots,u^n)$ of a system of QVI (quasi-variational inequalities) : this system was first considered by L.C. Evans and A. Friedman [15] - see also P.L. Lions and J.L. Menaldi [59], Belbas and S. Lenhart [2] . We fix $k \in (0,1]$, we consider $(u^1,\ldots,u^n)$ solution of :

$$\text{(QVI)}\quad A_i u^i \le f^i,\ \ u^i \le k+u^{i+1},\ (A_i u^i - f^i)(u^i - k - u^{i+1}) = 0 \text{ in } \mathcal{O};\ \ u^i = 0 \text{ on } \Gamma$$

where we set : $u^{n+1} \equiv u^1$ $(n+1 \equiv 1)$.

Because of the non-degeneracy and the smoothness assumptions, it is easy to show - cf. the above references or apply the arguments of the next section to the iterative scheme :

$$A_i u_p^i \le f^i,\ \ u_p^i \le k+u_{p-1}^{i+1},\ (A_i u_p^i - f^i)(u_p^i - k - u_{p-1}^{i+1}) = 0 \text{ in }\ ;\ u_p^i = 0 \text{ on } \Gamma$$

and let $p \to \infty$ - that there exists a unique solution $(u^1,\ldots,u^n)$ of (QVI) in $W^{2,\infty}(\mathcal{O})$. Furthermore, since $k > 0$ and all u^i vanish on Γ, we have on an open neighborhood of Γ (depending on i,k) and more generally

$$A_i u^i = f^i \quad \text{a.e. in } \{u^i < k+u^{i+1}\}$$

and thus by the local regularity theory for elliptic second order equations we see that u^i *is smooth in a neighborhood of* Γ *and more generally in the open set* $\{u^i < k+u^{i+1}\}$ (u^i is of class C^3). In addition $(u^i)_i$ corresponds to an optimal stochastic control with pay-off at switching times : more precisely let $i \in \{1,\dots,n\}$ and let us consider only admissible systems $\mathcal{A}_\theta^i$ where there exists a sequence $(\theta_j)_{j \geq 1}$ of increasing stopping times such that :

$$\theta_1 < \theta_2 < \dots < \theta_j < \theta_{j+1} \quad \text{a.s.}, \quad \theta_j \to +\infty \quad \text{a.s.};$$

$$\alpha_t(\omega) = \alpha_i \quad \text{if} \quad 0 \leq t < \theta_1(\omega), \ \alpha_t(\omega) = \alpha_{i+1} \quad \text{if} \quad \theta_1(\omega) \leq t < \theta_2(\omega), \dots,$$

$$\alpha_t(\omega) = \alpha_{i+j} \quad \text{if} \quad \theta_j(\omega) \leq t < \theta_{j+1}(\omega) \quad \text{for} \quad j \geq 1 \ ;$$

- where we set $\alpha_p = \alpha_q$ if $p \equiv q$ mod.n -

where $J(x,\mathcal{A}_\theta^i) = E\left\{\int_0^\tau f(X_t,\alpha_t)e^{-\lambda t}\,dt + \sum_{j\geq 1} 1_{(\theta_j \leq \tau)}\, k e^{-\lambda\theta_j}\right\}$.

It is clear from this interpretation that, as $k \to 0_+$, $u^i = u^i(k)$ converge to the same function $\tilde{u}$ which is the infimum over all controls α_t which are constant between stopping times and it is clear enough that : $\tilde{u} \equiv u$ - see also [15] , [59] ,[60] , [2] for related arguments.

We summarize all these arguments by the following Proposition which also indicates what is the mathematical formulation of the dynamic programming principle for the control problem defining $(u^i)_i$:

Proposition I.1 : *With the above assumptions, there exists a unique solution* $(u^1,\dots,u^n)$ *of* (QVI) *in* $W^{2,\infty}(\mathcal{O})$. *Furtheremore* $u^i \in C^3(\omega_i)$ *where* ω_i *is the open set defined by* $\{x/u^i(x) < k+u^{i+1}(x)\}$ *and* ω_i *contains an open neighborhood of* Γ. *Finally, if for each admissible system* σ *is a stopping time, we have :*

$$\begin{cases} u^i(x) = \inf\limits_{\mathcal{A}^i_\theta} J(x,\mathcal{A}^i_\theta) \ , \quad \forall\, i \in \{1,\dots,n\} \ , \quad \forall\, x \in \bar{\mathcal{O}} \ ; \\ u^i(x) = \inf\limits_{\mathcal{A}^i_\theta} \{E \int_0^{\tau\wedge\sigma} f(X_t,\alpha_t)e^{-\lambda t}\, dt + \sum\limits_{j\geq 1} 1_{(\theta_j \leq \tau\wedge\sigma)} k e^{-\lambda\theta_j} + \\ \qquad + \sum\limits_{j\geq 0} 1_{(\theta_j \leq \tau\wedge\sigma < \theta_{j+1})} u^{i+j}(X_{\sigma\wedge\tau}) e^{-\lambda\sigma\wedge\tau}\} \end{cases} \tag{13}$$

where we set : $u^p = u^q$ *if* $p \equiv q$ *mod.* n, $\theta_0 = 0$.

Observe finally that we only have to prove uniform bounds in $W^{1,\infty}(\mathcal{O})$ for $u^i = u^i(k)$ and that (7) holds uniformly with constants depending only on ν and $\sup\limits_{\alpha\in A} \|\phi(\cdot,\alpha)\|_{W^{2,\infty}(\mathbb{R}^N)}$ for $\phi = \sigma_{ij}, b_i, c, f$. This is achieved in the next section. In everything that follows, C denotes various constants depending only on ν, , $\sup\limits_{\alpha\in A} \|\phi(\cdot,\alpha)\|_{W^{2,\infty}(\mathbb{R}^N)}$ for $\phi = \sigma_{ij}, b_i, c, f$.

1.3. Proof of Theorem I.1.

As explained in the preceding section - keeping the same notations - we just have to prove (7) for $u^1,\dots,u^n$. This will be achieved in two steps :

1) we prove an estimate of the following form :

$$\| D^2 u^i \|_{L^\infty(\Gamma)} \leq C \ , \quad \text{for } 1 \leq i \leq n \ ; \tag{14}$$

2) we deduce (7) from (14) and from an argument inspired by a method due to Freidlin [18] and considered in P.L. Lions and J.L. Menaldi [59],[60]; M.V. Safonov [68],[69].

Let us only mention now the basic idea behind the proof : first if $\mathcal{O} = \mathbb{R}^N$

then because of (11) and of the assumption ($\lambda > \lambda_1$) we see that

$$J(x,\mathcal{A}) \in W^{2,\infty}(\mathbb{R}^N) \quad \text{and} \quad \sup_{\mathcal{A}} \|J(.,\mathcal{A})\|_{W^{2,\infty}(\mathbb{R}^N)} < \infty .$$

Then (7) follows since by a simple argument due to Krylov [25] we have :

$$\partial_\xi^2 u = \lim_{h\to 0} \frac{u(.+h\xi)+u(.-h\xi)-2u(.)}{h^2}$$

$$= \lim_{h\to 0} \frac{1}{h^2} \{\inf_{\mathcal{A}}(J(.+h\xi,\mathcal{A})) + \inf_{\mathcal{A}}(J(.-h\xi,\mathcal{A})) - 2\inf_{\mathcal{A}}(J(.,\mathcal{A}))\}$$

$$\leq \lim_{h\to 0} \sup_{\mathcal{A}} \{\frac{1}{h^2} [J(.+h\xi,\mathcal{A}) + J(.-h\xi,\mathcal{A}) - 2J(.,\mathcal{A})]\}$$

$$\leq \sup \|J(.,\mathcal{A})\|_{W^{2,\infty}(\mathbb{R}^N)} .$$

Of course since τ depends on x, this argument fails if $\mathcal{O} \equiv \mathbb{R}^N$ and the main idea is to localize the preceding argument by the use of the dynamic programming principle (see step 2 below); then some extra term appears involving the second-derivatives of u^i on Γ and if (14) is available we are then able to conclude.

Step 1 : Proof of (14)

Actually we first show that we have : $\|u^i\|_{W^{1,\infty}(\mathcal{O})} \leq C$. Of course in view of the stochastic interpretation, we immediately have : $\|u^i\|_{L^\infty(\mathcal{O})} \leq C$. Next, we may apply (5) and the arguments of Part 1 [36] to deduce the desired bound. Indeed it is easy to deduce from (5) (using $d(x) = \text{dist}(x,\Gamma)$) the existence of $w \in C^2(\bar{\mathcal{O}})$ such that : $A_i w \geq 1$ in $\bar{\mathcal{O}}$, $w \geq 0$ in $\bar{\mathcal{O}}$ and $w = 0$ on Γ (see also below). Then either from maximum principle or from Itô's formula , we deduce : $|u^i| \leq C\, w$ in $\bar{\mathcal{O}}$, $\forall\, i$.
Next, we argue as follows (cf Part 1 [36] and recall that $\lambda_1 \geq \lambda_o$) : let

$x,x' \in \bar{\mathcal{O}}$, we denote for each $\mathcal{A}_\theta^i$ by $\tau = \tau(x)$, $\tau' = \tau(x')$, $X_t = X_t(x)$, $X'_t = X_t(x')$. Then applying (13) with $\sigma = \tau \wedge \tau'$, we deduce easily :

$$u^i(x) - u^i(x') \leq C \quad E\int_0^\sigma |x-x'| e^{\lambda t} e^{-\lambda t} dt + \sup_j E\{(u^j(x)-u^j(x'))e^{-\lambda\sigma}\}$$

$$\leq C|x-x'| + C\, E[w(X_\sigma)e^{-\lambda\sigma}] \leq C|x-x'| + C\, E[d(X_\sigma)e^{-\lambda\sigma}]$$

$$\leq C|x-x'| + C\, E[|X_\sigma - X'_\sigma|\, e^{-\lambda\sigma}] \leq C|x-x'| \ .$$

It is also possible to give an analytic argument.

We now turn to the proof of (14) : (14) is obtained via a p.d.e. argument using barrier functions.
We first remark that because of (5) there exist $d_o > 0$, and smooth vector fields τ, n such that if x_o is fixed on Γ :

— $\tau, n, d(x) = \mathrm{dist}(x,\Gamma) \in C^2(\bar{\Gamma}_o)$ where $\Gamma_o = \{x \in \mathcal{O}\ ,\ d(x) < d_o\}$;

— $|n(x)| = |\nabla d(x)| = 1$ in $\bar{\Gamma}_o$; $-\nabla d(x) = n(x)$ is the unit out-normal on Γ ;

— $\exists \nu > 0$, $\forall (x,\alpha) \in \bar{\Gamma}_o \times A$; $a_{ij}(x,\alpha) n_i(x) n_j(x) \geq \frac{\nu}{2}$, $a_{ij}(x,\alpha)\partial_i d(x) \partial_j d(x) \geq \frac{\nu}{2}$

— $(n(x),\tau(x)) = 0$ on $\bar{\Gamma}_o$; $\tau(x_o)$ is a prescribed tangent vector τ_o to Γ; at x_o ; finally $\|\tau\|_{C^2}$, $\|n\|_{C^2}$ depend only on $\mathcal{O}$.

We next introduce the following test function, reminiscent of Bernstein methods for estimating derivatives of nonlinear elliptic equations :

$$w^i(x) = |Du^i(x)|^2 - \left|\frac{\partial u^i}{\partial n}(x)\right|^2 + \left(1 + \frac{\partial u^i}{\partial \tau}(x)\right)^2 - 1.$$

If we denote by $\mathcal{O}_i = \{x \in \mathcal{O} / u^i(x) < k + u^{i+1}(x),\ d(x) < d_o\}$ we claim that we have on $\mathcal{O}_i$:

$$A_i w^i \leq C \quad \text{in} \quad \mathcal{O}_i \ , \quad \forall\, i \tag{15}$$

Let us assume temporarily that (15) is proved : we deduce from (15) that we have : $|w^i(x)| \le Cd(x)$ in $\overline{\Gamma}_0$, $\forall i$.

Indeed let us consider the following barrier function w :

$$w(x) = C(e^{\alpha d(x)}-1)$$

then we have on $\overline{\Gamma}_0$: $A_i w \ge Ce^{\alpha d}\ [\alpha^2 a_{ij}\partial_i d\partial_j d - C\alpha]$ and choosing α and C large enough, we find :

$$A_i w \ge C \text{ in } \overline{\Gamma}_0 ,\quad w \ge w^i \text{ if } d(x) = d_0 ,\quad 0 \le w \le Cd(x) \text{ in } \overline{\Gamma}_0 ,$$

$$w = 0 \text{ on } \Gamma .$$

We now claim that (15) implies : $|w^i| \le w$. Indeed if for example we had : $\max_{i,x\in\Gamma_0} (w^i(x)-w(x)) > 0$, then choosing i_0 , x_0 realizing the maximum we would argue as follows : first $x_0 \in \Gamma_0$ and if $x_0 \notin \mathcal{O}_{i_0}$, then $u^{i_0}(x_0) = k+u^{i_0+1}(x_0)$. But since $u^{i_0}, u^{i_0+1} \in W^{2,\infty}(\mathcal{O})$ and $u^{i_0} \le k+u^{i_0+1}$; this would yield : $Du^{i_0}(x_0) = Du^{i_0+1}(x_0)$ and thus $w^{i_0}(x_0) = w^{i_0+1}(x_0)$. Since we cannot have for all i $u^i(x_0) = k+u^{i+1}(x_0)$, repeating this argument we find i_1 such that (i_1,x_0) is still a maximum and $x_0 \in \mathcal{O}_{i_1}$. Therefore in all cases we may assume that $x_0 \in \mathcal{O}_{i_0}$. But from (15) and the above inequalities satisfied by w, we deduce using the maximum principle

$$0 \le A_{i_0}(w-w^{i_0})(x_0) \le \lambda(w-w^{i_0})(x_0)$$

and this contradiction shows : $|w^i| \le w$ in $\overline{\Gamma}_0$ and thus :

$$|w^i(x)| \le Cd(x) \text{ in } \overline{\Gamma}_0 .$$

But this inequality implies :

$$C \geq \frac{\partial w^i}{\partial n}(x_o) = 2 \frac{\partial^2 u^i}{\partial n \partial \tau}(x_o) + 2\partial_k u^i \partial_j \tau_k n_j \geq 2 \frac{\partial^2 u^i}{\partial n \partial \tau}(x_o) - C$$

and since x_o is arbitrary on Γ and τ is an arbitrary unit tangent vector to Γ at x_o, (14) will be proved if we prove : $\left|\frac{\partial^2 u^i}{\partial n^2}\right| \leq C$ on Γ; and this is obtained from the equation which holds on Γ :

$$A_i u^i = f^i \quad \text{on } \Gamma .$$

We next turn to the proof of (15); to simplify notations, we skip the superscript i and we denote by ϕ_i the derivative with respect to x_i of ϕ. We first compute the first and second derivatives of w :

$$w_k = 2u_i u_{ik} - 2(u_i n_i)(u_{ik} n_i + u_i n_{ik}) + 2(1+u_i \tau_i)(u_{ik}\tau_i + u_i \tau_{i,k})$$

$$w_{k\ell} = 2u_i\ u_{ik} + 2u_i u_{ik\ell} - 2(u_{i\ell} n_i + u_i n_{i,\ell})(u_{ik} n_i + u_i n_{i,k}) +$$

$$- 2(u_i n_i)(u_{ik\ell} n_i + u_{ik} n_{i,\ell} + u_{i\ell} n_{i,k} + u_i n_{i,k\ell})$$

$$+ 2(u_{i\ell}\tau_i + u_i \tau_{i,\ell})(u_{ik}\tau_i + u_i \tau_{i,k})$$

$$+ 2(1+u_i\tau_i)(u_{ik\ell}\tau_i + u_{ik}\tau_{i,\ell} + u_{i\ell}\tau_{i,k} + u_i \tau_{i,k\ell}) \quad .$$

If we fix $x_o \in \mathcal{O}_i$, rotating the axis, we may assume $n(x_o) = e_N, \tau(x_o) = \beta e_1$. We now compute Aw at the point x_o and we find - recall that u is bounded in $W^{1,\infty}$:

$$Aw \leq C - a_{k\ell}(x_o) \quad w_{k\ell}(x_o) - b_k(x_o) w_k(x_o) \leq C - 2 \sum_{i<N} a_{k\ell} u_{i\ell} u_{ik}$$

$$+ 4a_{k\ell}\ u_j n_{j,\ell} u_{Nk} + 4a_{k\ell} u_N u_{ik} n_{i,\ell} - 4(1+\beta u_1) a_{k\ell} u_{ik} \tau_{i,\ell}$$

$$+ 2u_i(-a_k\ u_{ik\ell} - b_k u_{ik} + \lambda u_i) - 2u_N(-a_{k\ell} u_{Nk\ell} - b_k u_{Nk} + \lambda u_N)$$

$$+ 2(1+\beta u_1)(-a_{k\ell} u_{1k\ell} - b_k u_{1k} + \lambda u_1) .$$

On the other hand differentiating the equation satisfied by u on $\mathcal{O}$ we find :

$$- a_{k\ell} u_{ik\ell} - b_k u_{ik} + \lambda u_i = f_i + a_{k\ell,i} u_{k\ell} + b_{k,i} u_k$$

and this implies, using Cauchy-Schwarz inequalities :

$$Aw \leq C - 2 \sum_{i<N} a_{k\ell} u_{i\ell} u_{ik} + C \sup_{1\leq i\leq N} (a_{k\ell} u_{ik} u_{i\ell})^{1/2} + C \sup_{1\leq i\leq N} |a_{k\ell,i} u_{k\ell}|$$

Next, in view of a general lemma due to O. Oleinik [65] - see also D.W. Stroock and S.R.S. Varadhan [72] - we have :

$$\sup_{1\leq i\leq N} |a_{k\ell,i} u_{k\ell}| \leq C_o (a_{k\ell} u_{ik} u_{i\ell})^{1/2}$$

where C_o depends only on $\|\sigma\|_{W^{2,\infty}(\mathbb{R}^N)}$. Hence we deduce :

$$Aw \leq C - \sum_{i<N} a_{k\ell} u_{i\ell} u_{ik} + C(a_{k\ell} u_{Nk} u_{N\ell})^{1/2} \quad .$$

But because of (5) and the choice of n, we have :

$$\forall \xi \in \mathbb{R}^N \quad a_{k\ell} \xi_k \xi_\ell \geq \frac{\nu}{2} \xi_N^2$$

and we obtain in this way :

$$Aw \leq C - \frac{1}{2} \sum_{i<N} a_{k\ell} u_{i\ell} u_{ik} - \frac{\nu}{4} \sum_{i<N} u_{iN}^2 + C|u_{NN}| \ .$$

We now use the equation and (5) to deduce :

$$|u_{NN}| \leq C + C \sum_{i<N} |u_{iN}| + C \Big| \sum_{i,j<N} a_{ij} u_{ij} \Big|$$

and we conclude since :

$$\Big| \sum_{i,j<N} a_{ij} u_{ij} \Big| \leq C \Big\{ \sum_{i,k,\ell<N} a_{k\ell} u_{i\ell} u_{ik} \Big\}^{1/2} \leq C \Big\{ \sum_{i<N} a_{k\ell} u_{i\ell} u_{ik} \Big\}^{1/2} .$$

Step 2 : Proof of (7)

The reasoning now follows [59], [60], [68] . Let $\xi \in \mathbb{R}^N$, $|\xi| = 1$ and let $h > 0$. We denote by $\mathcal{O}_h = \{x \in \mathcal{O}, d(x) > h\}$. For each $\mathcal{A}_\theta$ we denote by $\sigma_h = \tau(x) \wedge \tau(x+h\xi) \wedge \tau(x-h\xi)$; applying (13) and using the same argument as in the beginning of this section we find :

$$\forall i, \forall x \in \mathcal{O}_h , \frac{1}{h^2} \{u^i(x+h\xi) + u^i(x-h\xi) - 2u^i(x)\}$$

$$\leq \sup_{\mathcal{A}_\theta} \frac{1}{h^2} \{J(x+h\xi,\sigma_h,\mathcal{A}_\theta) + J(x-h\xi,\sigma_h,\mathcal{A}_\theta) - 2J(x,\sigma_h,\mathcal{A}_\theta)\} +$$

$$+ \sup_{j,\mathcal{A}_\theta} \frac{1}{h^2} E\{(u^j(X^{+h}_{\sigma_h}) + u^j(X^{-h}_{\sigma_h}) - 2u^j(X_{\sigma_h}))e^{-\lambda\sigma_h} ;$$

where $J(x,\sigma,\mathcal{A}_\theta) = E \int_0^\sigma f(X_t,\alpha_t)e^{-\lambda t}\, dt$ and $X_t = X_t(x), X^{+h}_t = X_t(x+h\xi)$, $X^{-h}_t = X_t(x-h\xi)$.

As explained before, since $\lambda > \lambda_1$, $J(.,\sigma_h,\mathcal{A}_\theta) \in W^{2,\infty}(\mathbb{R}^N)$ (uniformly in $\sigma_h,\mathcal{A}_\theta$) and thus the first term in the right-hand side is bounded. Next, if we denote by $\Gamma_h = \{x \in \bar{\mathcal{O}} , d(x) \leq 2h^{3/4}\}$ and if $\Omega_h = \{\omega/\sigma_h < \infty, d(X_{\sigma_h}) \leq h^{3/4}$, $|X^{\pm h}_{\sigma_h} - X_{\sigma_h}| \leq h^{3/4}\}$, we deduce from the above inequalities for all i and for all $x \in \bar{\mathcal{O}}_h$:

$$\frac{1}{h^2} \{u^i(x+h\xi) + u^i(x-h\xi) - 2u^i(x)\}$$

$$\leq C + C \sup_i ||D^2u^i||_{L^\infty(\mathcal{O})} \sup_{\mathcal{A}_\theta} E [1_{\Omega^c_h} e^{-\alpha\sigma_h}] +$$

$$+ C \sup_j ||D^2u^j||_{L^\infty(\Gamma_h)} ;$$

for some $\alpha > 0$ $(\alpha < \lambda - \lambda_1)$. If we prove that

$$\sup_{\mathcal{A}} E [1_{\Omega^c_h} e^{-\alpha\sigma_h}] \to 0 \text{ as } h \to 0_+ , \qquad (16)$$

then letting $h \to 0_+$ and remarking that for h small enough, $u^i \in C^2(\Gamma_h)$, we deduce :

$$\forall i , \quad \partial_\xi^2 u^i \leq C + C \sup_i \|D^2 u^i\|_{L^\infty(\Gamma)} \quad \text{in } \mathcal{O} \; ; \text{ and (7) then follows}$$

from (14).

To prove (16), we consider

$$F_h^T = \{\omega / \sigma_h \leq T , \quad |X_{\sigma_h}^{+h} - X_{\sigma_h}| > h^{3/4} \quad \text{or} \quad |X_{\sigma_h}^{-h} - X_{\sigma_h}| > h^{3/4}\}$$

and we observe :

$$\sup_{\mathcal{A}} E\,[1_{\Omega_h^c}\, e^{-\alpha\sigma_h}] \leq e^{-\alpha T} + \sup_{\mathcal{A}} P(F_h^T) \quad \text{for all} \quad T < \infty \, .$$

Since $E\{\sup_{0 \leq t \leq T} |X_t^{\pm h} - X_t|\} \leq C(T)\, h$, we deduce :

$$\sup_{\mathcal{A}} P(F_h^T) \leq 2C(T)\, h^{1/4} \, .$$

Therefore we have :

$$\limsup_{h \to 0_+} \sup_{\mathcal{A}} E\,[1_{\Omega_h^c}\, e^{-\alpha\sigma_h}] \leq e^{-\alpha T} \quad , \text{ for all} \quad T < \infty$$

and we conclude letting $T \to +\infty$. □

I.4. Variants

In this section we present some variants of Theorem I.1 : of course since the conclusion of all the results below is always that (7) holds, we could each time repeat the consequences of (7), namely Theorem I.2, Corollary I.1., but we will not do so.

First of all we indicate how it is possible to relax (5) : to this end we assume that $\Gamma = \Gamma_o \cup \Gamma_1 \cup \Gamma_2$ where $\Gamma_o, \Gamma_1, \Gamma_2$ are closed, disjoint, smooth (possibly empty). We will consider the following extension of (5) :

$$\begin{cases} \forall (x,\alpha) \in \Gamma_o \times A\ , \quad \sigma(x,\alpha) = 0\ , \quad b(x,\alpha).n(x) \leq 0\ ; \\ \exists \nu > 0\ ,\ \forall (x,\alpha) \in \Gamma_1 \times A\ , \quad a_{ij}(x,\alpha)n_i(x)n_j(x) \geq \nu > 0; \\ \exists \nu > 0\ ,\ \forall (x,\alpha) \in \Gamma_2 \times A\ , \quad \sigma(x,\alpha) = 0\ ,\ b(x,\alpha).n(x) \geq \nu > 0\ . \end{cases} \qquad (5')$$

Theorem I.3 : *Let* $\mathcal{O}$ *be a smooth open set and let us assume* (9), (5') *(if* $\mathcal{O} \neq \mathbb{R}^N$*) and* $\lambda > \lambda_1$. *Then* u *satisfies :* $u \in W^{1,\infty}(\mathcal{O})$, $u = 0$ *on* $\Gamma_1 \cup \Gamma_2$ *and* (7) *holds.*

Remark I.4 : It is possible to relax a little bit the assumptions on Γ_o and Γ_2 . For example, an appropriate modification of the proofs below yields that we only need to assume : $a_{ij}(x,\alpha)n_i(x)n_j(x) = 0$ and $b_i(x,\alpha)n_i(x) - a_{ij}(x,\alpha)\partial_{ij}d(x) \geq \nu > 0$ on Γ_2 ; $a_{ij}(x,\alpha)n_i(x)n_j(x) = 0$ and $b_i(x,\alpha)n_i(x) - a_{ij}(x,\alpha)\partial_{ij}d(x) \leq 0$ on Γ_o .

Remark I.5 : As we already explained, (7) is a regularity result which is the analogue of $W^{2,\infty}$ regularity for linear degenerate second order elliptic operators : thus it is clear that one needs some assumption like (5) or (5') in order to obtain such a regularity.
On the other hand one could ask if $\lambda > \lambda_1$ is really necessary : we will see in the next section that in the uniformly non-degenerate case we may get rid of $\lambda > \lambda_1$. But in general, this assumption is necessary. Let us point out that this assumption is used only to insure some regularity of cost-functions like quantities - where the use of (11) is fundamental. In Genis and N.V. Krylov [20] , one example is given where $\mathcal{O} = \mathbb{R}$ (N = 1) , the "best" λ_1 insuring (11) is computed explicitly and where, if $\lambda \leq \lambda_1$, u is shown to have some regularity but does not satisfy (7) (or (8)).

Let us also remark that we admit in Theorem I.3 the case of an unbounded domain : in this case it is reasonable to ask if it is possible to relax the assumptions on σ,b,c,f. More precisely we indicate now a result where we relax the boundedness assumptions in x of σ,b,c,f and their derivatives (but we still assume some uniform boundedness in α).

We will assume :

$$\left\{\begin{array}{l} \phi(.,\alpha) \in W^{2,\infty}_{loc}(\mathbb{R}^N) \text{ for } \alpha \in A, \text{ for } \phi = \sigma_{ij}, b_i, c, f;\ \sigma(x,.) \in C(A) \text{ for } x \in \mathbb{R}^N \\ \forall \varepsilon > 0, \exists K_\varepsilon > 0, \quad \|\sigma(x,\alpha)\| + |b(x,\alpha)| \le \varepsilon |x| + K_\varepsilon, \quad \forall (x,\alpha) \in \mathbb{R}^N \times A \\ \exists C > 0 , \ \|D_x\sigma(x,\alpha)\| + |D_x b(x,\alpha)| \le C \quad \text{on} \quad \mathbb{R}^N \times A \\ \exists C, p \ge 0, \ |f| + |D_x f| + |D_x^2 f| + \|D_x^2\sigma\| + |D_x^2 b| + |c| + |D_x c| + |D_x^2 c| \\ \qquad \le C + C\,|x|^p \quad \text{on } \mathbb{R}^N \times A \end{array}\right. \tag{17}$$

We then have

<u>Theorem I.4</u> : *Let* $\mathcal{O}$ *be a smooth open set in* $\mathbb{R}^N$ *and let* $\lambda > \lambda_1$. *We assume in addition* (5') *(if* $\mathcal{O} \neq \mathbb{R}^N$*)* *and* (17). *Then for all* $R < \infty$, *if we denote by* $\mathcal{O}_R = \mathcal{O} \cap B_R$, $u \in W^{1,\infty}(\mathcal{O}_R)$, $u = 0$ *on* $\Gamma_1 \cup \Gamma_2$ *and*

$$\exists C_R < \infty, \ \forall \xi \in \mathbb{R}^N : |\xi| = 1 , \quad \partial_\xi^2 u \le C_R \ \text{in}\ \mathcal{D}'(\mathcal{O}_R) .$$

<u>Remark I.4</u> : Actually one can prove that one has :

$$|Du(x)| \le C + C|x|^q \quad \text{in } \mathcal{O}$$

$$\partial_\xi^2 u \le C + C|x|^q \quad \text{in } \mathcal{O} , \quad \forall \xi \quad \mathbb{R}^N : |\xi| = 1$$

for some $C, q \ge 0$. □

We will only prove Theorem I.3. in the case when $\Gamma_0 = \Gamma$ and in the case when $\Gamma_2 = \Gamma$: the general case is obtained via a straightforward combination of the various arguments. Roughly speaking, the method is to deduce from boundary estimates (like (14)) an estimate over $\mathcal{O}$ (like (7)). Now in order to obtain boundary estimates, one argues on Γ like in the preceding section; in addition Γ_o does not count since the processes X_t never reach Γ_o in finite time and thus (14) on Γ_o is not necessary. Finally the reason why on Γ_2 we still have some boundary estimates is due to the assumption (5') : indeed let u be a smooth solution near Γ_2 of : $A_\alpha u = f_\alpha$, $u = 0$ on Γ_2 ; where α is arbitrary in A. Then we deduce from the equation :

$$\nabla u = \frac{\partial u}{\partial n} n(x), \quad \frac{\partial u}{\partial n} = f(x,\alpha)\ (-b_i(x,\alpha)\, n_i(x))^{-1} \text{ on } \Gamma_2.$$

And differentiating the equation we get on Γ_2 for all $k \in \{1,\ldots,N\}$:

$$b_i(x,\alpha)\partial_{ik}u(x) + \partial_k b_i(x,\alpha)\partial_i u(x) = 0 \quad \text{on } \Gamma_2$$

- indeed remark that $\partial_k a_{ij}(x,\alpha) = \partial_k \sigma_{i\ell}(x,\alpha)\sigma_{j\ell}(x,\alpha) + \sigma_{i\ell}(x,\alpha)\partial_k\sigma_{j\ell}(x,\alpha)$ $= 0$ on Γ_2 - . And this implies an obvious bound on $\frac{\partial}{\partial n}(\partial_k u)$ on Γ_2 , that is a bound on $|D^2 u|$ on Γ_2 .

In order to make rigorous this type of estimate, we have to argue in a slightly different way - see the proof below.

After proving the cases $\Gamma_o = \Gamma$, $\Gamma_2 = \Gamma$ of Theorem I.3; we will prove Theorem I.4 in the case when $\mathcal{O} = \mathbb{R}^N$.

Before going into the proof, let us mention that, with appropriate adaptations of the general method of proof we present, it is possible to treat more general cases than the case of assumption (5'). It is of course impossible to give the most general statement and we will not try to do so; instead let us indicate a few examples in which directions one can look for similar regularity results.

Example I.1 : We will see in section II.2. that the optimal cost function in problems with an additional optimal stopping time problem still satisfies (7) or (7') . Now the optimal cost function u can be interpreted also as the optimal cost function associated with the following Hamilton-Jacobi-Bellman equation :

$$\max \{\sup_{\alpha \in A} \{A_\alpha u(x) - f(x,\alpha)\}, u(x) - \psi(x)\} = 0 \text{ in } \mathcal{O} , \quad u = 0 \text{ on } \Gamma$$

and thus we are in the same situation as above (with (5) satified by the operators A_α for example) except for an additional totally degenerate operator $B = \lambda$ and an additional cost $g(x) = \lambda\psi(x)$. We will see in section II.2 that if $\psi > 0$ or if $\psi = 0$ on Γ, regularity results like those obtained in Theorems I.1-3 are preserved. The idea is, when, for example, $\psi > 0$ on Γ , then nearby Γ we have $u < \psi$ and thus we are back to the situation we already treated. Therefore we see that by similar methods we can treat cases when $A = A_1 \cup A_2$ (with A_1, A_2 disjoint), (5') holds for $\alpha \in A_1$, and for $\alpha \in A_2$ we have : $\sigma = \nabla\sigma = 0$ on Γ , $b = \nabla b = 0$ on Γ and either $f \geq \nu > 0$ or $f \equiv 0$ on Γ. Actually one could assume that this holds only on some closed part of Γ , while analogues of (5') hold for $\alpha \in A_2$ on the complementary in Γ also assumed to be closed. More generally, if we know that for $\alpha \in A_2$ we have in a neighborhood of Γ : $A_\alpha u < f(x,\alpha)$, then we need to assume (5') only for $\alpha \in A_1$.

Example I.2 : Assume for example that nearby Γ we have $\sigma \equiv 0$, $c \equiv \lambda$, $f(x,\alpha) = f(x)$ and $\sup_{\alpha \in A} \{-b_i(x,\alpha)p_i\} = |p|$ for all $p \in \mathbb{R}^N$.
Assume in addition, to simplify, that $f(x,\alpha) \geq 0$ on $\bar{\mathcal{O}} \times A$. Then clearly $u \geq 0$ and if $\lambda > \lambda_0$ by the results of Part 1 [36] : $u \in W^{1,\infty}(\mathcal{O})$, $u = 0$ on Γ. In addition - see [37] ,[58] - u is a viscosity solution of :

$$|Du| + \lambda u = f(x) \quad \text{in} \quad \tilde{\Gamma}$$

where $\tilde{\Gamma}$ is on open neighborhood of Γ ($\tilde{\Gamma} = \{x \in \mathcal{O}, d(x) < \varepsilon_0\}$ for some $\varepsilon_0 > 0$). Now if $f \equiv 0$ or if $f > 0$ on Γ, one shows, as in P.L. Lions [58], that u is also the solution of the above first-order Hamilton-Jacobi equation built by the method of characteristics. Therefore $u \in C^2(\Gamma')$ where $\Gamma' = \{x \in \bar{\mathcal{O}},\ d(x) \leq \varepsilon_1\}$ for some $\varepsilon_1 > 0$. Then if $\lambda > \lambda_1$, this implies that (7) holds. It is possible to extend this type of argument by assuming that on Γ (to simplify) $\sigma \equiv 0$ and in a neighborhood of Γ, the Hamiltonian $H(x,t,p) = \sup_{\alpha \in A} \{-b_i(x,\alpha).p_i + c(x,\alpha)t - f(x,\alpha)\}$ is such that - roughly speaking - the method of characteristics for the Hamilton-Jacobi equation : $H(x,u,Du) = 0$ in $\mathcal{O}$, $u = 0$ on Γ "yields a smooth solution nearby Γ ".

Proof of Theorem I.3 in the case when $\Gamma_0 = \Gamma$: We introduce $u^i = u^i(k) = \inf_{\mathcal{A}^i_\theta} J(x, \mathcal{A}^i_\theta)$, with the same notations as in Sections I.2-3. Now we remark that (5') implies that for all $\mathcal{A}$ and in particular for all $\mathcal{A}^i_\theta$ and for all $x \in \mathcal{O}$: $\tau = +\infty$ a.s. Indeed let $\tilde{d}$ be in $C^2(\bar{\mathcal{O}})$ such that : $\tilde{d} = d$ in a neighborhood of Γ and $\tilde{d}(x) > 0$ in $\mathcal{O}$. Then if we set $Y_t = \text{Log}\, \tilde{d}(X_t)$, for $t < \tau' = \inf(t \geq 0, X_t \notin \mathcal{O})$ we have from Itô's rule :

$$dY_t = \frac{\nabla \tilde{d}(X_t)}{\tilde{d}(X_t)} \cdot \sigma(X_t,\alpha_t) dB_t + a_{ij}(X_t,\alpha_t) \Big\{ \frac{\partial_{ij}\tilde{d}(X_t)}{\tilde{d}(X_t)} - \frac{\partial_i \tilde{d}(X_t) \partial_j \tilde{d}(X_t)}{(\tilde{d}(X_t))^2} \Big\} dt + b_i(X_t,\alpha_t) \frac{\partial_i \tilde{d}(X_t)}{\tilde{d}(X_t)} dt .$$

But because of (5') :

$$\Big| \frac{\nabla \tilde{d}(X_t)}{\tilde{d}(X_t)} \sigma(X_t,\alpha_t) \Big| \leq C , \quad \Big| \frac{a_{ij}(X_j,\alpha_t)}{\tilde{d}(X_t)^2} \Big| \leq C$$

and

$$b_i(X_t,\alpha_t)\partial_i\tilde{d}(X_t) \geq - C\, \tilde{d}(X_t) \text{ ; and we get :}$$

$$dY_t = \phi_t \, dB_t + \psi_t \, dt \; ,$$

where ϕ_t is a bounded progessively measurable process and ψ_t is a progressively measurable process bounded from below. This implies that for all $T < \infty$:

$$\inf_{[0,T]} Y_t \geq - CT + \inf_{[0,T]} \left(\int_0^t \phi_s dB_s\right)$$

and thus for all $T < \infty$, $P(\tau' \leq T) = 0$ and thus $\tau = \tau' = +\infty$ a.s. Therefore we may consider that u , u^i are defined on $\mathbb{R}^N$ by :

$$u(x) = \inf_{\mathcal{A}} J(x,\mathcal{A}) \; ; \quad u^i(x) = \inf_{\mathcal{A}^i_\theta} J(x,\mathcal{A}^i_\theta)$$

where

$$J(x,\mathcal{A}) = E \int_0^\infty f(X_t,\alpha_t) \exp\left(- \int_0^t c\right)dt$$

and

$$J(x,\mathcal{A}^i_\theta) = E \int_0^\infty f(X_t,\alpha_t) \exp\left(- \int_0^t c\right)dt + \sum_{i\geq 1} k\left(\exp - \int_0^{\theta_i} c\right) .$$

Next since $\lambda > \lambda_1$, the cost functions $J(x,\mathcal{A})$ are bounded in $W^{2,\infty}(\mathbb{R}^N)$ uniformly in $\mathcal{A}$. This immediately yields on one hand that $u, u^i \in W^{1,\infty}(\mathbb{R}^N)$ (with bounds independent of k) and that we have :

$$\exists\, C > 0, \forall \xi \in \mathbb{R}^N : |\xi| = 1 \; , \; \partial_\xi^2 u \leq C \; , \; \partial_\xi^2 u^i \leq C \text{ in } \mathcal{D}'(\mathbb{R}^N) \; . \; \square$$

<u>Remark I.5</u> : The above proof shows that it is enough to assume on Γ_0 :

$$\sigma^T(x,\alpha).n(x) = 0 \quad \text{on} \quad \Gamma_o \times A$$

$$b_i(x,\alpha).n_i(x) - a_{ij}(x,\alpha)\partial_{ij}d(x) \leq 0 \quad \text{on} \quad \Gamma_o \times A \ .$$

Proof of Theorem I.3 in the case when $\Gamma_2 = \Gamma$: Exactly as in section I.2, it is enough to consider the case of smooth coefficients σ,b,c,f and of a finite set $A = \{\alpha_1,\ldots,\alpha_n\}$. To simplify the notation we take $c \equiv \lambda$. Furthermore as in section I.2, considering $\sigma_\varepsilon(x,\alpha) = (\sigma(x,\alpha),\ \varepsilon\chi(x)I_N)$ where $\chi \in C_b^2(\mathbb{R}^N)$, $\chi > 0$ in $\mathcal{O}$, $\chi = 0$ on Γ , we may assume without loss of generality that we have on each compact K of $\mathcal{O}$:

$$\exists\ \nu = \nu(K) > 0\ , \qquad a(x,\alpha) \geq \nu\ I_N \qquad \forall x \in K\ ,\ \forall\ \alpha \in A\ .$$

We next introduce $u^i = u^i(k) = \inf\limits_{\mathcal{A}_\theta^i} J(x,\mathcal{A}_\theta^i)$ with the same notation as before

Because of (5') - if $\lambda > \lambda_o$ (see Part 1 [36]) and in particular if $\lambda > \lambda_1$ - u^i is bounded in $W^{1,\infty}(\mathcal{O})$, $u^i = 0$ on Γ : this is proved exactly as in Part 1 and in section I.3. Remark also that the dynamic programming principle still holds : make a direct proof or deduce it from the nondegenerate case by tedious approximation argument. Then this implies that there exists ε_o (depending on k) such that :

$$u^i < k + u^{i+1} \quad \text{if} \quad x \in \{z \in \bar{\mathcal{O}},\ d(z) \leq \varepsilon_o\}\ .$$

Denoting by $\Gamma' = \{z \in \mathcal{O},\ d(z) < \varepsilon_o\}$, one proves easily that one has :

$$A_i u^i = f^i \quad \text{in} \quad \mathcal{D}'(\Gamma'),$$

make an approximation argument using the nondegenerate case or argue directly from (13) . Since $u^i \in W^{1,\infty}(\mathcal{O})$, this shows, from local regularity theory of uniformly elliptic linear equations, that $u^i \in C^2(\Gamma')$. And this implies as in [15] , [59] that $u^i \in W^{2,\infty}_{loc}(\mathcal{O})$ - recall that the operators A_i are uniformly elliptic "inside" $\mathcal{O}$.

Next let u_o^i be the solution of :

$$- b_j \partial_j u_o^i + \lambda u_o^i = f^i , \qquad u_o^i = 0 \text{ on } \Gamma$$

built nearby Γ by the method of characteristics. Such a function u_o^i exists and is smooth say C^2 - on $\overline{\Gamma}'$ choosing possibly ε_o smaller than above. Let us emphasize that u_o^i is independent of k. We claim that there exists a constant $C = C(k) > 0$ such that :

$$|u^i - u_o^i| \leq C(k)d^3 , \qquad \forall x \in \overline{\mathcal{O}} . \tag{18}$$

Indeed remark first that choosing ε_o smaller we may assume that $d \in C^2(\overline{\Gamma}')$ and :

$$|\nabla d| = 1 \text{ on } \Gamma' , \quad b_j(x,\alpha_i).\partial_j d(x) \geq \frac{\nu}{2} > 0$$

on Γ'. Then we have on Γ' :

$$A_i(u_o^i+Cd^3) \geq - (K+CKd) \sup_{k,\ell} |a_{k\ell}(x,\alpha_i)| + f^i + 3C\, d^2 \frac{\nu}{2}$$

$$\geq - (Kd^2+CKd^3)+f^i + 3C\, d^2 \frac{\nu}{2}$$

where K denotes various constants independent of k. Then if we choose $\varepsilon_o \leq \frac{\nu}{K}$ and $C \geq \frac{2K}{\nu}$, we deduce :

$$A_i(u_o^i+Cd^3) \geq f^i \text{ in } \Gamma' , \quad u_o^i + Cd^3 = 0 \text{ on } \Gamma .$$

Choosing C even larger, we may assume : $u_o^i + Cd^3 \geq u^i$ on $\partial\Gamma'$.
Then, arguing in the same way for $u_o^i - Cd^3$, we deduce (18) applying the maximum principle.

We now prove (7) and we keep the notation of step 2 in section I.3; we find for all i and for all $x \in \mathcal{O}_h$:

$$\frac{1}{h^2}\{u^i(x+h\xi) + u^i(x-h\xi) - 2u^i(x)\} \leq C + \delta(h) + $$
$$+ \sup_{j,\mathcal{A}_\theta^i} [\frac{1}{h^2} E \; 1_{\Omega_h}\{u^j(X_{\sigma_h}^{+h}) + u^j(X_{\sigma_h}^{-h}) - 2u^j(X_{\sigma_h})\} e^{-\lambda\sigma_h}]$$

where $\delta(h) \to 0$ as $h \to 0_+$. Replacing in the above inequality u^j by u_o^j (take h such that $2h^{3/4} \leq \varepsilon_o$) we find using (18) :

$$\frac{1}{h^2}\{u^i(x+h\xi) + u^i(x-h\xi) - 2u^i(x)\} \leq C + \delta(h)$$
$$+ \sup_{j,\mathcal{A}_\theta^i} [\frac{1}{h^2} E \; 1_{\Omega_h} \{u_o^j(X_{\sigma_h}^{+h}) + u_o^j(X_{\sigma_h}^{-h}) - 2u_o^j(X_{\sigma_h})\} e^{-\lambda\sigma_h}]$$
$$+ \frac{1}{h^2} 3C(k)\,(2h^{3/4})^3$$

indeed on Ω_h , $d(X_{\sigma_h}) \leq h^{3/4}$ and $d(X_{\sigma_h}^{\pm}) \leq 2h^{3/4}$.

Since we assumed $\lambda > \lambda_1$, the second term in the upper bound is bounded by $K \| D^2 u_o^j \|_{L^\infty(\Gamma_h)}$ that is by a quantity bounded uniformly in k and in h small. If we then let h go to 0_+ , we find (7) .

<u>Proof of Theorem I.4 in the case when $\mathcal{O} = \mathbb{R}^N$</u> : In view of the arguments given above, we only have to prove that if the coefficients σ, b, f (again we take $c \equiv \lambda$ to simplify) are smooth and satisfy (17) and if $\lambda > \lambda_1$ then we have :

$$\exists\, C \geq 0 , \quad \exists\, q \geq 0 , \;\forall\, x \in \mathbb{R}^N, \;\forall \mathcal{A} \quad |D_x J(\cdot,\mathcal{A})| \leq C + C|x|^q \qquad (19)$$

$$\exists\, C \geq 0 , \quad \exists\, q \geq 0 , \;\forall\, x \in \mathbb{R}^N, \;\forall \mathcal{A} \quad \|D_x^2 J(\cdot,\mathcal{A})\| \leq C + C|x|^q . \qquad (20)$$

We first remark that because of (17), we obtain by standard arguments on solutions of stochastic differential equations the following bounds :

$$\forall\varepsilon > 0\ ,\ \forall m \geq 2\ ,\ \exists\, K_\varepsilon > 0\ ,\ \forall x \in \mathbb{R}^N,\ \forall \mathcal{A}\quad E[|X_t|^m] \leq K_\varepsilon(1+|x|^m)e^{\varepsilon t};\tag{21}$$

$$\exists\, K > 0\ ,\ \forall x \in \mathbb{R}^K,\ \forall \mathcal{A}\quad \sup\ E\ \{|\frac{\partial X_t}{\partial x_i}|^4\} \leq K\ e^{2\lambda_1 t}\quad ;\tag{22}$$

$$\forall\varepsilon > 0\ ,\exists K_\varepsilon, \exists m > 0,\ \forall\ x \in \mathbb{R}^N\ ,\ \forall \mathcal{A}\quad \sup_{i,j}\ E\{|\frac{\partial^2 X_t}{\partial x_i \partial x_j}|^2\} \leq K_\varepsilon(1+|x|^m)e^{2(\lambda_1+\varepsilon)t}\tag{23}$$

This yields for all $x \in \mathbb{R}^N$ and for all $\mathcal{A}$:

$$|D\ J(x,\mathcal{A})| \leq \int_0^\infty E\{(C+C|X_t|^m)\ \sum_i |\frac{\partial X_t}{\partial x_i}|\}\ e^{-\lambda t}\ dt$$

for some $C, m \geq 0$. Using (21), (22) and the Cauchy-Schwarz inequality, we easily deduce (19) since $\lambda > \lambda_1$.
In the same way we have for all $x \in \mathbb{R}^N$ and for all $\mathcal{A}$:

$$\|D_x^2 J(x,\mathcal{A})\| \leq \int_0^\infty E\{(C+C|X_t|^m)\ \sum_{i,j} |\frac{\partial^2 X_t}{\partial x_i \partial x_j}| + (C+C|X_t|^m)\ \sum_i |\frac{\partial X_t}{\partial x_i}|^2\}\ e^{-\lambda t} dt$$

for some $C, m \geq 0$. And we deduce easily (20) from (21)-(22)-(23) and the fact that $\lambda > \lambda_1$.

We conclude this section by mentioning a result concerning optimal control problems with a pay-off at the exit from $\bar{\mathcal{O}}$: we consider now cost functions $J(x,\mathcal{A})$ of the form :

$$J(x,\mathcal{A}) = E\ \{\int_0^\infty f(X_t,\alpha_t)\exp(-\int_0^t c(X_s,\alpha_s)ds)\ dt + \phi(X_\tau)\ \exp(-\int_0^\tau c(X_t,\alpha_t)dt)\}$$

and the optimal cost function is still given by (3). We will give here only the analogue of Theorem I.3 (but Theorem I.4 extends to this case as well) and since its proof is totally similar to those given above, we will skip it.

Theorem I.5 : *Let* $\phi \in W^{3,\infty}(\mathbb{R}^N)$, *let* $\lambda > \lambda_1$, *let* $\mathcal{O}$ *be a smooth open set and let us assume* (9) , (5') *(if* $\mathcal{O} \neq \mathbb{R}^N$ *). Then we have :*

i) $u \in W^{1,\infty}(\mathcal{O})$, $u = \phi$ *on* $\Gamma_1 \cup \Gamma_2$

ii) (7) *and* (8) *hold :*

$$\exists C > 0 , \ \forall \xi \in \mathbb{R}^N : |\xi| = 1 , \quad \partial_\xi^2 u \le C \text{ in } \mathcal{D}'(\mathcal{O}) ; \tag{7}$$

$$\sup_{\alpha \in A} ||A_\alpha u||_{L^\infty(\mathcal{O})} < \infty. \tag{8}$$

iii) *The* HJB *equation holds :*

$$\sup_{\alpha \in A} \{A_\alpha u(x) - f(x,\alpha)\} = 0 \quad \text{a.e. in } \mathcal{O} . \tag{4}$$

I.5. The non-degenerate case

In this section, we consider the very particular case when all the matrices $a(x,\alpha)$ are definite positive uniformly in x,α, that is, when we assume :

$$\exists \nu > 0 , \ \forall (x,\alpha) \in \bar{\mathcal{O}} \times A , \quad a(x,\alpha) \ge \nu I_N . \tag{24}$$

To simplify we will consider only the case of a bounded smooth domain $\mathcal{O}$ and we will denote by $\underline{\lambda}_1$, $\bar{\lambda}_1$ the following quantities :

$$\begin{cases} \underline{\lambda}_1 = \sup \{\lambda \ge 0 ; \ \sup_{x \in \bar{\mathcal{O}}} \ \sup_{\mathcal{A}} \ E[e^{\lambda\tau}] < +\infty \} \\ \\ \bar{\lambda}_1 = \sup \{\lambda \ge 0 ; \ \sup_{x \in \bar{\mathcal{O}}} \ \inf_{\mathcal{A}} E[e^{\lambda\tau}] < +\infty . \end{cases} \tag{25}$$

Those two quantities were introduced in P.L. Lions [56] and for reasons detailed below were called *demi-eigenvalues*. In addition it was proved in [56] that one has :

$$0 < \underline{\lambda}_1 \leq \inf_{\alpha \in A} \lambda_1(B_\alpha) \leq \sup_{\alpha \in A} \lambda_1(B_\alpha) \leq \overline{\lambda}_1 < +\infty$$

where B_α is the operator defined by : $B_\alpha = -a_{ij}(x,\alpha)\partial_{ij} - b_i(x,\alpha)\partial_i$ and $\lambda_1(B)$ is the first eigenvalue (for Dirichlet conditions) of the operator B.

We then have the :

<u>Theorem I.6</u> : *We assume* (24), (9) , *Next, if* $\lambda = \inf_{x,\alpha} c(x,\alpha) > -\underline{\lambda}_1$ *(resp. if* $f \geq 0$ *on* $\mathbb{R}^N \times A$ *and* $\lambda > -\overline{\lambda}_1$*) , the optimal cost function is the unique solution (resp. the unique nonnegative solution) in* $W^{2,\infty}(\mathcal{O})$ *of :*

$$\sup_{\alpha \in A} \{A_\alpha u(x) - f(x,\alpha)\} = 0 \quad \textit{a.e. in } \mathcal{O}, \quad u = 0 \quad \textit{on} \quad \Gamma .$$

In addition there exists $\theta \in (0,1)$ *depending only on* a *and* $\mathcal{O}$ *such that* $u \in C^{2,\theta}(\mathcal{O})$.

<u>Remark I.6</u> : We will not prove this result taken from P.L. Lions [56] . Let us remark that *the regularity result* ($C^{2,\theta}(\mathcal{O})$) *is due to L.C. Evans* [13], [14] (see also N.S. Trudinger [74]). We conjecture that $u \in C^{2,\theta}(\overline{\mathcal{O}})$ and we hope to come back on this point in a future publication.

<u>Remark I.7</u> : In particular if $\lambda \geq 0$, $u \in W^{2,\infty}(\mathcal{O})$. Let us also mention that this result still holds if $\mathcal{O}$ is unbounded (or $\mathcal{O} = \mathbb{R}^N$) but of course $\underline{\lambda}_1$ and $\overline{\lambda}_1$ may vanish in this case (if $\mathcal{O} = \mathbb{R}^N$, $\underline{\lambda}_1 = \overline{\lambda}_1 = 0$) .

<u>Remark I.8</u> : The solvability of (4) for uniformly elliptic operators has been studied by many authors : H. Brezis and L.C. Evans [6] solved the case when

$A = \{\alpha_1,\alpha_2\}$ by a variational approach and proved a $C^{2,\theta}$ regularity result (this was slightly extended in P.L. Lions [54]) ; at the same time L.C. Evans and A. Friedman [15] and J.L. Menaldi and the author [59], [60], [61] solved the case when the matrices $a(x,\alpha)$ do not depend on x by p.d.e. techniques; the general case was settled by P.L. Lions [53],[55] by p.d.e. arguments but still with the assumption $(\lambda > \lambda_1)$ which was relaxed by L.C. Evans and P.L. Lions [16] by slight modifications of the arguments introduced in [53],[55].

Similar results - with similar proofs - hold in the case of non-homogeneous boundary conditions, i.e. the case of a terminal pay-off at the exit from $\bar{\mathcal{O}}$. The following result which extends a result due to S. Lenhart [35] is proved by appropriate modifications of the methods of [53],[55],[56],[16] :

Corollary I.2 : *We assume* (24) *and* (9) . *Let* $\phi \in C(\bar{\mathcal{O}})$ *and let*

$$u(x) = \inf_{\mathcal{A}} E\left\{\int_0^\tau f(X_t,\alpha_t)\exp\left(-\int_0^t c(X_s,\alpha_s)ds\right) + \phi(X_\tau)\exp\left(-\int_0^\tau c(X_t,\alpha_t)dt\right)\right\}$$

If $\lambda > -\underline{\lambda}_1$ *(resp. if* $\lambda > -\bar{\lambda}_1$ *and if* $f \geq 0$ *on* $\mathbb{R}^N \times A$*) then* u *is the unique solution (resp. the unique nonnegative solution) in* $W^{2,\infty}_{loc}(\mathcal{O}) \cap C(\bar{\mathcal{O}})$ *of :*

$$\sup_{\alpha \in A}\{A_\alpha u(x) - f(x,\alpha)\} = 0 \quad a.e. \text{ in } \mathcal{O}, \quad u = \phi \quad \text{on } \Gamma .$$

In addition there exists $\theta \in (0,1)$ *depending only on* a , $\mathcal{O}$ *such that* $u \in C^{2,\theta}(\mathcal{O})$.

Let us also mention the following result concerning optimal markovian controls which is an immediate consequence of the results of N.V. Krylov concerning the solvability of stochastic differential equations with measurable coefficients, see [28] , [24] and [72].

Corollary I.3 : *Under the assumptions of Theorem I.6 or Corollary I.2, then for each* $\varepsilon > 0$ *, there exist a Borel function* $\alpha_\varepsilon(x)$ *and a strong Markov process* X_t *solution (for some* $\mathcal{A}$*) of*

$$dX_t = \sigma(X_t,\alpha_\varepsilon(X_t))dB_t + b(X_t,\alpha_\varepsilon(X_t))dt, \quad X_0 = x$$

such that for all $x \in \bar{\mathcal{O}}$:

$$u(x) \le E\int_0^\tau f(X_t,\alpha_\varepsilon(X_t))\exp(-\int_0^t c)dt + \phi(X_\tau)\exp(-\int_0^\tau c) \le u(x) + \varepsilon .$$

In addition if A *is compact, then we may take* $\varepsilon = 0$ *above.*

We now conclude this section by a few considerations on $\underline{\lambda}_1$, $\bar{\lambda}_1$. First we recall a few results from P.L. Lions [56] - where below we assume that (24) and (9) hold and that $\mathcal{O}$ is connected - : there exist $\phi_1,\psi_1 \in C^2(\mathcal{O}) \cap W^{2,\infty}(\mathcal{O})$ satisfying :

$$\sup_{\alpha\in A} \{B_\alpha\phi_1(x)\} = \underline{\lambda}_1\phi_1(x) , \quad \phi_1 < 0 \quad \text{in } \mathcal{O}, \quad \phi_1 = 0 \quad \text{on } \Gamma$$

$$\sup_{\alpha\in A} \{B_\alpha\psi_1(x)\} = \bar{\lambda}_1\psi_1(x) , \quad \psi_1 > 0 \quad \text{in } \mathcal{O}, \quad \psi_1 = 0 \quad \text{on } \Gamma.$$

In addition let $\mu \in \mathbb{R}, \chi \in W^{2,\infty}(\mathcal{O})$ be such that :

$$\sup_{\alpha\in A} \{B_\alpha \chi(x)\} = \mu\chi(x) \quad \text{in } \mathcal{O} , \quad \chi = 0 \quad \text{on } \Gamma .$$

Then if $\chi \ge 0$ in $\bar{\mathcal{O}}$, $\mu = \bar{\lambda}_1$ and $\chi = k\psi_1$ for some $k \ge 0$;

while if $\chi \le 0$ in $\bar{\mathcal{O}}$, $\mu = \underline{\lambda}_1$ and $\chi = k\phi_1$ for some $k \ge 0$.

In addition - see [56] for more details - $\underline{\lambda}_1$ and $\bar{\lambda}_1$ also possess the properties "of first eigenvalues " concerning bifurcation theory.

We conclude by showing that $\underline{\lambda}_1$, $\overline{\lambda}_1$ also arise when considering optimization questions on first eigenvalues of elliptic second-order operators. Indeed let $\mathcal{C}$ be the set of all elliptic operators B such that there exist $m \geq 1$, $\alpha_1,\ldots,\alpha_m \in A$, $\chi_1,\ldots,\chi_m \in C(\overline{\mathcal{O}})$ satisfying :

$$0 \leq \chi_i \leq 1 \quad \forall i, \quad \sum_{i=1}^{m} \chi_i = 1 \text{ on } \overline{\mathcal{O}} ;$$

and

$$B = \sum_{i=1}^{m} \chi_i A_{\alpha_i} .$$

We now consider the minimization problems : find $\underline{\lambda}$, $\overline{\lambda}$ given by

$$\begin{cases} \overline{\lambda} = \sup_{B \in \mathcal{C}} \lambda_1(B) \\ \underline{\lambda} = \inf_{B \in \mathcal{C}} \lambda_1(B) . \end{cases} \tag{26}$$

<u>We claim that we have</u> : $\underline{\lambda}_1 = \underline{\lambda}$, $\overline{\lambda}_1 = \overline{\lambda}$.

Indeed let $B \in \mathcal{C}$; if ψ_1 is the "eigenfunction" (positive) associated with $\overline{\lambda}_1$, we have : $B_\alpha \psi_1 \leq \overline{\lambda}_1 \psi_1$ in $\mathcal{O}$, $\psi_1 = 0$ on Γ for all $\alpha \in A$ and therefore we have :

$$B\psi_1 \leq \overline{\lambda}_1 \psi_1 \text{ in } \mathcal{O} , \quad \psi_1 > 0 \text{ in } \mathcal{O} , \quad \psi_1 = 0 \text{ on } \Gamma .$$

But this implies : $\overline{\lambda}_1 \geq \lambda_1(B)$. And we proved : $\overline{\lambda}_1 \geq \overline{\lambda}$. To prove the converse, we argue as follows : let $\varepsilon > 0$, we denote by

$$\mathcal{O}^\varepsilon = \{x \in \mathbb{R}^N , \operatorname{dist}(x,\overline{\mathcal{O}}) < \varepsilon\}$$

Of course, corresponding to $\mathcal{O}^\varepsilon$ there exist $\overline{\lambda}_1^\varepsilon, \psi_1^\varepsilon$ and it is easily checked that $\overline{\lambda}_1^\varepsilon \uparrow \lambda_1$ as $\varepsilon \downarrow 0_+$. We are going to show that we can find $B \in \mathcal{C}$ such that

$$\lambda_1(B) \geq \overline{\lambda}_1^\varepsilon - \varepsilon .$$

First let $(\alpha_i)_{i\geq 1}$ be a dense family in A : clearly we have - ε being fixed, we skip it -

$$\sup_{1\leq i\leq m} \{B_{\alpha_i}\psi_1\} = \bar{\lambda}_1\psi_1 + f_m \quad \text{in } \mathcal{O}^\varepsilon$$

where f_m is bounded in $L^\infty(\mathcal{O}^\varepsilon)$ and $f_m \underset{m}{\to} 0$ in $L^p(\mathcal{O}^\varepsilon)$ for all $p < \infty$. Next, we observe that there exist $\chi_1,\ldots,\chi_m \in L^\infty(\mathcal{O}^\varepsilon)$ such that :

$$0 \leq \chi_i \leq 1 \text{ a.e. in } \mathcal{O}^\varepsilon \ \forall i\ , \ \sum_{i=1}^m \chi_i = 1 \text{ a.e. in } \mathcal{O}^\varepsilon$$

and

$$\sum_{i=1}^m \chi_i A_{\alpha_i}\psi_1 = \sup_{1\leq i\leq m} \{A_{\alpha_i}\psi_1\} \text{ a.e. in } \mathcal{O}^\varepsilon .$$

We then introduce $\chi_i^\delta \in C(\bar{\mathcal{O}}^\varepsilon)$ satisfying : $0 \leq \chi_i^\delta \leq 1$ in $\mathcal{O}^\varepsilon$,

$$\sum_{i=1}^m \chi_i^\delta = 1 \text{ in } \mathcal{O}^\varepsilon, \quad \| \chi_i^\delta - \chi_i \|_{L^p(\mathcal{O}^\varepsilon)} \underset{\delta\to 0}{\longrightarrow} 0 \ \ \forall i\ , \forall p < \infty .$$

We have thus found $B\ (= B(m,\delta)) \in \mathcal{C}$: $B = \sum_{i=1}^m \chi_i^\delta A_{\alpha_i}$, such that :

$$B\psi_1 = \bar{\lambda}_1\psi_1 + f_m^\delta \quad \text{in } \mathcal{O}^\varepsilon;$$

where $f_m^\delta \to 0$ in $L^p(\mathcal{O}^\varepsilon)$ $\forall p < \infty$ as $\delta \to 0$, and then $m \to \infty$. We next consider ϕ_m^δ the solution of :

$$B\phi_m^\delta = f_m^\delta \text{ in } \mathcal{O}\ , \quad \phi_m^\delta = 0 \text{ on } \Gamma .$$

Because of the Alexandrov-Pucci-Krylov estimate (see [67] ,[24] for example) one has : $\|\phi_m^\delta\|_{L^\infty(\mathcal{O})} \to 0$ as $\delta \to 0$ and then $m \to \infty$. On the other hand, we have on $\mathcal{O}$:

$$\begin{cases} B(\psi_1 - \phi_m^\delta) = \bar{\lambda}_1\psi_1 \geq (\bar{\lambda}_1-\varepsilon)(\psi_1-\phi_m^\delta) & \text{in } \mathcal{O} \\ \psi_1 - \phi_m^\delta \geq 0 & \text{in } \bar{\mathcal{O}} \end{cases}$$

choosing δ small and m large so that : $\|\phi_m^\delta\|_{L^\infty(\mathcal{O})} \le \max(1, \frac{\bar{\lambda}_1 - \varepsilon}{\varepsilon}) \inf_{\mathcal{O}} \psi_1$.

And this yields : $\lambda_1(B) \ge \bar{\lambda}_1^\varepsilon - \varepsilon$, and we conclude letting ε go to 0 . One proves that $\underline{\lambda}_1 = \underline{\lambda}$ in a similar way.

Example I.3. : Let $0 < \nu < \mu$. Consider the set $\mathcal{B}$ of elliptic operators B given by : $B = - a_{ij}(x)\partial_{ij}$, where $a_{ij} \in C(\bar{\mathcal{O}})$ $(1 \le i,j \le N)$ and (a_{ij}) satisfies :

$$\mu I_N \ge a(x) \ge \nu I_N \quad \text{in } \bar{\mathcal{O}} .$$

We still denote by : $\underline{\lambda} = \inf_{B \in \mathcal{B}} \lambda_1(B)$, $\bar{\lambda} = \sup_{B \in \mathcal{B}} \lambda_1(B)$.

We next introduce $A = \{a = a^T , \ \nu I_N \le a \le \mu I_N\}$, to this control set correspond $\underline{\lambda}_1$, $\bar{\lambda}_1$, ϕ_1, ψ_1 satisfying :

$$\begin{cases} \sup_{a \in A} \{- a_{ij}\, \partial_{ij}\psi_1\} = \bar{\lambda}_1 \psi_1 \ \text{in } \mathcal{O} , \ \psi_1 > 0 \ \text{in } \mathcal{O} , \ \psi_1 = 0 \text{ on } \Gamma . \\ \inf_{a \in A} \{- a_{ij}\, \partial_{ij}\phi_1\} = \underline{\lambda}_1 \phi_1 \ \text{in } \mathcal{O} , \ \phi_1 > 0 \ \text{in } \mathcal{O} , \ \phi_1 = 0 \text{ on } \Gamma ; \\ \phi_1, \psi_1 \in C^2(\mathcal{O}) \cap W^{2,\infty}(\mathcal{O}) . \end{cases}$$

By the above result, we then have : $\underline{\lambda} = \underline{\lambda}_1$, $\bar{\lambda} = \bar{\lambda}_1$.

Indeed it is easy to approximate any $B \in \mathcal{B}$ by a sequence of operators lying in $\mathcal{C}$ (built as above from A) - with equicontinuous coefficients.

Remark I.9. : The result and the arguments above are to be compared with the results in P.L. Lions [45] .

Remark I.10. : Let us finally remark that one can replace in all the results above the assumption (9) by assuming (9) for $\phi = \sigma_{ij}$, b_i , c and

$$\sup_{\alpha\in A} \|f(.,\alpha)\|_{W^{1,\infty}(\mathbb{R}^N)} < \infty \ , \ f(x,.) \in C(A) \quad \text{for all} \quad x \in \mathbb{R}^N$$

$$\exists C > 0 \ , \ \forall \alpha \in A, \ \forall \xi \in \mathbb{R}^N : |\xi| = 1 \ , \ \partial_\xi^2 f(.,\alpha) \le C \quad \text{in } \mathcal{D}'(\mathbb{R}^N) \ .$$

II - RELATED QUESTIONS

II.1. Local regularity

We have seen in section I that, at least if $\lambda > \lambda_1$, one could deduce the regularity of u inside $\mathcal{O}$ (in the form of (7)) from the knowledge of the regularity of u nearby or on Γ . But clearly this argument may be localized and we find the :

Theorem II.1. : *Let* ω *be a bounded smooth open set included in* $\mathcal{O}$. *We assume that* (9) *hold, that* $\lambda > \lambda_1$ *and that we have either :*

$$\exists \nu > 0 \ , \ \forall (x,\alpha) \in \partial\omega \times A \ , \quad a(x,\alpha) \ge \nu I_N \tag{27}$$

or

$$\forall (x,\alpha) \in \partial\omega \times A \ , \quad \sigma^T(x,\alpha).n(x) = 0, \ \ b_i(x,\alpha)n_i(x) - a_{ij}(x,\alpha)\partial_{ij}d(x) \le 0$$

(where $n(x)$ *is the unit outward normal to* $\partial\omega$ *and* $d(x) = \text{dist}(x,\partial\omega)$*) .* *In case we assume* (27) *, we assume in addition that* u *is u.s.c. on* $\bar{\mathcal{O}}$. *Then* $u \in W^{1,\infty}(\omega)$ *and we have :*

$$\exists C > 0 \ , \quad \forall \xi \in \mathbb{R}^N : |\xi| = 1 \ , \quad \partial_\xi^2 u \le C \quad \textit{in } \mathcal{D}'(\omega) \ .$$

Remark II.1. : Of course this regularity yields the same consequences as Theorem I.1. (cf. Theorem I.2., Corollary I.1). It is also possible to relax the assumptions on ω and to consider the optimal cost function for the control problem with pay-off at the exit from $\mathcal{O}$.

Remark II.2. : We conjecture that in case we assume (27), the upper-semi-continuity of u on $\mathcal{O}$ is not needed. Let us recall that in Part 1 [36], we gave general results insuring that u is u.s.c. on $\bar{\mathcal{O}}$.

Proof of Theorem II.1. : First, if we assume (28), then by the proof of Theorem I.3. in the case when $\Gamma_0 = \Gamma$, we see that for all $\mathcal{A}$ and for all $x \in \omega$: $\tau = +\infty$ a.s. It is then very easy to conclude since $\lambda > \lambda_1$ (cf. the proof of Theorem I.3).

Next, if we assume (27), we first show that u is smooth in a neighborhood of $\partial\omega$ and this implies Theorem II.1. by the same method as in step 2 of Theorem I.1. (section I.3). We recall (see Part 1 [36]) that u satisfies :

$$u(x) = \inf_{\mathcal{A}} E \left\{ \int_0^\sigma f(X_t,\alpha_t)e^{-\lambda t} + u(X_\sigma)e^{-\lambda\sigma} \right\}, \quad \forall x \in \bar{\gamma}$$

where γ is a smooth open set containing $\partial\omega$ and such that (27) holds on $\bar{\gamma}$; and where σ is the first exit time of X_t from γ (or $\bar{\gamma}$) . Since u is u.s.c. on $\bar{\mathcal{O}}$, there exist $u^\varepsilon \in C(\bar{\mathcal{O}})$ such that :

$$\|u^\varepsilon\|_{L^\infty(\mathcal{O})} \le C \text{ (ind. of } \varepsilon), \quad u^\varepsilon(x) \downarrow u(x) \text{ as } \varepsilon \downarrow 0_+$$

for all $x \in \bar{\mathcal{O}}$. Denoting by v^ε the function defined by :

$$v^\varepsilon(x) = \inf_{\mathcal{A}} E \left\{ \int_0^\sigma f(X_t,\alpha_t)e^{-\lambda t} + u^\varepsilon(X_\sigma)\, e^{-\lambda\sigma} \right\}, \quad \forall x \in \bar{\gamma}$$

we deduce that : $v^\varepsilon \downarrow u$ as $\varepsilon \downarrow 0_+$, for all $x \in \bar{\gamma}$.

On the other hand, in view of Corollary I.2., $v^\varepsilon \in W^{2,\infty}_{loc}(\gamma) \cap C(\overline{\gamma})$ (or $C^{2,\theta}(\gamma) \cap C(\overline{\gamma})$) is the solution of :

$$\sup_{\alpha \in A} \{A_\alpha v^\varepsilon(x) - f(x,\alpha)\} = 0 \quad \text{in} \quad \gamma, \quad v^\varepsilon = u^\varepsilon \quad \text{on} \quad \partial\gamma.$$

But since the estimates shown in S. Lenhart [35], L.C. Evans [14] yield for each $\gamma' \subset\subset \gamma$:

$$\| v^\varepsilon \|_{W^{2,\infty}(\gamma')} \le C(\gamma') \, \| v^\varepsilon \|_{L^\infty(\gamma)}$$

$$\| v^\varepsilon \|_{C^{2,\theta}(\gamma')} \le C(\gamma') \, \| v^\varepsilon \|_{W^{2,\infty}(\gamma)}$$

we see that v^ε is bounded in $W^{2,\infty}_{loc}(\gamma)$ (or $C^{2,\theta}(\gamma)$) and thus u belongs to $C^{2,\theta}(\gamma)$; and we conclude.

II.2. Other problems

We give in this section the analogues of the results proved in the preceding sections for stochastic control problems of the following kinds : optimal stopping and continuous control problems, time dependent problems and problems with switching costs.

We begin by the optimal stopping problems : we keep the notation used before concerning admissible systems and we define for $x \in \overline{\mathcal{O}}$, $\mathcal{A}$ admissible system and θ stopping time (adapted to the σ-fields F_t of $\mathcal{A}$) a cost function :

$$J(x,\mathcal{A},\theta) = E\{\int_0^{\tau \wedge \theta} f(X_t, \alpha_t)\exp(-\int_0^t c)dt + 1_{(\theta \le \tau)} \psi(X_\theta)\exp(-\int_0^\theta c)\} \;;$$

and the optimal cost function is now defined by :

$$u(x) = \inf_{\mathcal{A},\theta} J(x,\mathcal{A},\theta)$$

where we consider in the infimum all admissible systems $\mathcal{A}$ and all stopping times θ. We will assume that ψ satisfies

$$\psi \in W^{1,\infty}(\mathcal{O}); \quad \exists C>0, \ \forall \xi \in \mathbb{R}^N : |\xi| = 1 , \quad \partial_\xi^2 \psi \le C \ \text{ in } \mathcal{D}'(\mathcal{O}) \tag{29}$$

where $\mathcal{O}$ is a bounded smooth domain.

Exatly as in section I, one proves

Theorem II.2. : *Let us assume* (9), (5') *(if* $\mathcal{O} \neq \mathbb{R}^N$*)* , (29) *and*

$$\psi > 0 \quad \text{on } \Gamma_i \quad \text{or} \quad \psi = 0 \text{ on } \Gamma_i \quad \text{for } i = 1,2. \tag{30}$$

Then, if $\lambda > \lambda_1$ *and if* ψ *has bounded second derivatives nearby* $\Gamma_1 \cup \Gamma_2$ *, we have :*

i) $u \in W^{1,\infty}(\mathcal{O})$, $u = 0$ *on* $\Gamma_1 \cup \Gamma_2$

ii) (7) *and* (8) *hold :*

$$\exists C > 0 , \quad \forall \xi \in \mathbb{R}^N : |\xi| = 1 , \quad \partial_\xi^2 u \le C \text{ in } \mathcal{D}'(\mathcal{O}) \tag{7}$$

$$\sup_\alpha \|A_\alpha u\|_{L^\infty(\mathcal{O})} < +\infty . \tag{8}$$

iii) *The* HJB *equation holds :*

$$\max \{u-\psi, \sup_{\alpha \in A} \{A_\alpha u(x) - f(x,\alpha)\}\} = 0 \quad \text{a.e. in } \mathcal{O} .$$

Remark II.3. : Of course exactly as in section I.4., we can consider variants of (5') and we may give general results including both Theorem I.3. and II.2. In addition, as in Theorem I.4., one could consider unbounded domains and unbounded coefficients. Let us also mention that if $\phi > 0$ on Γ_i then it is enough to assume (29) only on $\mathcal{O}_i^\delta$ (for all $\delta > 0$) where $\mathcal{O}_i^\delta = \{x \in \mathcal{O},\ \text{dist}(x,\Gamma_i) > \delta\}$, we do not need to assume that ψ has bounded second derivatives near $\Gamma_1 \cup \Gamma_2$.

Remark II.4. : Even in the non-degenerate case, we do not know if the assumption : $\psi \geq 0$ on $\Gamma_1 \cup \Gamma_2$ is enough to insure that (7) holds.
In the particular case where $A = \{\alpha\}$ and $a(x,\alpha)$ is definite positive, this was proved by R. Jensen [23] : using the results and methods of [23] , we were only able to treat the case when the matrices $a(x,\alpha)$ are, uniformly on $\Gamma \times A$, definite positive and when $(a(x,\alpha).n(x))$ $(a_{ij}(x,\alpha)n_i(x)n_j(x))$ is independent of α on Γ.

In the uniformly elliptic case we may relax the assumptions on λ exactly as in section I.5 : we will now consider

$$u(x) = \inf_{\mathcal{A},\theta} \quad E \left\{ \int_0^{\tau\wedge\theta} f(X_t,\alpha_t)\exp\left(-\int_0^t c\right) + 1_{(\theta < \tau)}\psi(X_\theta)\exp\left(-\int_0^\theta c\right) \right.$$
$$\left. + 1_{(\tau \leq \theta)}\,\phi(X_\tau)\exp\left(-\int_0^\tau c\right)\right\}$$

where $\phi \in C(\Gamma)$ and ψ satisfies now :

$$\psi \in C(\bar{\mathcal{O}});\ \forall \delta > 0, \quad \exists C_\delta > 0\ ,\ \forall \xi \in \mathbb{R}^N : |\xi| = 1\ ,\ \partial_\xi^2\psi \leq C_\delta \quad \text{in } \mathcal{D}'(\mathcal{O}_\delta) \tag{29'}$$

where $\mathcal{O}_\delta = \{x \in \mathcal{O},\ \text{dist}(x,\Gamma) > \delta\}$.

Theorem II.3. : *Let us assume that* $\psi \geq \phi$ *on* Γ *and that* (9), (29') *and* (24) *hold :*

$$\exists \nu > 0\ ,\ \forall (x,\alpha) \in \bar{\mathcal{O}} \times A\ , \quad a(x,\alpha) \geq \nu\, I_N\ .$$

i) *If* $\lambda > -\underline{\lambda}_1$ *then* $u \in W^{2,\infty}_{loc}(\mathcal{O}) \cap C(\overline{\mathcal{O}})$ *is the unique solution of*

$$\max \{u-\psi \,,\ \sup_{\alpha\in A} \{A_\alpha u - f(.,\alpha)\}\} = 0 \ \text{a.e. in } \mathcal{O} \,,\ u = \phi \ \text{on } \Gamma .$$

ii) *If* $f(.,\alpha),\phi,\psi$ *are nonnegative* $(\forall\alpha)$ *and if* $\lambda \in \mathbb{R}$ *then* $u \in W^{2,\infty}_{loc}(\mathcal{O}) \cap C(\overline{\mathcal{O}})$ *and* u *satisfies :* $\max \{u-\psi, \sup_{\alpha\in A} \{A_\alpha u-f(.,\alpha)\}\} = 0$ *a.e. in* $\mathcal{O}$, $u = \phi$ *on* Γ. *In addition if* A *is compact,* u *is the minimum element of the set* S *of super solutions* v *satisfying (for example) :*

$$v \in W^{2,\infty}_{loc}(\mathcal{O}) \cap C(\overline{\mathcal{O}}) \,,\quad v \geq 0 \text{ in } \mathcal{O} \,,\ \max \{u-\psi, \sup_{\alpha\in A}\{A_\alpha u-f(.,\alpha)\}\} \geq 0$$
$$\text{a.e. in } \mathcal{O}, v \geq \phi \text{ on } \Gamma$$

Finally, if $\lambda > -\overline{\lambda}_1$, u *is the unique nonnegative solution of the HJB equation in* $W^{2,\infty}_{loc}(\mathcal{O}) \cap C(\overline{\mathcal{O}})$.

<u>Remark II.5.</u> : Most of this theorem is identical to the results of section I.5 and is proved by straightforward adaptations of the proofs in [55], [35], [16], [56]. The case $\lambda \geq 0$ was considered by S. Lenhart [35]. The main new point is part ii) : let us emphasize that we only assume that $\lambda \in \mathbb{R}$ i.e. :

$$\inf_{x,\alpha} c(x,\alpha) > -\infty .$$

<u>Proof of Part ii)</u> : To simplify we take $c \equiv \lambda$ and we may assume $\lambda < 0$. Then we have : $0 \leq u(x) \leq J(x,\mathcal{A},0) = \psi(x)$ in $\mathcal{O}$ and thus $u \in L^\infty(\mathcal{O})$.
Next, we observe that in view of an easy argument (for example a martingale argument as in N.V. Krylov [24]) we also have :

$$u(x) = \inf_{\mathcal{A},\theta} E \{\int_0^{\tau\wedge\theta} f(X_t,\alpha_t) - \lambda u(X_t) \ dt + 1_{(\theta \leq \tau)}\psi(X_\theta) + 1_{(\tau\leq\theta)}\phi \ (X_\tau)\}.$$

But in view of Part 1, for smooth functions of x $g(x,\alpha) = f(x,\alpha) - \lambda w(x)$ one has :

$$v(x)=\inf_{\mathcal{A},\theta} E\left\{\int_0^{\tau\wedge\theta} g(X_t,\alpha_t)dt + 1_{(\theta<\tau)}\psi(X_\theta) + 1_{(\tau\le\theta)}\phi(X_\tau)\right\} \in C(\bar{\mathcal{O}})$$

and thus :

$$|(u-v)(x)| \le \sup_{\mathcal{A}} E\{\int_0^{\tau} |\lambda||u-w| (X_t)dt\} \le C \|u-w\|_{L^N(\mathcal{O})}$$

where we used (24) and the Alexandrov-Pucci-Krylov inequality (see N.V. Krylov [24]). Thus approximating u by a sequence of smooth w converging to u in $L^N(\mathcal{O})$, we see that $u \in C(\bar{\mathcal{O}})$.

Next, we observe that we may define $\underline{\lambda}_1$, $\bar{\lambda}_1$, as functions of $\mathcal{O}$ and clearly one has : $\underline{\lambda}_1(\mathcal{O}) \le \underline{\lambda}_1(\mathcal{O}')$ if $\mathcal{O}' \subset \mathcal{O}$, $\bar{\lambda}_1(\mathcal{O}) \le \bar{\lambda}_1(\mathcal{O}')$ if $\mathcal{O}' \subset \mathcal{O}$. Furthermore it is an easy exercise to show that :

$$\underline{\lambda}_1(\mathcal{O}) \uparrow +\infty \text{ if } \operatorname{diam}(\mathcal{O}) \downarrow 0 .$$

Therefore for ω smooth open set such that $\underline{\lambda}_1(\omega) > -\lambda$ and $\omega \subset \mathcal{O}$, we may apply Part i) of the above theorem. Since u also satifies (cf. Part 1 [36]):

$$u(x) = \inf_{\mathcal{A},\theta} E\left\{\int_0^{\sigma\wedge\theta} f(X_t,\alpha_t)e^{-\lambda t} dt + 1_{(\sigma\le\theta)}u(X_\sigma) + 1_{(\sigma>\theta)}\psi(X_\theta)\right\}$$

where σ is the first exit time of X_t from $\bar{\omega}$; we deduce : $u \in W^{2,\infty}_{loc}(\omega)$. And from the above remarks we finally deduce $u \in W^{2,\infty}_{loc}(\mathcal{O})$. In addition the equation holds in ω and thus in $\mathcal{O}$.

Finally, let us prove that : $u(x) = \min_{v\in S} v(x) \quad \forall\, x \in \bar{\mathcal{O}}$.

Let $\omega = \{x \in \mathcal{O},\ u(x) < \psi(x)\}$, ω is an open set. We denote by σ the first entrance time of X_t in ω . Using for example Corollary I.3. and Itô's formula, it is easy to show that if $v \in S$ then we have :

$$v(x) \ge \inf_{\mathcal{A}} E\left\{\int_0^{\sigma\wedge T} f(X_t,\alpha_t)e^{-\lambda t}dt + v(X_{\sigma\wedge T})\, e^{-\lambda(\sigma\wedge T)}\right\}$$

for all $T < \infty$, $x \in \bar{\mathcal{O}}$. But since A is compact, one can find $\alpha(x)$ and X_t

such that for all $T < \infty$, $x \in \bar{\mathcal{O}}$:

$$v(x) \geq E\{ \int_0^{\sigma \wedge T} f(X_t,\alpha_t)e^{-\lambda t}dt + v(X_{\sigma \wedge T})e^{-\lambda(\sigma \wedge T)}\}$$

and taking $T \to +\infty$, we deduce since f,v are nonnegative :

$$v(x) \geq E\{\int_0^{\sigma} f(X_t,\alpha_t)e^{-\lambda t}dt + v(X_\sigma)e^{-\lambda\sigma}\}$$

$$\geq E\{\int_0^{\sigma} f(X_t,\alpha_t)e^{-\lambda t}dt + 1_{(\sigma<\tau)}\psi(X_\sigma)e^{-\lambda\sigma} + 1_{(\sigma=\tau)}\phi(X_\sigma)e^{-\lambda\sigma}\}$$

$$\geq \inf_{\mathcal{A},\theta} J(x,\mathcal{A},\theta) = u(x) . \quad \square$$

Remark II.6. : We do not know if the assumption on A is necessary. On the other hand as soon as $\lambda \leq -\bar{\lambda}_1$, there is no uniqueness of nonnegative solutions of the HJB equation : indeed take $f \equiv 0$, $\psi \equiv 1$ and $\lambda = \bar{\lambda}_1$. Clearly $u \equiv 0$ in $\bar{\mathcal{O}}$ if $\phi \equiv 0$. But (cf. section I.5) there exists $\psi_1 > 0$ in $\mathcal{O}$ such that :

$$\psi_1 = 0 \text{ on } \Gamma, \quad \sup_{\alpha\in A}\{A_\alpha\psi_1(x)\} = 0 \text{ in } \mathcal{O}, \quad \psi_1 \in C^2(\mathcal{O}) \cap W^{2,\infty}(\mathcal{O})$$

and $v = \Theta\psi_1$ satisfies, for $0 \leq \Theta \leq \|\psi_1\|_\infty^{-1}$:

$$\max\{u(x)-\psi(x), \sup_{\alpha\in A}\{A_\alpha v(x)-f(x,\alpha)\}\} = 0 \text{ in } \mathcal{O}, \quad v = 0 \text{ on } \Gamma .$$

To motivate the assumption on A , we may also modify the above example by assuming : $\inf_{\alpha\in A} f(x,\alpha) \equiv 0$ in $\bar{\mathcal{O}}$, $\psi \equiv 1$, $\lambda = -\bar{\lambda}_1$, $\phi \equiv 0$. Then clearly $\min_{v\in S} v(x) \equiv 0$ in $\bar{\mathcal{O}}$, but it is not clear that $u \equiv 0$ in $\bar{\mathcal{O}}$. $\square$

We next consider *time-dependent problems* and we will keep the notation of Parts 1-2 [36],[37] : let $T > 0$ be fixed, the state process X_s is now given by:

$$dX_s = \sigma(X_s,t+s,\alpha_s)dB_s + b(X_s,t+s,\alpha_s)ds, \quad X_0 = x \in \bar{\mathcal{O}}$$

where $t \in [0,T]$ is fixed, and thus $X_s = X_s\ (x,t)$. We consider cost functions of the following type :

$$J(x,t,\mathcal{A}) = E\{\int_0^{\tau\wedge(T-t)} f(X_s,t+s,\alpha_s)\exp\{-\int_0^s c(X_\mu,t+\mu,\alpha_\mu)d\mu\} +$$

$$+ 1_{((T-t)\le\tau)}\ u_o(X_{T-t})\exp\{-\int_0^{T-t} c(X_s,t+s,\alpha_s)ds\}\}$$

and the optimal cost function is given by :

$$u(x,t) = \inf_{\mathcal{A}} J(x,t,\mathcal{A}) \quad , \qquad \text{for}(x,t) \in \overline{Q} = \overline{\mathcal{O}} \times [0,T] \ ;$$

where $\mathcal{O}$ is a smooth open set in $\mathbb{R}^N$ and $Q = \mathcal{O}\times(0,T)$.

We will assume that the coefficients $\sigma(x,t,\alpha), b(x,t,\alpha), c(x,t,\alpha)$ and the data $f(x,t,\alpha), u_o(x)$ satisfy :

$$\begin{cases} \sup_{t,\alpha} \|\phi(.,t,\alpha)\|_{W^{2,\infty}(\mathbb{R}^N)} < +\infty; \quad \phi(x,.,.) \in C([0,T]\times A), \ \ \forall\ x \in \mathbb{R}^N ; \\ \\ \text{for all } \phi = \sigma_{ij} \quad (1 \le i \le N,\ 1 \le j \le m),\ b_i(1 \le i \le N),\ c\ . \end{cases} \tag{31}$$

$$u_o \in W^{1,\infty}(\mathbb{R}^N) \ ;\ \exists\, C > 0\ ,\quad \forall\ \xi \in \mathbb{R}^N : |\xi| = 1\ ,\ \ \partial_\xi^2 u_o \le C \quad \text{in } \mathcal{D}'(\mathbb{R}^N) \tag{32}$$

Exactly as in section I, one can prove part i) of the

<u>Theorem II.4.</u> : *We assume* (31),(32) *and the analogue of* (5') *(if* $\mathcal{O} \ne \mathbb{R}^N$*)*

$$\begin{cases} \forall(x,t,\alpha) \in \Gamma_o\times[0,T]\times A,\ \sigma = 0\ ,\ \ b.n(x) \le 0 \\ \exists\nu > 0\ ,\ \ \forall(x,t,\alpha) \in \Gamma_1\times[0,T]\times A,\ \ a_{ij}(x,t,\alpha)n_i(x)n_j(x) \ge \nu > 0 \\ \exists\nu > 0\ ,\ \ \forall(x,t,\alpha) \in \Gamma_2\times[0,T]\times A,\ \sigma = 0\ ,\ \ b.n(x) \ge \nu > 0\ . \end{cases} \tag{5''}$$

Finally we assume that u_o *has bounded second derivatives nearby* $\Gamma_1 \cup \Gamma_2$

and $u_0 = 0$ *on* $\Gamma_1 \cup \Gamma_2$. *(if* $\mathcal{O} \neq \mathbb{R}^N$ *).*

Then we have :

i) $u \in C(\bar{Q})$, $u = 0$ *on* $(\Gamma_1 \cup \Gamma_2) \times [0,T]$, $u = u_0$ *on* $\bar{\mathcal{O}} \times \{T\}$,

$D_x u \in L^\infty(Q)$ *and :*

$$\exists C > 0 \ , \ \forall \xi \in \mathbb{R}^N : |\xi| = 1 \ , \ \partial_\xi^2 u \le C \text{ in } \mathcal{D}'(Q) \ .$$

ii) $u \in W^{2,1,2}(Q \cap B_R)(\forall R < \infty)$ *and there exists* $C > 0$ *such that*

$$\begin{cases} -C \le A_\alpha u \le C + \dfrac{\partial u}{\partial t} & \text{in } \mathcal{D}'(Q) \text{ for all } \alpha \in A \\ -\dfrac{\partial u}{\partial t} - C\Delta u \ge -C & \text{in } \mathcal{D}'(Q) \ . \end{cases} \tag{34}$$

iii) *Furthermore the* HJB *equation holds :*

$$-\frac{\partial u}{\partial t} + \sup_{\alpha \in A} \{A_\alpha u - f(x,t,\alpha)\} = 0 \quad \text{a.e. in } Q \ . \tag{35}$$

Remark II.7. : The same extensions as in section I.4 : hold various variants of (5") are possible that we will not give. Let us also remark that considering the process $Y_s = (X_s, t+s)$, one sees that the above result is "almost" a consequence of Theorem I.3. taking $\Gamma_0' = (\Gamma_0 \times [0,T]) \cup (\bar{\mathcal{O}} \times \{0\})$, $\Gamma_1' = ((\Gamma - \Gamma_0) \times [0,T]) \cup (\bar{\mathcal{O}} \times \{T\})$ - except for the regularity conditions assumed on the coefficients. It is possible to give one general result including both theorems but we will skip such a heavy and straightforward extension. Let us also mention that it is possible to treat the case of different boundary conditions than $u = 0$ on $(\Gamma_1 \cup \Gamma_2) \times [0,T]$ and one has results analogous to Theorems I.5.

Remark II.8. : Exactly as in section I.4, we can treat the case of unbounded coefficients in x . More precisely, we only need to assume :

$$
\begin{cases}
\phi(\cdot,t,\alpha) \in W^{2,\infty}_{loc}(\mathbb{R}^N) \quad \text{for } \alpha \in A,\ t \in [0,T], \quad \text{for } \phi = \sigma_{ij}, b_i, c, f ; \\
\phi(x,\cdot,\cdot) \in C([0,T] \times A) \quad \text{for } x \in \mathbb{R}^N \ ; \ u_0 \in W^{2,\infty}_{loc}(\mathbb{R}^N) \ ; \\
\exists C > 0 \ , \ \|\sigma\| + |b| \leq C|x| + C \ , \ \forall (x,t,\alpha) \in \mathbb{R}^N \times [0,T] \times A \qquad (17') \\
\exists C > 0 \ , \ \|D_x\sigma\| + |D_x b| \leq C \ , \qquad \forall (x,t,\alpha) \in \mathbb{R}^N \times [0,T] \times A \\
\exists C > 0 \ , \ \exists p > 0 \ , \ \sup_{|\beta| \leq 2} \{|D_x^\beta f| + |D_x^\beta c| + |D_x^\beta u_0|\} + \|D_x^2\sigma\| + \|D_x^2 b\| \leq \\
\qquad \leq C + C|x|^p \ , \ \forall (x,t,\alpha) \in \mathbb{R}^N \times [0,T] \times A.
\end{cases}
$$

As we already indicated i) is proved exactly as in section I. Furthermore we will see below that ii) is deduced from i) and that iii) follows from the regularity implied by i), ii) and from the fact that u is the viscosity solution of the HJB equation, using the results of Part 2 [37] exactly as Theorem I.2. follows from Theorem I.1. Therefore we just have to prove ii) : but, recall (cf. Part 1 [36]) that u satisfies :

$$
-\frac{\partial u}{\partial t} + A_\alpha u \leq f(\cdot,\alpha) \quad \text{in } \mathcal{D}'(Q) \ , \quad \forall \alpha \in A \ .
$$

Since (33) implies easily : $A_\alpha u \geq - C$ in $\mathcal{D}'(Q)$, $\forall \alpha \in A$; we just have to prove that $\frac{\partial u}{\partial t} \in L^2(Q)$ and we already know that :

$$
\frac{\partial u}{\partial t} \geq - C \quad \text{in } \mathcal{D}'(Q) \ . \tag{34'}
$$

We now prove that $\frac{\partial u}{\partial t} \in L^2(Q)$: to this end we will assume that u is smooth, the argument being easily justified by a tedious approximation method. Because of (5") , we can find $\xi \in C^2(\bar{\mathcal{O}})$ (we will assume that $\mathcal{O}$ is bounded to simplify) such that : $\xi = 0$ on Γ_0 , $\xi > 0$ in $\bar{\mathcal{O}} - \Gamma_0$ and there exists $C > 0$:

$$
a(x,\alpha) \leq C\, \xi^2(x)\, I_N \ , \qquad \forall x \in \bar{\mathcal{O}} \ , \quad \forall \alpha \in A \ .
$$

But this inequality combined with (35) yields :

$$0 = -\frac{\partial u}{\partial t} + \sup_{\alpha} \{A_\alpha u - f(x,t,\alpha)\} \leq -\frac{\partial u}{\partial t} - C\xi^2 \Delta u + C \quad \text{in} \quad Q ,$$

and we obtain :

$$-\frac{\partial u}{\partial t} - C\xi^2 \Delta u \geq -C \quad \text{in} \quad Q , \quad u = 0 \quad \text{on} \quad \Gamma_1 \cup \Gamma_2 .$$

Therefore we deduce :

$$-\frac{\partial u}{\partial t} - C \operatorname{div}(\xi^2 \nabla u) \geq -C \quad \text{in} \quad Q , \quad u = 0 \quad \text{on} \quad \Gamma_1 \cup \Gamma_2 .$$

Next, in view of (34'), $C + \frac{\partial u}{\partial t} \geq 0$ in Q ; and multiplying the above inequality by $C + \frac{\partial u}{\partial t}$ and integrating by parts over Q we find (recall that $D_x u \in L^\infty(Q)$ and $\frac{\partial u}{\partial t} = 0$ on $\Gamma_1 \cup \Gamma_2$) :

$$\left|\frac{\partial u}{\partial t}\right|^2_{L^2(Q)} + \frac{C}{2} \int_0^T \frac{d}{dt} \left\{\int_{\mathcal{O}} \xi^2 \, |\nabla u|^2 \, dx\right\} dt \leq C \left|\frac{\partial u}{\partial t}\right|_{L^2(Q)} + C$$

and we conclude easily.

Remark II.9. : Let us mention that from i), ii) one may deduce :

$$|u(x,t) - u(y,s)|^2 \leq C \{|x-y|^2 + |t-s|^{\varepsilon/(N+2)}\} , \quad \forall (x,t),(y,s) \in \overline{Q} .$$

Indeed, we have :

$$\begin{aligned}
|u(x,t) - u(y,s)|^2 &\leq C|x-y|^2 + C|u(x,s) - u(x,t)|^2 \\
&\leq C|x-y|^2 + \frac{C}{\delta^N} \int_{B(x,\delta)} |u(\xi,s) - u(\xi,t)|^2 d\xi + C\delta^2 \\
&\leq C|x-y|^2 + C\delta^2 + \frac{C}{\delta^N} \int_{B(x,\delta)} \left\{\int_s^t \left|\frac{\partial u}{\partial \lambda}(\xi,\lambda)\right| d\lambda\right\}^2 d\xi \\
&\leq C|x-y|^2 + C\delta^2 + \frac{C|t-s|}{\delta^N}
\end{aligned}$$

and we conclude choosing $\delta = |t-s|^{1/(N+2)}$.

Let us give a few additionnal results :

Corollary II.1. : *Under the assumptions of Theorem II.4. and if we assume in addition :* $u_o \in W^{2,\infty}(\mathbb{R}^N)$ *and*

$$\sup_{x,\alpha} \left\| \frac{\partial \phi}{\partial t}(x,.,\alpha) \right\|_{L^\infty(0,T)} < \infty, \text{ for } \phi = \sigma_{ij},\ b_i,\ c,\ f\ ; \tag{36}$$

then $\frac{\partial u}{\partial t} \in L^\infty(Q)$ *and thus :* $\sup_{\alpha \in A} \|A_\alpha u\|_{L^\infty(Q)} < +\infty.$

This follows easily from the fact that (36) implies :

$$E\,[|X_s(x,t_1) - X_s(x,t_2)|^2] \le C_T|t_1-t_2|^2 \quad \text{if} \quad x \in \bar{\mathcal{O}}\ , \quad s \in [0,T]\ ;$$

for all $T < \infty$.

Corollary II.2. : *Under the assumptions of Theorem II.4. and if there exist* ω *bounded open set included in* $\mathcal{O}$, $p \in \{1,\dots,N\}$, $k \ge 1$, $\nu > 0$ *such that :*

$$\forall\, x \in \omega,\ \exists \alpha_1,\dots,\alpha_k \in A\ ,\ \forall \xi \in \mathbb{R}^N\ ,\ \sum_{\ell=1}^{k} a_{ij}(x,\alpha_\ell)\xi_i\xi_j \ge \nu \sum_{i=1}^{p} \xi_i^2\ .$$

Then $\partial_{ij}\, u \in L^2(\omega \times (0,T))$ *for* $1 \le i,j \le p$. *If in addition,* $u_o \in W^{2,\infty}(\mathbb{R}^N)$ *and* (36) *holds, then* $\partial_{ij}\, u \in L^\infty(\omega \times (0,T))$ *for* $1 \le i,j \le p$.

Remark II.10. : In particular if the matrices $a(x,t,\alpha)$ are definite positive uniformly on $\bar{Q} \times A$, if (31), (36) hold and if $u_o \in W^{2,\infty}(\mathbb{R}^N)$ with $u_o = 0$ on Γ . We find that $u \in W^{2,1,\infty}(Q)$. This result was shown by P.L. Lions [46] by p.d.e. methods.

Remark II.11. : Let us finally point out that one could have similar results for "combination problems" where we combine optimal stopping and time dependent problems.

We conclude this section by a few results on *optimal stochastic control problems with switching costs* : we take $A = \{\alpha_1,\ldots,\alpha_n\}$ and we denote by $\sigma^i(x) = \sigma(x,\alpha_i)$, $a^i = a(x,\alpha_i)$, $b_i(x) = b(x,\alpha_i)$, $c^i(x) = c(x,\alpha_i)$, $f^i(x) = f(x,\alpha_i)$, $A_i = A_{\alpha_i}$ for $1 \le i \le n$. Let $k(i,j) \ge 0$ be given (for $1 \le i \ne j \le n$). We consider admissible systems given by $\mathcal{A}^i = (\Omega,F,F^t,P,\Theta,\ell(t,\omega))$ where Θ is a sequence $(\Theta_j)_{j\ge 0}$ of F^t-stopping times such that : $\Theta_0 = 0 < \Theta_1 < \ldots < \Theta_j < \Theta_{j+1}$ a.s. and $\lim_{j\uparrow+\infty} \Theta_j = +\infty$ a.s. Finally $\ell(t,\omega)$ is an integer-valued process, constant on intervals $[\Theta_j,\Theta_{j+1}[$ (for $j \ge 0$) and such that : $\ell(t,\omega) = i$ if $0 \le t < \Theta_1(\omega)$. We will also use the notations : ℓ_t for $\ell(t,.)$, ℓ_{t-} for $\lim_{s\uparrow t-} \ell_s$.

Then the state process is given by the solution of :

$$dX_t = \sigma^{\ell_t}(X_t)dB_t + b^{\ell_t}(X_t)dt \ , \quad X_0 = x$$

that is a process which "switches" at times Θ_i from one diffusion process to another one. We finally give the cost functions and the optimal cost function :

$$J(x,\mathcal{A}^i) = E\,\{\int_0^\tau f^{\ell_t}(X_t)\exp(-\int_0^t c^{\ell_s}(X_s)ds) + $$

$$+ \sum_{j\ge 1} k(\ell_{\Theta_j^-},\ell_{\Theta_j})\exp(-\int_0^{\Theta_j} c^{\ell_t}(X_t)dt\} \ ;$$

$$u^i(x) = \inf_{\mathcal{A}^i} J(x,\mathcal{A}^i) \ ;$$

for $x \in \bar{\mathcal{O}}$. We assume that $\mathcal{O}$ is a smooth open set in $\mathbb{R}^N$. Then we prove as in section I the following :

<u>Theorem II.5.</u> : *Under the assumptions* (9), (5') *(if* $\mathcal{O} \ne \mathbb{R}^N$) *and if* $\lambda > \lambda_1$ *then we have :*

i) $u^i \in W^{1,\infty}(\mathcal{O})$, $u^i = 0$ *on* $\Gamma_1 \cup \Gamma_2$;

ii) *the following inequality holds :*

$$\exists C > 0 \ , \ \forall \xi \in \mathbb{R}^N : |\xi| = 1 \ , \quad \partial_\xi^2 u^i \leq C \ \text{ in } \mathcal{D}'(\mathcal{O}) \ ;$$

and thus $A_i u^i \in L^\infty(\mathcal{O})$.

iii) *the HJB equations hold : for all* $i \in \{1,\ldots,n\}$ *we have*

$$\max \{A_i u^i(x) - f^i(x) \ , \ \max_{j \neq i} \{u^i(x) - k(i,j) - u^j(x)\}\} = 0 \quad \text{a.e. in } \mathcal{O} \ .$$

iv) *Let* $i \in \{1,\ldots,n\}$, *let* ω *be an open set included in* $\mathcal{O}$ *and let* $p \in \{1,\ldots,N\}$. *We assume there exists* $\nu > 0$ *such that*

$$\forall \ x \in \omega, \ \ a^i_{k\ell}(x)\xi_k\xi_\ell \geq \nu \sum_{k=1}^{p} \xi_k^2 \quad \text{for all } \ \xi \in \mathbb{R}^N \ .$$

Then $\partial_{ij} u \in L^\infty(\omega)$ *for* $1 \leq i, \ j \leq p$.

Remark II.12. : Exactly as in section I.4. we could give extensions of this result where (5') is relaxed, we could treat the case of unbounded coefficients in x and the case of different boundary conditions. We could also treat optimal stopping or time-dependent problems. We may treat more general problems where for the system i, the state is not allowed to switch to some particular dynamics, i.e. for all $i \in \{1,\ldots,N\}$, we take $K(i) \subset \{1,\ldots,N\}-\{i\}$ (K_j can be empty) and in the preceding description we add the restrictions :

$$\ell_{\Theta_j} \in K(\ell_{\Theta_{j-}}) \quad \text{for all } \ j \geq 1 \ .$$

This corresponds to the case $k(i,j) = +\infty$ if $j \notin K(i), j \neq i$. For this kind of problem, the same result holds except that the HJB equation is now :

$$\max_i \{A_i u^i - f^i \ , \ \max_{j\in K(i)} \{u^i - k(i,j) - u^j\}\} = 0 \quad \text{a.e. in } \mathcal{O} .$$

Finally let us point out that we could combine all these variants or extensions.

Remark II.13. : In the uniformly elliptic case, i.e. if $a^i(x)$ are definite positive uniformly in x, then we may take $\lambda > -\underline{\lambda}_1$ and then $u^i \in W^{2,\infty}(\mathcal{O})$, while if $f^i \geq 0$ then we may take $\lambda > -\overline{\lambda}_1$ and then $u^i \in W^{2,\infty}(\mathcal{O})$ is the unique nonnegative solution of the HJB equations.

Remark II.14. : Observe that if for some distinct integers $i_1,\dots,i_p$ in $\{1,\dots,N\}$ we have : $k(i_1,i_2) = k(i_2,i_3) = \dots = k(i_{p-1},i_p) = k(i_p,i_1) = 0$ then : $u^{i_1} = u^{i_2} = \dots = u^{i_p}$ in $\bar{\mathcal{O}}$. We will say that if this occurs there is a cycle $(i_1,\dots,i_p)$. In particular if $k(i,j) = 0$ for all $i \neq j$ $u^i \equiv u (\forall i)$ is the optimal cost function associated with the HJB equation :

$$\sup_{1\leq i\leq n} \{A_i u - f^i\} = 0 \quad \text{a.e. in } \mathcal{O}, \qquad u = 0 \text{ on } \Gamma_1 \cup \Gamma_2 .$$

Remark II.15.: One can show that, if there is no cycle, $(u^1,\dots u^n)$ is the unique solution of the HJB equations in the following sense :

$$\begin{cases} u_i \in C(\bar{\mathcal{O}}) \ , & u^i = 0 \text{ on } \Gamma_1 \cup \Gamma_2 \quad \text{for all } i \\ A_i u^i \leq f^i \text{ in } \mathcal{D}'(\mathcal{O}) \ , & u^i \leq k(i,j) + u^j \quad \text{for all } j \neq i \ , \\ & \qquad \text{for all } i \\ A_i u^i = f^i \text{ in } \mathcal{D}'(\mathcal{O}_i) \text{ where } & \mathcal{O}_i = \{x \in \mathcal{O} / u^i < \operatorname{Min}_{j\neq i} (k(i,j)+u^j)\} \ , \forall i . \end{cases}$$

This is proved by a simple verification argument using the result of D.W. Stroock and S.R.S. Varadhan [70] and those of the appendix of Part 2 [37] .

But, as soon as there exists a cycle, such a uniqueness result is false and the situation is then identical to the one described in section III.

II.3. Other boundary conditions

First we remark that if $\lambda > 0$, the case of periodic boundary conditions is totally similar to the case of $\mathcal{O} = \mathbb{R}^N$ and that we will consider the case $\lambda = 0$ with periodic boundary conditions and the case of state constraints in some future publications (see also J.M. Lasry and P.L. Lions [34]). Therefore in this section we will only consider the case of *reflected diffusion processes*. We keep the notations of Part 1 [36] : we consider a smooth bounded open set $\mathcal{O}$ in $\mathbb{R}^N$, a smooth vector-field γ (of class C^3 for instance) on $\mathbb{R}^N$ such that :

$$\exists \nu > 0 , \quad \gamma(x).n(x) \geq \nu > 0 \quad , \quad \forall x \in \Gamma . \tag{37}$$

For each admissible system the state process X_t is now given by the solution of the following problem :

$$dX_t = \sigma(X_t,\alpha_t)dB_t + b(X_t,\alpha_t)dt - 1_{(X_t \in \Gamma)} \gamma(X_t)dA_t$$

$$X_o = x \in \overline{\mathcal{O}} , \quad A_o = 0$$

where X_t , A_t are continuous processes and A_t is non-decreasing (adapted to F_t of course). Since we will always assume (9) , the solvability of this problem is insured by standard results on the stochastic differential equations with reflection - cf. Ikeda and Watanabe [22], D.W. Stroock and S.R.S. Varadhan [71] ; see also H. Tanaka [72], A.S. Sznitman and the author [62] . The cost functions and the optimal cost function are given by :

$$J(x,\mathcal{A}) = E\left\{\int_0^\infty f(X_t,\alpha_t)\exp\left(-\int_0^t c(X_s,\alpha_s)ds\right)\right\}$$

$$u(x) = \inf_{\mathcal{A}} J(x,\mathcal{A}) .$$

Let (Γ_o,Γ_1) be a partition of Γ into two smooth, closed parts, possibly empty. We have the :

Theorem II.6. : *We assume that* (9) *holds, that* $\lambda > \lambda_1$ *and that we have :*

$$\left|\begin{array}{l} \exists\varepsilon_o > 0 \ , \ \forall\, x \in \{z \in \mathcal{O},\ \mathrm{dist}(x,\Gamma_1) \in (0,\varepsilon_o)\} , \\ \qquad\qquad\qquad\qquad \inf_{\alpha\in A}\ \inf_{|\xi|=1} (a_{ij}(x,\alpha)\xi_i\xi_j) > 0 \ ; \\ \forall(x,\alpha) \in \Gamma_o \times A \ , \ \sigma^T.n = 0 \ , \ b_i n_i - a_{ij}\partial_{ij}\, d \le 0 \ . \end{array}\right. \qquad (38)$$

Then we have :

i) $u \in C^{o,\alpha}(\bar{\mathcal{O}})$ *for some* $0 < \alpha \le 1$, $u \in W^{1,\infty}_{loc}(\mathcal{O} \cup \Gamma_o)$.

ii) *The following inequalities hold :*

$$\forall\,\delta > 0 \ , \ \exists C_\delta > 0 \ , \ \forall\,\xi \in \mathbb{R}^N : |\xi| = 1 \ , \ \partial^2_\xi u \le C_\delta \ \textit{in}\ \mathcal{D}'(\mathcal{O}_\delta) \ ;$$

$$\forall\,\delta > 0 \ , \ \exists C_\delta > 0 \ , \ \sup_{\alpha\in A} \|A_\alpha u\|_{L^\infty(\mathcal{O}_\delta)} \le C_\delta \ ;$$

where $\mathcal{O}_\delta = \{x \in \mathcal{O},\ \mathrm{dist}(x,\Gamma_1) > \delta\}$.

iii) *The HJB equation holds :*

$$\sup_{\alpha\in A} \{A_\alpha u(x) - f(x,\alpha)\} = 0 \quad \textit{a.e. in}\ \mathcal{O} .$$

Remark II.16. : Let us point out that (38) means that the matrices $a(x,\alpha)$ are definitive positive, uniformly in α, in a neighborhood of Γ_1 but not necessarily on Γ_1 . Let us also mention that we could treat as well unbounded domains with unbounded coefficients, optimal-stopping problems, time-dependent problems, problems with switching costs... Finally, exactly as in section I, the above estimates yield some extra regularity if some matrices (or convex combinations) are nondegenerate in some directions. In particular, if (38) holds on $\mathcal{O}$ i.e. if *the matrices are definite positive uniformly in* α , *for all* $x \in \mathcal{O}$, then we do not need to assume $\lambda > \lambda_1$ and *if* $\lambda > 0$, $u \in W^{2,\infty}_{loc}(\mathcal{O})$.

To prove Theorem II.6., we only have to remark that by the results of Part 1 [36] , we already know that $u \in C^{0,\alpha}(\bar{\mathcal{O}})$ for some $\alpha \leq 1$ and thus u is also given by :

$$u(x) = \inf_{\mathcal{A}} E \left\{\int_0^{\tau_\delta} f(X_t,\alpha_t)\exp\left(-\int_0^t c(X_s,\alpha_1)ds\right) + u(X_{\tau_\delta})\exp\left(-\int_0^{\tau_\delta} c(X_t,\alpha_t)dt\right)\right\}$$

where τ_δ is the first exit time from $\bar{\mathcal{O}}_\delta$ of X_t and where δ is arbitrary in $(0,\varepsilon_0)$. Applying Theorem II.1., we conclude easily. □

Let us point out that we do not know, even in the uniformly elliptic case, if C_δ in ii) can be taken independent of δ : the only partial answer in this direction is given by :

Theorem II.7. : *We assume that* (9) *holds, that* $\lambda > \lambda_1$ *and that we have :*

$$\begin{cases} \exists \nu > 0 \ , \ \forall (x,\alpha) \in \Gamma_1 \times A \ , \quad a(x,\alpha) \geq \nu \quad I_N \\ \forall (x,\alpha) \in \Gamma_0 \times A \ , \quad \sigma^T.n = 0 \ , \quad b_i n_i - a_{ij}\partial_{ij} d \leq 0 \ ; \end{cases} \tag{38'}$$

$$\forall\ x \in \Gamma_1\ ,\ \ (a.\gamma)(a_{ij}\gamma_i\gamma_j)^{-1}\ \ \textit{is independent of}\ \ \alpha \in A\ . \qquad (39)$$

Then we have :

i) $u \in W^{1,\infty}(\mathcal{O})$, $u \in W^{2,\infty}(\tilde{\Gamma})$ (*where* $\tilde{\Gamma}$ *is a neighborhood of* Γ_1 *in* $\bar{\mathcal{O}}$).

ii) *The estimates* (7) *and* (8) *hold.*

iii) *The HJB equation holds :* $\sup_{\alpha\in A} \{A\ u(x)-f(x,\alpha)\} = 0$ *a.e. in* $\mathcal{O}$; *and we have :*

$$\frac{\partial u}{\partial \gamma} = 0 \ \ \textit{on}\ \ \Gamma_1\ . \qquad (40)$$

Remark II.17. : Again if the matrices $a(x,\alpha)$ definite positive, uniformly in (x,α) , then it is enough to assume $\lambda > 0$ and (39) and the above result still holds (and $u \in W^{2,\infty}(\mathcal{O})$) . We conjecture that at least in this case (39) is not necessary.

Proof of Theorem II.7. : The main new point consists in proving that $u \in W^{2,\infty}(\tilde{\Gamma})$: to explain the new argument we will assume $\Gamma_0 = \emptyset$ and thus $\Gamma_1 = \Gamma$. In view of (38') , there exists $\varepsilon_0 > 0$ such that :

$$a(x,\alpha) \geq \nu\ I_N \ \text{ for all } \ \alpha \in A\ , \text{ for } \ x \ \text{ such that } \ d(x) \leq \varepsilon_0\ .$$

We next introduce $\varepsilon > 0$ and we assume that A is finite : $A = \{\alpha_1,\ldots,\alpha_n\}$- the general case is obtained by remarking that the bounds below do not depend on n . Finally we assume that the coefficients and the data are smooth and that the matrices are definitive positive uniformly in (x,α)-the general case is then obtained by an approximation argument. We consider as in [15],[59], [55] the solution $(u_\varepsilon^1,\ldots,u_\varepsilon^n)$ of the following penalized system :

$$A_i u_\varepsilon^i + \beta_\varepsilon(u_\varepsilon^i - u_\varepsilon^{i+1}) = f^i \quad \text{in } \mathcal{O} \ , \quad \frac{\partial u_\varepsilon^i}{\partial \gamma} = 0 \quad \text{on } \Gamma$$

where $A_i = A_{\alpha_i}$, $f^i = f(.,\alpha_i)$ - we denote by $a^i(x) = a(x,\alpha_i)$, $b^i(x) = b(x,\alpha_i)$ and we assume, to simplify $c^i \equiv \lambda > 0$ - and β_ε satisfies

$$\beta_\varepsilon(t) = \frac{1}{\varepsilon}\beta(t), \quad \beta \in C^\infty(\mathbb{R}) \ , \quad \beta(t) = 0 \ \text{ if } \ t \le 0 \ , \quad \beta''(t) \ge 0 \ \text{ on } \ \mathbb{R} \ ,$$
$$\beta'(t) > 0 \ \text{ if } \ t > 0 \ .$$

We will first prove that, for some appropriate choice of a neighborhood $\tilde{\Gamma}$ of Γ in $\mathcal{O}$, we have : $\|u\|_{W^{2,\infty}(\Gamma)} \le C$ (independently of i,n,ε ...).
To this end we introduce $\xi \in \mathcal{D}_+(\bar{\mathcal{O}})$ such that $0 \le \xi \le 1$ in $\bar{\mathcal{O}}$, $\frac{\partial \xi}{\partial \gamma} = 0$ on Γ, $\xi \equiv 1$ in a neighborhood of a given point $\bar{x} \in \Gamma$ and Supp ξ is contained in an open neigborhood ω of $\bar{x}$ satisfying : there exist, for all i C^4-diffeomorphisms $\emptyset_i$ sending ω onto an open neighborhood ω^i of 0 and $\emptyset_i(\bar{x}) = 0$, $\emptyset_i(\omega \cap \mathcal{O}) = \omega^i \cap (x_N > 0)$, $\emptyset_i(\omega \cap \mathcal{O}^c) = \omega^i \cap (x_N < 0)$. In addition, we require $\Phi = \emptyset_i$ to satisfy :

$$a_{\ell m}^i \Phi_\ell^N \Phi_m^k = 0 \quad \text{in } \omega \cap \mathcal{O}, \quad \Phi_i^k \gamma_i = 0 \quad \text{on } \Gamma \cap \omega \quad \text{for } k < N$$

where Φ_j^k means $\frac{\partial}{\partial x_j}(\Phi^k)$. These conditions mean that the change of unknown $\tilde{u}(y) = u(x)$ (for $y = \emptyset_i(x)$) transforms the operator $(a_{k\ell}^i \partial_{k\ell})$ into $(\tilde{a}_{k\ell}^i \partial_{k\ell})$ where $\tilde{a}_{Nk}^i = 0$ on $\omega^i \cap (y_N \ge 0)$ and the boundary condition : $\frac{\partial u}{\partial \gamma} = 0$ on $\Gamma \cap \omega$ into $\frac{\partial \tilde{u}}{\partial y_N} = 0$ on $\omega^i \cap (y_N = 0)$.

It is clearly enough to estimate : $\bar{K}^\varepsilon = \sup_{i,x} \xi(x)|D^2 u_\varepsilon^i(x)|$. Let i_0, x_0 be a point of maximum; we transform ω into ω^{i_0} , u_ε^j into $\tilde{u}_\varepsilon^j$, ξ into $\tilde{\xi}$: the operator A_i becomes $\tilde{A}_i$ and we have :

$$\begin{cases} \tilde{A}_i \tilde{u}^i_\varepsilon + \beta_\varepsilon(\tilde{u}^i_\varepsilon - \tilde{u}^{i+1}_\varepsilon) = \tilde{f}^i & \text{in } \omega^{i_o} \cap (y_N > 0) \\ \dfrac{\partial \tilde{u}^i_\varepsilon}{\partial y_N} = 0 & \text{in } \omega_{i_o} \cap (y_N = 0) . \end{cases}$$

In addition : $\tilde{a}^{i_o}_{nk} = 0$ on $\omega^{i_o} \cap (y_N \geq 0)$ for $k < N$; and because of (39) we have : $\tilde{a}^j_{Nk} = 0$ on $\omega^{i_o} \cap (y_N = 0)$ for all j and for $k < N$.

If $y_o = \emptyset^{i_o}(x_o)$, we have clearly : $\overline{K}^\varepsilon \leq C + C_o \tilde{\xi}(y_o) \, |D^2 \tilde{u}^{i_o}_\varepsilon(y_o)|$ where (here and below) C denotes various constants independent of i,n,ε and C_i depends only on $\mathcal{O}$,γ and a. We then denote by

$K^\varepsilon = \sup_{i,x} \tilde{\xi}(x) \, |D^2 \tilde{u}^i_\varepsilon(x)|$. We also have : $K^\varepsilon \leq C + C_1 \overline{K}^\varepsilon$.

We are going to estimate K , to this end we skip everywhere below the superscript ε and we consider the following auxiliary functions (as in [16], [35]) :

$$w^i(x) = \xi^2(x) \, |D^2 u^i(x)|^2 + 2N^2 \nu \ \ K\xi \, \alpha_{k\ell} \, \partial_{k\ell} u^i + \lambda |\nabla u^i|^2$$

(to simplify notation we write $\tilde{\phi} = \phi$ for all ϕ and we set $y = x$) , where $\alpha_{k\ell} = \tilde{a}^{i_o}_{k\ell}(y_o)$, and $\lambda > 0$ will be determined below. We may without loss of generality assume that u^i are bounded in $W^{2,\infty}(\mathcal{O})$. We next claim that we have :

$$- \frac{\partial w^i}{\partial x_N} \leq C_2 K^2 \qquad \text{on} \quad \omega^{i_o} \cap (x_N = 0) . \qquad (41)$$

Indeed, recall that $\dfrac{\partial \xi}{\partial x_N} = 0$, $\dfrac{\partial u^i}{\partial x_N} = 0$ and $\alpha_{Nk} = 0$ if $k < N$, therefore

$$\frac{\partial w^i}{\partial x_N} = 2\xi^2(x)\,\partial_{NN}u^i\,\partial_{NNN}u^i + 2N^2\nu K\xi\alpha_{NN}\,\partial_{NNN}u^i \quad \text{on} \quad \omega^{i}{}_{o} \cap (x_N = 0)\ .$$

Now we deduce from the equation and the fact that on $\omega^{i}{}_{o} \cap (x_N = 0)$, $a^i_{Nk} = 0$ for $k < N$:

$$|\partial_{NNN}u^i| \le C + C|D^2u^i|$$

and (41) is proved. We then introduce

$$\overline{w}^i(x) = e^{\mu x_N}\, w^i(x)\ , \text{ where } \mu > 0 \text{ is determined below.}$$

Without loss of generality we may assume that $\overline{w}^i$ attains its maximum over i and x for $i = 1$, $x = x^1 \in \text{Supp}\ \xi$.

The same argument as in [16],[35] yields that if $x^1_N > 0$ and if $\delta, \mu > 0$ are fixed, there exists $\lambda_o = \lambda_o(\delta,\mu)$ large enough such that :

$$\overline{w}^i(x^1) \le \delta K^2 + C(\delta,\mu).$$

On the other hand if $x^1_N = 0$, then $\frac{\partial \overline{w}^i}{\partial x_N}(x^1) \le 0$

and this implies in view of (41) :

$$\overline{w}^i(x^1) \le \frac{C_2}{\mu} K^2\ .$$

Therefore in all cases we have, if $\lambda = \lambda_o(\delta,\mu)$,

$$w^i(x) \le e^{-\mu x_N}\, w^i(x) \le \max\left(\frac{C_2}{\mu} K^2, \delta K^2 + C(\delta,\mu)\right)\ ;$$

in particular, we may apply this inequality for $i = i_o$, $x = y_o$ and we find, using the same argument as in [16],[35] :

$$\bar{K}^2 \leq C + C_3 \xi^2(y_o)|D^2 u^{i_o}(y_o)|^2 \leq C + C_3 w^{i_o}(y_o) + CK + C(\delta,\mu)$$

$$\leq \max(\frac{C_4}{\mu} K^2, C_3\delta K^2) + CK + C(\delta,\mu) \leq \max(\frac{C_6}{\mu}\bar{K}^2, C_5\delta\bar{K}^2) + C(\delta,\mu)(1+\bar{K})$$

and we conclude choosing $\delta = \frac{1}{2C_5}$, $\mu = 2C_6$.

This shows that for some neighborhood of Γ we have :

$$\|u_\varepsilon^i\|_{W^{2,\infty}(\tilde{\Gamma})} \leq C \quad \text{(ind.of } i,\varepsilon,n\ldots) .$$

It is now easy to conclude, since in view of the local estimates proved in [35] , we deduce : $\|u_\varepsilon^i\|_{W^{2,\infty}(\mathcal{O})} \leq C'$ (where C' now depends on the uniform ellipticity constant of $a(x,\alpha)$ over $\bar{\mathcal{O}}$); and exactly as in [15] , [55] u_ε^i converges uniformly to $u \in W^{2,\infty}(\mathcal{O})$ satisfying :

$$\sup_{1\leq i\leq n} \{A_i u - f^i\} = 0 \text{ in } \mathcal{O} , \quad \frac{\partial u}{\partial \gamma} = 0 \text{ on } \Gamma \text{ and } \|u\|_{W^{2,\infty}(\tilde{\Gamma})} \leq C$$

(independently of $n,\ldots$). By an easy verification result we deduce that u is the optimal cost function of the associated stochastic control problem and we conclude.

II.4. Unbounded coefficients

In this section we consider problems where the coefficients σ,b,c,f are not bounded uniformly with respect to $\alpha \in A$. Two cases may occur :

1st case : the quantity $H(\xi,p,t,x) = \sup_{\alpha\in A} \{-a_{ij}(x,\alpha)\xi_{ij} - b_i(x,\alpha)p_i + c(x,\alpha)t - f(x,\alpha)\}$ is finite for all symmetric $N \times N$ matrices ξ (we will denote by S^N the space of such matrices), vectors p, reals t and points x in $\mathbb{R}^N$.

Example : $\sigma \equiv 0$, $A = \mathbb{R}^N$, $b_i(x,\alpha) = \alpha$, $c(x,\alpha) \equiv \lambda$, $f(x,\alpha) = f(x) + \frac{1}{2}|\alpha|^2$.
Then : $H(\xi,p,t,x) = \frac{1}{2}|p|^2 + \lambda t - f(x)$.

In this case the results we obtain are somewhat similar to those obtained in section I.

2nd case : H is not always finite. In this case, following N.V. Krylov [24], [26], one introduces :

$$p(x,\alpha) = \{1 + (a_{ij}(x,\alpha))^2 + (b_i(x,\alpha))^2 + (c(x,\alpha))^2 + (f(x,\alpha))^2\}^{1/2}$$

We will see that under very general assumptions, the optimal cost function satisfies a renormalized HJB equation :

$$\sup_{\alpha\in A} \{p(x,\alpha)^{-1} \{A_\alpha u(x) - f(x,\alpha)\}\} = 0 \ .$$

Of course many choices of p are possible and in general all these choices give equivalent equations (see N.V. Krylov [24] for a general discussion of this point).

We begin with the study of the first case : we will thus assume that

$$H(\xi,p,t,x) = \sup_{\alpha\in A} \{-a_{ij}(x,\alpha)\xi_{ij} - b_i(x,\alpha)p_i + c(x,\alpha)t - f(x,\alpha)\} \tag{42}$$

is finite and continuous on $S^N \times \mathbb{R}^N \times \mathbb{R} \times \mathbb{R}^N$;

and that the coefficients satisfy :

$$\begin{cases} \varphi(.,\alpha) \in W^{2,\infty}(\mathbb{R}^N) \text{ for all } \alpha \in A,\ \varphi(x,.) \in C(A) \text{ for all } x \in \mathbb{R}^N \\ \sup_{\alpha\in A} \|D^\beta\varphi(.,\alpha)\|_{L^\infty(\mathbb{R}^N)} < +\infty, \text{ for } |\beta| = 1,2 \text{ for } \varphi = \sigma_{ij}, b_i, c, f \\ \sup_{\alpha\in K} \|\varphi(.,\alpha)\|_{L^\infty(\mathbb{R}^N)} \leq C(K) < \infty, \text{ for any compact set } K \subset A, \\ \text{for } \varphi = \sigma_{ij}, b_i, c, f \ ; \end{cases} \tag{43}$$

and we may define λ_1 as before. We will also assume :

$$\begin{cases} \sup_{\alpha,\Gamma} \|\sigma(x,\alpha)\| < \infty \\ \exists\nu > 0,\ \forall(x,\alpha) \in \Gamma \times A,\ a_{ij}(x,\alpha)n_i n_j \geq \nu. \end{cases} \tag{44}$$

We have then

Theorem II.8 : *Let* $\lambda > \lambda_1$, $\mathcal{O}$ *be an open smooth set in* $\mathbb{R}^N$ *and let us assume that* (42), (43) *and* (44) *(if* $\mathcal{O} \neq \mathbb{R}^N$*) hold. In addition, we assume if*

$\mathcal{O} \neq \mathbb{R}^N$ *that there exists* $\underline{u} \in W^{1,\infty}(\mathcal{O})$ *such that* $\underline{u} = 0$ *on* Γ *and :*

$$\forall \alpha \in A, \quad A_\alpha \underline{u} \leq f(.,\alpha) \text{ in } \mathcal{D}'(\mathcal{O}) . \tag{45}$$

Then $u \in W^{1,\infty}(\mathcal{O})$, $u = 0$ *on* Γ *and* (7) *holds :*

$$\exists C > 0, \forall \xi \in \mathbb{R}^N \; ; |\xi| = 1, \quad \partial_\xi^2 u \leq C \text{ in } \mathcal{D}'(\mathcal{O}) . \tag{7}$$

Thus $A_\alpha u \in L^\infty(\mathcal{O})$ *(for all* $\alpha \in A$*) and the HJB equation holds :*

$$H(D^2u, Du, u, x) = 0 \quad \text{a.e. in } \mathcal{O}.$$

Remark II.18 : As in the preceding section various extensions of this result are possible (unbounded coefficients in x ; non-zero boundary conditions, extensions of (44)) and we might treat as well optimal stopping or time-dependent problems. In addition if we assume, instead of (44) :

$$\begin{cases} \sup_{\alpha,\mathcal{O}} \|\sigma(x,\alpha)\| < \infty \\ \\ \exists \nu > 0, \forall (x,\alpha) \in \bar{\mathcal{O}} \times A, \; a_{ij}(x,\alpha)\, n_i(x) n_j(x) \geq \nu \end{cases} \tag{44'}$$

then it is enough to assume $\lambda \geq 0$ (or $\lambda > 0$ if $\mathcal{O}$ is unbounded). Finally, results similar to Corollary I.1 are deduced from (7).

Remark II.19 : To motivate (45), let us point out that if $A = \mathbb{R}^N$, $\sigma = I_N$, $b_i(x,\alpha) = -\alpha$, $c(x,\alpha) \equiv \lambda$, $f(x,\alpha) = f(x) + L(\alpha)$, where $L(\alpha)$ is a convex continuous function such that :

$\lim_{|\alpha|\to\infty} L(\alpha)|\alpha|^{-1} = +\infty$, then we have :

$$H(\xi,p,t,x) = -\frac{1}{2}\,\mathrm{Tr}(\xi) + H(p) + \lambda t - f(x)$$

with $H = L^*$. Therefore, the HJB equation reduces to :

$$-\frac{1}{2}\,\Delta u + H(\nabla u) + \lambda u = f(x) \text{ in } \mathcal{O},\ u = 0 \text{ on } \Gamma$$

and $\frac{H(p)}{|p|} \to +\infty$ as $|p| \to \infty$. It is known (cf. P.L. Lions [47]) that (45) is then in general necessary for the solvability of this equation, indeed (45) in this case reduces to

$$-\frac{1}{2}\,\Delta\underline{u} + H(\nabla\underline{u}) + \lambda\underline{u} \leq f \quad \text{in } \mathcal{O}\ ,\ \underline{u} \in W^{1,\infty}(\mathcal{O}),\ \underline{u} = 0 \text{ on } \Gamma.$$

Remark II.20 : An important example of applications is the following framework :

$$H = \sup_{\alpha\in A}\{-a_{ij}(x,\alpha)\xi_{ij} + H^{\alpha}(x,\nabla u) + \lambda u\}$$

(we took to simplify $c(x,\alpha) \equiv \lambda$ and we could treat $H^{\alpha}(x,u,\nabla u)$) ; where a_{ij}, H^{α} satisfy :

$$\sup_{\alpha\in A} \|\sigma(.,\alpha)\|_{W^{2,\infty}(\mathbb{R}^N)} < \infty,\ \sigma(x,.) \in C(A) \text{ for } x \in \mathbb{R}^N \qquad (46)$$

$$\begin{cases} H^{\alpha} \text{ is convex in } p \text{ for all } x \in \mathbb{R}^N, \ H^{\alpha} \in C(\mathbb{R}^N \times \mathbb{R}^N \times A) \\ \sup_{\alpha} \|H^{\alpha}\|_{W^{1,\infty}(\mathbb{R}^N \times B_R)} < \infty, \text{ for all } R < \infty \end{cases} \tag{47}$$

Writing, at least formally,

$$H^{\alpha}(x,p) = \sup_{\xi \in \mathbb{R}^N} \{-b(x,\alpha,\xi).p - f(x,\alpha,\xi)\}$$

- we have many possible choices of such decompositions - we see that with a new control set given by $A \times \mathbb{R}^N$ this is a special case of the above situation. For example, if we choose $b(x,\alpha,\xi) = \xi$, $f(x,\alpha,\xi) = H^{\alpha}(x,.)^* = L^{\alpha}(x,\xi)$, the corresponding cost function is now if $\mathcal{O} = \mathbb{R}^N$:

$$J(x,\mathcal{A}) = E \int_0^{\infty} L^{\alpha_t}(X_t,\xi_t)\, e^{-\lambda t}\, dt$$

while the state process X_t solves :

$$dX_t = \sigma(X_t,\alpha_t)\, dB_t + \xi_t dt \ , \ X_o = x \ .$$

In this case (i.e. if we assume (46)-(47)) we can obtain more general results insuring that $u \in W^{1,\infty}(\mathcal{O})$. Assume that $\lambda > \lambda_1$ (depending only on σ,α), that (45), (44) hold if $\mathcal{O} \neq \mathbb{R}^N$ and that we have : $\forall \varepsilon > 0$, $\exists R_o > 0$, $\forall \alpha \in A$:

$$- p.\frac{\partial H^{\alpha}}{\partial x}(x,p) \leq \frac{(H^{\alpha})^2}{2\mu N} + (\lambda - \lambda_1)|p|^2 + \varepsilon|p|^2 \{\frac{\partial H^{\alpha}}{\partial p_i} p_i - H^{\alpha}\}$$

for all $x \in \bar{\mathcal{O}}$ and for all $|p| \geq R_0$; where $\mu = \operatorname{Tr} a^{\alpha}$ (if $\mu = 0$, the inequality holds). Then $u \in W^{1,\infty}(\mathcal{O})$. One proves this claim with the use of p.d.e. techniques in order to obtain the Lipschitz estimates, which are identical to those of P.L. Lions [47], [48], [58].

Another type of result that one can obtain is the following : assume that (45), (46), (47) hold and that we have : σ does not depend on x and $H^{\alpha}(x,p)\ |p|^{-1} \to +\infty$ as $|p| \to \infty$ uniformly in $(x,\alpha) \in \bar{\mathcal{O}} \times A$

$$\begin{cases} \forall R < \infty,\ \exists \nu_R > 0,\ \forall (x,p,q) \in \bar{\mathcal{O}} \times \bar{B}_R \times \bar{B}_R, \\ (\frac{\partial H^{\alpha}}{\partial p}(x,p) - \frac{\partial H^{\alpha}}{\partial p}(x,q), p-q) \geq \nu_R\ |p-q|^2 . \\ \\ \forall \varepsilon > 0,\ \exists k \in (0,1),\ \exists R_0 > 0,\ \forall \alpha \in A,\ \forall |p| \leq R_0,\ \forall x \in \bar{\mathcal{O}} \\ \\ -\ p.\frac{\partial H^{\alpha}}{\partial x}(x,p) \leq k\ \frac{(H^{\alpha})^2}{\mu N} + \varepsilon\ |p|^2 (\frac{\partial H^{\alpha}}{\partial p}.p - H^{\alpha}) ; \end{cases}$$

then $u \in W^{1,\infty}(\mathcal{O})$, $u = 0$ on Γ and we have :

$$\forall \delta > 0,\ \exists C_{\delta} > 0,\ \forall \xi \in \mathbb{R}^N : |\xi| = 1,\ \partial_{\xi}^2 u > \leq C_{\delta} \text{ in } \mathcal{D}'(\mathcal{O}_{\delta}) .$$

This type of result is proved exactly as in P.L. Lions [58] where the case $\sigma = 0$ is treated.

Remark II.21 : The above results and remarks may be extended a lot when we assume that all the matrices $a(x,\alpha)$ are definite positive uniformly in $\alpha \in A$. Instead of giving such extensions, we prefer to indicate that in section II.6, we will give existence results for p.d.e. which, when they can be interpreted as HJB equations, immediately give by standard verification methods that the optimal cost function of the associated control problem has the regularity of the solution of the equation.

We will not prove Theorem II.8 since it is, as is the following result, a consequence of the following method : one takes a dense sequence $(\alpha_j)_{j\geq 1}$ in A and one considers u_n the optimal cost function corresponding to admissible systems where the control process α_t takes its values in $\{\alpha_1,\ldots,\alpha_n\}$. We may apply Theorem I.1 to u_n and we remark that : $u_n \geq \underline{u}$ in $\bar{\mathcal{O}}$ (cf. Part 1 [36]). Because of (43), (44) one deduces, as in Part 1, that u_n is bounded in $W^{1,\infty}(\mathcal{O})$. From the proof of Theorem I.1, we deduce that "D^2u_n on Γ" is bounded (to justify this claim, one first approximates the problem as in section I.3) and this yields easily that (7) holds uniformly in n. Taking $n \to \infty$, we conclude easily : indeed u_n is a viscosity solution of :

$$\sup_{1\leq j\leq n} (A_{\alpha_j} u_n(x) - f(x,\alpha_j)) = 0 \text{ in } \mathbb{R}^N$$

and $H_n(\xi,p,t,x) = \sup_{1\leq j\leq n} (-a_{k\ell}(x,\alpha_j)\xi_{k\ell} - b_k(x,\alpha_j)p_k + c(x,\alpha_j)t - f(x,\alpha_j))$ increases to H as $n \uparrow +\infty$ and thus, by Dini's Lemma, the convergence is uniform on compact sets. This shows that u is a viscosity solution of the HJB equation and because of the results of Part 2 [37] and the regularity already proved we deduce that the HJB equation holds a.e.

With the same arguments, we prove :

Theorem II.9 : *Let* $\mathcal{O} = \mathbb{R}^N$, $\lambda > \lambda_1$. *We assume* (43) *and :*

$$\exists C > 0,\ \forall(x,\alpha) \in \mathbb{R}^N \times A,\ \ f(x,\alpha) \geq -C .$$

Then $u \in W^{1,\infty}(\mathbb{R}^N)$, (7) *holds and we have :*

$$\sup_{\alpha \in A} \|p(.,\alpha)^{-1} A_\alpha u\|_{L^\infty(\mathbb{R}^N)} < \infty . \tag{49}$$

Therefore the renormalized HJB equation holds :

$$\sup_{\alpha \in A} \{p(x,\alpha)^{-1} \{A_\alpha u(x) - f(x,\alpha)\}\} = 0 \text{ a.e. in } \mathbb{R}^N . \tag{50}$$

Remark II.22 : We could as well treat the case of coefficients unbounded in x or optimal stopping problems.

Remark II.23 : This result extends those due to N.V. Krylov [26] where some non-degeneracy hypothesis is assumed.

Example : Let us consider the case when $A = \{\alpha \in S^N, \alpha \geq I_N\}$, $a(x,\alpha) = \alpha$, $b(x,\alpha) \equiv 0$, $c(x,\alpha) = \lambda \operatorname{Tr}(\alpha)$, $f(x,\alpha) = f(x) \operatorname{Tr}(\alpha)$ with $f \in W^{2,\infty}(\mathbb{R}^N)$. Then applying Theorem II.9 (and Corollary I.1) we have : $u \in W^{2,\infty}(\mathbb{R}^N)$ solves :

$$\begin{cases} \sup_{\alpha \geq I_N} \{(-\alpha_{ij}\, \partial_{ij} u + \lambda \operatorname{Tr}(\alpha_{ij}) u - f \operatorname{Tr}(\alpha_{ij})) \dfrac{1}{(1+\sum_{i>j} \alpha_{ij}^2)^{1/2}}\} = 0 \\ \\ \text{a.e. in } \mathbb{R}^N . \end{cases} \tag{50}$$

It is an easy exercise to check that (50) holds if and only if :

$$\begin{cases} (D^2u - (\lambda u+f)I_N) \geq 0 \text{ a.e. in } \mathbb{R}^N, \\ \\ \inf\limits_{1\leq i\leq N} \lambda_i(D^2u-\lambda uI_N) = f(x) \text{ a.e.} \end{cases} \tag{50'}$$

where $\lambda_1,\ldots,\lambda_N$ denote the eigenvalues of the symmetric matrix $D^2u - \lambda uI_N$. It is worth pointing out that the unnormalized HJB equation is, in this case, equivalent to : $(D^2u - (\lambda u+f)I_N) \geq 0$ a.e. in $\mathbb{R}^N$ and $-\Delta u + N\lambda u = f$ and this is in general false !

II.5. Monge-Ampère equations.

In this section we consider the classical Monge-Ampère equations :

$$\det(D^2u) = g \quad \text{in } \mathcal{O},\ u \text{ convex on } \bar{\mathcal{O}},\ u = 0 \text{ on } \Gamma \tag{51}$$

where $\mathcal{O}$ is a smooth bounded open convex set satisfying, for example :

$$\exists p \in W^{2,\infty}(\mathcal{O}),\ p = 0 \text{ on } \Gamma,\ (D^2p) \geq I_N \text{ in } \mathcal{O}. \tag{52}$$

The Monge-Ampère equations (and similar equations) arise in differential geometry (cf. A.D. Alexandrov [1], N.V. Pogorelov [66], L. Nirenberg [63], S.Y. Cheng and S.T. Yau [8]) and it was remarked (independently) by B. Gaveau [19] and N.V. Krylov [26] that the Monge-Ampère equations are, in fact, HJB equations. This remark is based upon the following algebraic fact : let A be a symmetric matrix, denote by $V = \{B \in S^N, B > 0, \det B = \frac{1}{N^N}\}$, then we have : $\inf\limits_{B\in V} \mathrm{Tr}(AB) > -\infty$ if and only if $A \geq 0$. In addition, if this is the case, then we have :

$$\inf_{B\in V} \operatorname{Tr}(AB) = (\det A)^{1/N} . \tag{53}$$

Therefore (51) is equivalent to :

$$\sup_{B\in V} \{-b_{ij}\ \partial_{ij} u\} = -f \quad \text{in } \mathcal{O} \text{ , u convex on } \bar{\mathcal{O}}, \ u = 0 \quad \text{on } \Gamma$$

where $f = g^{1/N}$.

It is then natural to consider the optimal stochastic control problem where

$$A = V,\ \sigma(x,\alpha) = \sqrt{2}\ \alpha^{1/2} \text{ , } b(x,\alpha) = 0 \text{ , } c(x,\alpha) = 0 \text{ , } f(x,\alpha) = f(x) \text{ .}$$

We then have : $X_t = x + \int_0^t \sqrt{2}\ \alpha_s^{1/2} . dB_s$ and

$$u(x) = \inf_{\mathcal{A}} E \int_0^\tau f(X_t)dt \text{ .} \tag{54}$$

Let us recall that in view of Alexandrov-Pucci-Krylov inequalities we have :

$$\sup_{\mathcal{A}} |E \int_0^\tau f(X_t)dt| \le C \|f\|_{L^N(\mathcal{O})}$$

where C depends only on N and diam $\mathcal{O}$. And thus (54) has a meaning.

We have the

Theorem II.10 : *We assume* (52) *and*

$$\begin{cases} f \in W^{1,\infty}(\mathcal{O}) \ ; \\ \exists C > 0, \ \forall \xi \geq \mathbb{R}^N \ : |\xi| = 1, \ \partial_\xi^2 f \geq -C \text{ in } \mathcal{D}'(\mathcal{O}). \end{cases} \tag{55}$$

Then there exists a unique $u \in W^{2,\infty}(\mathcal{O})$ *solution of* (51) *and* u *is given by* (54). *In addition if* $\omega = \{x \in \mathcal{O}, f(x) > 0\}$, *then* $u \in W^{3,p}_{loc}(\omega)$ *for all* $p < \infty$.

Remark II.24 : Similar results hold for related equations including the general Monge-Ampère equations considered in P.L. Lions [51], and also if the zero boundary condition is replaced by :

$u = \varphi$ on Γ , where $\varphi \in C^{1,1}(\Gamma)$. Finally we can relax (52) and assume only

$$\exists R > 0, \ \forall y \in \Gamma, \ \exists y_0 \in \mathbb{R}^N, \ \mathcal{O} \subset B(y_0,R) \text{ and } |y-y_0| = R \ . \tag{52'}$$

Remark II.25 : The nondegenerate case, i.e. when $f > 0$ on $\bar{\mathcal{O}}$, has been studied by many authors : a first partial proof of the existence of a smooth solution $u \in W^{1,\infty}(\mathcal{O}) \cap C^\infty(\mathcal{O})$ was given by N.V. Pogorelov [66] and the proof was completed by S.Y. Cheng and S.T. Yau [8]. This proof uses geometrical arguments. Recently we gave an alternate proof using only p.d.e. arguments and in particular the relations of problems like (51) and the HJB equations, and a penalty method (similar to the one used in Part 1 [36]) : we refer to [48], [50], [51]. At the same time S.Y. Cheng and S.T. Yau [9] announced another analytical method. All these results concerned the existence of a smooth solution $u \in C^\infty(\mathcal{O}) \cap W^{1,\infty}(\mathcal{O})$; but the existence of $u \in C^\infty(\bar{\mathcal{O}})$ was settled only recently by L. Caffarelli, J. Spruck and L. Nirenberg [7]. We will use their result. Let us also indicate that once we know that $u \in C^2_{loc}(\omega)$, if f has some regularity, say $f \in C^{k,\alpha}_{loc}(\omega)$ for $k \geq 1$, $\alpha \in (0,1)$ then $u \in C^{k+2,\alpha}_{loc}(\omega)$.

Proof of Theorem II.10 : We first remark that there exists $f^n \in C^\infty(\bar{\mathcal{O}})$ such that : $f^n > 0$ in $\bar{\mathcal{O}}$, f^n is bounded in $W^{1,\infty}(\mathcal{O})$ and f^n converges uniformly on $\bar{\mathcal{O}}$ to f, and f^n satisfies :

$$\exists C > 0,\ \forall n \geq 1,\ \forall \xi \in \mathbb{R}^N : |\xi| = 1,\ \partial^2_\xi f^n \geq -C \text{ in } \mathcal{O}.$$

From the results of [7], we deduce that there exists a unique $u^n \in C^2(\bar{\mathcal{O}}) \cap C^\infty(\mathcal{O})$ solution of (51) (with $g = f^N$ replaced by $g^n = (f^n)^N$). We are going to prove that u^n is bounded in $W^{2,\infty}(\mathcal{O})$ therefore u^n (or appropriate subsequences) will converge to some u in $W^{2,\infty}(\mathcal{O})$, and for example, from the stability results of weak solutions (see [8]) we see that u solves (51). The uniqueness is an easy consequence of the maximum principle (in Bony's form, for example, see [5], [52]).

Now, in view of (52), we have :

$$\det(\lambda D^2 p) \geq \lambda^N \text{ in } \mathcal{O},\ p \text{ is convex on } \bar{\mathcal{O}},\ p = 0 \text{ on } \Gamma ;$$

therefore choosing λ large enough (independently of n) we see, using the maximum principle :

$$\lambda p \leq u^n \leq 0 \text{ in } \bar{\mathcal{O}} .$$

This implies not only that u^n is bounded in $L^\infty(\mathcal{O})$, but also that Du^n is bounded on Γ and u^n being convex, this yields : u^n is bounded in $W^{1,\infty}(\mathcal{O})$.

We next prove that D^2u^n is bounded on Γ. Let $x_o \in \Gamma$ and let τ be a C^2 tangent vector field to $\partial\Gamma$ (with C^2 norms bounded independently of x_o) and assume that $\tau(x_o) = \tau_o$ is a prescribed unit tangent vector to Γ at x_o. Differentiating (51) with respect to x_k we get :

$$a^o_{ij}\,\partial_{ijk}\,u^n = \partial_k\,f^n \text{ in } \mathcal{O} \quad 1 \le k \le N ,$$

with $a^o = \frac{1}{N}\,(D^2u^n)^{-1}\,\{\det(D^2u^n)\}^{1/N}\,(a^o(x) \in V,\ \forall x \in \bar{\mathcal{O}})$.

And this yields :

$$a^o_{ij}\partial_{ij}(\tau_k\partial_k u^n) = a^o_{ij}(\partial_{ij}\tau_k)\partial_k u^n + 2a^o_{ij}\partial_i\tau_k\partial_j u^n + \tau_k\partial_k f^n$$

$$= a^o_{ij}b_{ij} + \frac{2}{N}\,f^n(\partial_i\tau_i) + \tau_k\partial_k f^n \text{ in } \mathcal{O}$$

with $\|b\| \le C$, and : $\tau_k\partial_k u^n = 0$ on Γ.

Let λ be such that : $\frac{\lambda}{2}\,(D^2p) \ge b$ in $\mathcal{O}$, $\frac{\lambda}{2} \ge \frac{2}{N}\,f^n(\partial_i\tau_i) + \tau_k\partial_k f^n$; then we have :

$$a^o_{ij}\partial_{ij}(\lambda p) \ge |a^o_{ij}\partial_{ij}(\tau_k\partial_k u^n)|$$

and by the maximum principle we deduce : $|\tau_k\partial_k u^n| \le \lambda p$ in $\bar{\mathcal{O}}$.

This implies : $|\partial_{k\tau_o} u^n(x_o)| \le C$. And since $u^n = 0$ on Γ, we will have proved the estimate on $\|D^2u^n\|_{L^\infty(\Gamma)}$ if we show that : $|\frac{\partial^2 u^k}{\partial n^2}| \le C$ on Γ.

This is obtained from the equation (51) if we have estimates from below on $\frac{\partial^2 u^k}{\partial\tau_o^2}(x_o)$ for any tangent vector on Γ at x_o. Therefore this estimate will be proved if we show that there exists some $\nu > 0$ independent of k such that :

$$\frac{\partial u^k}{\partial n}(x_o) \geq \nu > 0 \ , \ \forall x_o \in \Gamma.$$

But remark that without loss of generality we may assume that $f^k \downarrow f$ on $\bar{\mathcal{O}}$ as $k \uparrow +\infty$ and thus $u^k \uparrow u$ as $k \uparrow +\infty$, where u is convex, Lipschitz on $\bar{\mathcal{O}}$ and in view of [8], u is the unique weak solution of :

$$\det(D^2 u) = g \text{ in } \ , \ u \text{ convex on } \bar{\mathcal{O}}, \ u = 0 \text{ on } \Gamma.$$

Since u is convex on $\bar{\mathcal{O}}$ and $u = 0$ on Γ, if u vanishes somewhere in $\mathcal{O}$ then $u \equiv 0$ and $g \equiv 0$ and the result is proved. On the other hand if $g \not\equiv 0$ then $u < 0$ in $\mathcal{O}$. Now from the convexity we deduce for $h \in (0,h_o)$

$$\frac{1}{h}\{u(x_o) - u(x_o - hn(x_o))\} \geq \frac{1}{h_o}\{u(x_o) - u(x_o - h_o n(x_o))\} \geq \nu > 0$$

and thus :

$$\frac{\partial u^k}{\partial n}(x_o) = \lim_{h \to 0_+} \frac{1}{h}\{u^k(x_o) - u^k(x_o - hn(x_o))\}$$

$$\geq \lim_{h \to 0_+} \frac{1}{h}\{u(x_o) - u(x_o - hu(x_o))\} \geq \nu > 0 \ .$$

We have proved : $\|D^2 u^k\|_{L^\infty(\Gamma)} \leq C$.

We may now conclude using either the method of proof introduced in section I.3 (using the stochastic representation) or an easy p.d.e. argument. Indeed let $\xi \in \mathbb{R}^N$, $|\xi| = 1$; differentiating twice (51) with respect to

ξ we obtain the following inequality :

$$a^o_{ij}\partial_{ij}(\partial^2_\xi u^n) \geq \partial^2_\xi f^n \text{ in } \mathcal{O},\ \partial^2_\xi u^n \geq -C \text{ on } \Gamma.$$

Observing that for λ large enough we have :

$$a^o_{ij}\partial_{ij}(\lambda-\lambda p) \leq -\lambda \operatorname{Tr} a^o_{ij} \leq -\lambda \leq \partial^2_\xi f^n,\ \lambda - \lambda p \geq C \text{ on } \Gamma$$

we deduce applying once more the maximum principle : $\partial^2_\xi u^n \leq \lambda-\lambda p \leq C$; and we conclude since u^n is convex : $\|D^2u^n\|_{L^\infty(\mathcal{O})} \leq C$.

To prove that u is given by (54), we only need to prove the corresponding formula for u^n and let $n \to \infty$ (use the Alexandrov-Pucci-Krylov estimate recalled above). In this case we remark (skipping the superscript n) that we have :

$$a^o_{ij}(x)\ \partial_{ij}\ u(x) = f(x) \text{ in } \bar{\mathcal{O}},\ u = 0 \text{ on } \Gamma,\ u \in C^2(\bar{\mathcal{O}}) \cap C^\infty(\mathcal{O}) \ ;$$

with $a^o(x) = \frac{1}{N}\,(D^2u(x))(\det D^2u)^{1/N} \in V,\ \forall x \in \bar{\mathcal{O}}$. Taking $\alpha(x) = \sqrt{2}(a^o(x))^{1/2} \in C(\bar{\mathcal{O}}) \cap C^\infty(\mathcal{O})$ and considering the diffusion process stopped at the exit from $\bar{\mathcal{O}}$ defined by :

$$dX_t = \alpha(X_t)dB_t\ ,\ X_o = x \in \bar{\mathcal{O}}\ ;$$

we deduce using Itô's formula :

$$u(x) = E \int_0^\tau f(X_t)dt = J(x,\mathcal{A}_x) \ , \ \ \forall x \in \bar{\mathcal{O}}$$

where $\mathcal{A}_x$ corresponds to the control $\alpha_t = a^o(X_t)$.
Since on the other hand, one shows easily using Itô's formula that

$$u(x) \leq J(x,\mathcal{A}) \ , \ \ \forall \mathcal{A}, \ \ \forall x \in \bar{\mathcal{O}} \ ;$$

we conclude.

II.6. Fully nonlinear elliptic degenerate equations.

Our purpose in this section is to deduce from the results and methods above existence and regularity results for solutions of fully nonlinear second-order elliptic equations :

$$H(D^2u,Du,u,x) = 0 \text{ in } \mathcal{O}$$

where $\mathcal{O}$ is, to simplify, a bounded smooth open set in $\mathbb{R}^N$.
We will also specify homogeneous Dirichlet boundary conditions to simplify the presentation but we could treat as well more general cases. It is quite clear that the above equation can be viewed as an HJB equation if and only if H is convex in $(\xi,p,t) \in S^N \times \mathbb{R}^N \times \mathbb{R}$.

The result which follows is proved exactly as in the preceding sections and concerns Hamiltonians H given by :

$$\cdot \ H(\xi,p,t,x) = \sup_{\alpha \in A} \{-a_{ij}(x,\alpha)\xi_{ij} - b_i(x,\alpha)p_i + f(t,x,\alpha)\} \qquad (56)$$

where $a = \frac{1}{2}\ \sigma\sigma^T$ and the coefficients σ,b satisfy (42)-(43)-(44) and f satisfies :

$$\left\{\begin{array}{l} f(.,.,\alpha) \in W^{2,\infty}_{loc}(\mathbb{R} \times \mathbb{R}^N) \text{ for all } \alpha \in A, \\ \\ f(t,x,.) \in C(A) \text{ for } (t,x) \in \mathbb{R} \times \bar{\mathcal{O}}. \\ \\ \sup_{\alpha\in A} \|D^\beta f(.,.,\alpha)\|_{L^\infty(-R,R)\times\bar{\mathcal{O}})} < +\infty, \text{ for } |\beta| = 1,2, \\ \text{for any } R < \infty. \\ \\ \sup_{\alpha\in K} \|f(.,.,\alpha)\|_{L^\infty((-R,R)\times\bar{\mathcal{O}})} < +\infty, \text{ for all } R < \infty, \\ \text{for any compact } K \\ \\ \frac{\partial f}{\partial t}(t,x,\alpha) \geq \lambda > 0 \text{ , for } (t,x,\alpha) \in \mathbb{R} \times \bar{\mathcal{O}} \times A. \end{array}\right. \tag{57}$$

Theorem II.11 : *Let* $\lambda > \lambda_1$; *we assume* (42), (43), (44), (45) *and* (57). *Then there exists a unique* $u \in W^{1,\infty}(\mathcal{O})$ *such that* (7) *holds,* $A_\alpha u \in L^\infty(\mathcal{O})$ *for all* $\alpha \in A$ *and* u *solves :*

$$H(D^2u,Du,u,x) = 0 \textit{ a.e. in } \mathcal{O}, \ u = 0 \textit{ on } \Gamma. \tag{58}$$

Remark II.26 : This result is only one example of the results we can obtain in this way : for example one could replace (44) by more general conditions. Let us also point out that the uniqueness result is proved exactly as the uniqueness results of next section and can be extended (to have a similar form to those of section III below).

We now turn to the uniformly elliptic case, that is we consider some Hamiltonian H satisfying :

$$\exists \nu,\mu > 0, \quad \mu I_N \geq \left(\frac{\partial H}{\partial \xi}\right) \geq \nu I_N \quad \text{on} \quad S^N \times \mathbb{R}^N \times \mathbb{R} \times \bar{\mathcal{O}} \; .$$

There are various results which can be obtained following the approach given in P.L. Lions [55], L.C. Evans and P.L. Lions [16]; all of them follow the same scheme of proof. One obtains various a priori estimates by differentiating the equation and applying the maximum principle to various auxiliary functions : we considered in [55], [16] the following auxiliary functions :

$$W_1(x) = |\nabla u|^2 + \lambda(C-u)^2 \quad , \quad \lambda, C > 0$$

$$W_2(x) = |D^2 u|^2 + \frac{2N^2}{\nu} K \, \mathrm{Tr}(\alpha_{ij}\partial_{ij}u) + \lambda|\nabla u|^2 \quad , \; \lambda > 0$$

where $K = \|D^2 u\|_{L^\infty(\mathcal{O})}$, $\alpha_{ij} = \frac{\partial H}{\partial \xi_{ij}}$ at some appropriate point ; W_1 enables us to obtain Lipschitz estimates, while W_2 yields $W^{2,\infty}$ estimates. One could give very general (and heavy !) results by looking carefully at the proofs in [55], [16] and making the necessary assumptions in order to apply the proofs of [55], [16] : let us at that stage point out that more general results are obtained by appropriate modifications of W_1, W_2 like : $\bar{W}_1 = (1 + \lambda e^{(C-u)^2})|\nabla u|^2$, $\bar{W}_2 = e^{\lambda|\nabla u|^2} \{|Du|^2 + \frac{2N^2}{\nu} K \, \mathrm{Tr}(\alpha_{ij}\partial_{ij}u)\}$. We will indicate only one result of this kind (obtained with the use of $\bar{W}_1$, W_2) which is meaningful from the control theory view-point : we will take H of the following form

$$H(\xi,p,t,x) = \sup_{\alpha\in A} \{-a_{ij}(x,\alpha)\xi_{ij} + H^{\alpha}(x,t,p)\} \tag{60}$$

where we assume (A is still a separable metric space) :

$$\exists\nu,\mu > 0,\ \mu I_N \geq a(x,\alpha) \geq \nu I_N \quad \forall(x,\alpha) \in \bar{\mathcal{O}} \times A\ . \tag{59'}$$

$$\sup_{\alpha\in A} \|a_{ij}(.,\alpha)\|_{W^{2,\infty}(\mathcal{O})} < \infty,\ a_{ij}(x,.) \in C(A) \quad \forall x \in \bar{\mathcal{O}}. \tag{61}$$

$$\begin{cases} \sup_{\alpha\in A} \|H^{\alpha}\|_{W^{2,\infty}(\mathcal{O}\times(-R,+R)\times B_R)} < \infty \quad \forall R < \infty, \\ \\ H^{\alpha}(x,t,p) \in C(A) \quad \forall x,t,p. \end{cases} \tag{62}$$

$$\begin{cases} \forall R < \infty,\ \exists\varepsilon > 0,\ \exists k \in (0,1) \text{ such that we have uniformly} \\ \\ \text{in } \alpha,\ x \in \bar{\mathcal{O}},\ |t| \leq R : \\ \\ \lim_{|p|\to\infty} \inf \frac{1}{\varepsilon}|p|^2 + \{\frac{\partial H^{\alpha}}{\partial p}.p - H^{\alpha}\} > 0 \\ \\ \lim_{|p|\to\infty} \inf \{\frac{1}{\varepsilon}|p|^4 + |p|^2\{\frac{\partial H^{\alpha}}{\partial p}.p - H^{\alpha}\}\}^{-1} \{\frac{k(H^{\alpha})^2}{a_{ii}} + \\ \\ \frac{\partial H^{\alpha}}{\partial x}.p + \frac{\partial H^{\alpha}}{\partial t}|p|^2\} \geq 0\ . \end{cases} \tag{63}$$

Theorem II.12 : *We assume* (59'), (61), (62), (63) *and that there exists* $\underline{u}$ *(resp.* $\bar{u}$*)* $\in W^{2,\infty}(\mathcal{O})$ *satisfying* : $\bar{u} = \underline{u} = 0$ *on* Γ, $\bar{u} \geq \underline{u}$ *in* $\bar{\mathcal{O}}$ *and*

$$\begin{cases} A_\alpha \underline{u} + H^\alpha(x,\underline{u},\nabla\underline{u}) \le 0 \quad \textit{in } \mathcal{O}\ , \textit{ for all } \alpha \in A \\ \\ \sup\limits_{\alpha\in A} \{A_\alpha \bar{u} + H^\alpha(x,\bar{u},\nabla\bar{u})\} \ge 0 \textit{ in } \mathcal{O}. \end{cases}$$

Then there exists $u \in W^{2,\infty}(\mathcal{O}) \cap C^{2,\theta}(\mathcal{O})$ *(for some* θ *depending only on* $\nu,\mu,\mathcal{O})$ *solution of :*

$$H(D^2u,Du,u,x) = 0 \textit{ in } \mathcal{O},\ u = 0 \textit{ on } \Gamma.$$

In the case when A is reduced to one point, this result is due to P.L. Lions [47]. The case of interest for control problems as shown in section II.4 is the case when H^α is convex in (t,p) : remark that in this case the first part of (63) immediately holds since we have

$$\frac{\partial H^\alpha}{\partial p} .p - H^\alpha \ge - H^\alpha(x,t,0) \ge - C_R \quad \text{if} \quad x \in \bar{\mathcal{O}}, \quad |t| \le R.$$

In addition if we assume $\frac{\partial H^\alpha}{\partial t} \ge 0$ (for example) then the existence of $\bar{u}$ is insured : indeed, assuming H to be locally smooth to simplify, we see that there exists $\bar{u} \in W^{2,\infty}(\mathcal{O})$ solution of : $\bar{u} = 0$ on Γ and

$$\sup_{\alpha\in A} \{-a_{ij}(x,\alpha)\partial_{ij}\bar{u} + \frac{\partial H^\alpha}{\partial x_i}(x,0,0)\partial_i\bar{u} + \frac{\partial H^\alpha}{\partial t}(x,0,0)\bar{u} + H^\alpha(x,0,0)\} = 0 \text{ in } \mathcal{O}\ .$$

And using the convexity of H^α, our claim is proved (one needs only $\frac{\partial H^\alpha}{\partial t} > -\underline{\lambda}_1$ where $\underline{\lambda}_1$ corresponds to $\sigma = \sqrt{2}\ a$, $b = - \frac{\partial H^\alpha}{\partial x_i}$).

Let us also point out that in the case of the so-called "natural assumptions" i.e. if we assume :

$$\left|\frac{\partial H^\alpha}{\partial p}.p\right| + |H^\alpha| \le C_R(1+|p|^2) \quad \text{for } p \in \mathbb{R}^N,\ x \in \bar{\mathcal{O}},\ |t| \le R$$

$$\left|\frac{\partial H^\alpha}{\partial x}\right| \le C_R(1+|p|) \quad \text{for } p \in \mathbb{R}^N,\ x \in \bar{\mathcal{O}},\ |t| \le R\ ;$$

then (63) is automatically satisfied.

III - UNIQUENESS RESULTS

Our goal in this section is to give various uniqueness results for solutions of the Hamilton-Jacobi-Bellman equation (4). We already explained why, besides (7), we cannot impose more regularity assumptions than :

$$u \in W^{1,\infty}(\mathcal{O})\ ,\quad \sup_{\alpha \in A} |A_\alpha u| \in L^\infty(\mathcal{O})$$

since in general the optimal cost function is not more regular. But as the following example shows there may be many solutions (with the above regularity) of (4) with prescribed boundary conditions.

Example : Take $A = [-1,+1]$, $\mathcal{O} = (-1,+1)$, $\sigma \equiv 0$, $b(x,\alpha) = -\alpha$, $c(x,\alpha) \equiv \lambda > 0$, $f(x,\alpha) \equiv 1$. Then it is easy to see that the optimal cost function is given by :

$$\bar{u}(x) = \frac{1}{\lambda}\,(1 - \exp(\lambda(x-1))) \quad \text{for } x \in [0,1],\ u(-x) = u(x) \text{ in } [-1,+1].$$

On the other hand, if we set :

$$\underline{u}(x) = \frac{1}{\lambda}(1 - \exp(-\lambda(x-1))) \quad \text{for} \quad x \in [0,1],\ u(-x) = u(x) \text{ in } [-1,+1] ;$$

or for $n \geq 1$:

$$u_n(x) = \frac{1}{\lambda}(1 - \exp(-\lambda(x - \frac{j}{2^{n+1}}))) \quad \text{for} \quad x \in [\frac{j}{2^{n+1}}, \frac{2j+1}{2^n}] ,$$

$$\text{for } 0 \leq j \leq 2^{n-1}-1$$

$$= \frac{1}{\lambda}(1 - \exp(\lambda(x - \frac{j+1}{2^{n-1}}))) \quad \text{for } x \in [\frac{2j+1}{2^n}, \frac{j+1}{2^{n-1}}] ,$$

$$\text{for } 0 \leq j \leq 2^{n-1}-1 ;$$

$$= u_n(-x) \quad \text{if} \quad x \in [-1,0] ;$$

then $u_n, \bar{u} \in W^{1,\infty}(-1,+1)$ (and thus have the required regularity) - they are even piecewise analytic - and satisfy the HJB equation :

$$|u'| + \lambda u = 1 \quad \text{a.e. in } \mathcal{O} . \qquad (4)$$

This example clearly shows that we need to impose some additional condition to guarantee the uniqueness. Remark also that, among the solutions given in the example, $\bar{u}$ is the only one to satisfy :

$$(\bar{u})'' \leq C \quad \text{in} \quad (-1,+1) \quad \text{for some} \quad C ,$$

indeed in all cases $\underline{u}''$, u_n'' are measures which contain some Dirac masses. Therefore, this suggests that (7) (or some generalization of (7)) should be the right condition to add. This will be proved in the next sections.

III.1. Main result

We will assume that the coefficients σ,b,c, f satisfy :

$$\left\{\begin{array}{l} \sup_{\alpha\in A} \|\varphi(.,\alpha)\|_{W^{1,\infty}(\mathbb{R}^N)} < \infty,\ \varphi(x,.) \in C(A) \text{ for all } x \in \mathbb{R}^N \\ \text{for } \varphi = \sigma_{ij}\ (1 \leq i \leq N,\ 1 \leq j \leq m),\ b_i (1 \leq i \leq N) \\ \sup_{\alpha\in A} \|\psi(.,\alpha)\|_{L^\infty(\mathbb{R}^N)} < \infty,\ \psi \text{ is continuous on } \bar{B}_R \text{ uniformly} \\ \text{in } \alpha \in A, \text{ for all } R < \infty,\ \psi \in C(\mathbb{R}^N \times A), \text{ for } \psi = c,\ f\ ; \end{array}\right. \tag{64}$$

and

$$\inf\{c(x,\alpha)/(x,\alpha) \in \mathbb{R}^N \times A\} = \lambda > 0\ . \tag{65}$$

Let $\mathcal{O}$ be a bounded open set in $\mathbb{R}^N$, we denote by $\mathcal{O}_\delta = \{x \in \mathcal{O}/\mathrm{dist}(x,\Gamma) > \delta\}$ and by τ_δ the first exit time of the state process X_t from $\bar{\mathcal{O}}_\delta$.

<u>Theorem III.1</u> : *We assume* (64) *and* (65). *Let* $u \in W^{1,\infty}_{loc}(\mathcal{O})$ *satisfy :*

$$\sup_{\alpha\in A} \|A_\alpha u\|_{L^\infty(\mathcal{O}_\delta)} \leq C_\delta, \quad \textit{for all } \delta > 0 \tag{8'}$$

$$\sup_{\alpha\in A} \{A_\alpha u(x) - f(x,\alpha)\} = 0 \quad \textit{a.e. in } \mathcal{O}\ . \tag{4}$$

If u *is SSH i.e. if* u *satisfies :*

$$\forall\delta > 0,\ \exists C_\delta > 0,\ \Delta u \le C_\delta \text{ in } \mathcal{D}'(\mathcal{O}_\delta) \quad ; \qquad (7")$$

then we have for all $\delta > 0$ *and for all* $x \in \bar{\mathcal{O}}_\delta$ *:*

$$u(x) = \inf_{\mathcal{A}} E \int_0^{\tau_\delta} f(X_t,\alpha_t) \exp(-\int_0^t c(X_s,\alpha_s)ds) + u(X_{\tau_\delta})\exp(-\int_0^{\tau_\delta} c).$$

Remark III.1 : Let us point out that here and everywhere below conditions like (7") may be modified by replacing Δ by any second - order uniformly elliptic operator with locally Lipschitz coefficients. We will indicate in section III.3 various extensions and variants of this result.

Theorem III.1 will be proved in the next section. Let us indicate immediately a few consequences : of course we could repeat such consequences for any of the variants and extensions considered in the following sections but we will not do so here.

Corollary III.1 : *We assume* (64) *and* (65) *and let* $u,v \in C(\bar{\mathcal{O}}) \cap W^{1,\infty}_{loc}(\mathcal{O})$ satisfy (8'), (4) *and* (7"). *Then we have :*

$$\max_{\bar{\mathcal{O}}}(u-v)^+ \le \max_{\Gamma_0}(u-v)^+$$

where Γ_0 *is closed, contained in* Γ *and such that :*

$$\forall x, \forall \mathcal{A} \quad P(\tau' < \infty,\ X_{\tau'} \notin \Gamma_0) = 0$$

where $\tau' = \inf(t \geq 0, X_t \notin \mathcal{O})$.

Remark III.2 : Of course if $u \equiv v$ on Γ_0, then $u \equiv v$ in $\bar{\mathcal{O}}$. Let us also point out that if u satisfies :

$$A_\alpha u \leq f(x,\alpha) \quad \text{in } \mathcal{D}'(\mathcal{O}) \quad \text{for all } \alpha \in A$$

and if v satisfies (8'), (7") and :

$$\sup_{\alpha \in A} \{A_\alpha v(x) - g(x,\alpha)\} \geq 0 \quad \text{a.e. in } \mathcal{O}$$

(where g satisfy the same conditions than f) ; then we have with similar arguments :

$$\max_{\bar{\mathcal{O}}} (u-v)^+ \leq \max \{\max_{\Gamma_0}(u-v)^+, \frac{1}{\lambda} \sup_{\alpha \in A} \|(f(.,\alpha) - g(.,\alpha))^+\|_{L^\infty(\mathcal{O})}\}$$

Indeed, in view of the results of Part 1 [36], we have :

$$u(x) \leq \inf_{\mathcal{A}} E \int_0^{\tau_\delta} f(X_t,\alpha_t) \exp(-\int_0^t c(X_t,\alpha_t)ds) + u(X_{\tau_\delta}) \exp(-\int_0^{\tau_\delta} c) ;$$

while the same proof as the one of Theorem III.1 yields :

$$v(x) \geq \inf_{\mathcal{A}} E \int_0^{\tau_\delta} f(X_t,\alpha_t) \exp(-\int_0^t c) + v(X_{\tau_\delta}) \exp(-\int_0^{\tau_\delta} c)$$

and we conclude easily taking $\delta \to 0_+$.

Remark III.3 : Of course Corollary III.1 still holds if $\mathcal{O}$ is unbounded, provided one assumes that $u,v \in C_b(\bar{\mathcal{O}})$. More generally, exactly as in Part 2 [37], we deduce that if $\mathcal{O}$ is unbounded and if we assume :

$$\begin{cases} \sup_{\alpha\in A} \|\varphi(.,\alpha)\|_{W^{1,\infty}(B_R)} < \infty \quad \text{for all } R < \infty,\ \varphi(x,.) \in C(A) \\ \qquad\qquad\qquad\qquad\qquad\qquad \text{for all } x \in \mathbb{R}^N \\ \forall\varepsilon > 0,\ \exists C_\varepsilon > 0,\ \forall\alpha \in A,\ |\varphi(x,\alpha)| \le \varepsilon|x| + C_\varepsilon,\ \text{for } \varphi = \sigma_{ij}, b_i\ ; \\ \exists C > 0,\ \exists m \ge 0,\ c + |f| \le C(1+|x|^m) \quad \text{on} \quad \mathbb{R}^N \times A \end{cases}$$

and (65) ; then (66) still holds provided we assume $u,v \in C(\bar{\mathcal{O}})$ and :

$$|u(x)| + |v(x)| \le C(1+|x|^p) \quad \text{in} \quad \bar{\mathcal{O}}\ ,\ \text{for some } C,p \ge 0.$$

Let us indicate another type of consequence which concerns stability results extending and simplifying various stability results due to N.V. Krylov [30], [31] : this result has to be viewed as one example of results which can be obtained by the same method. We consider sequences of coefficients σ, $(\sigma_n)_{n\ge1}$; b, $(b_n)_{n\ge1}$; c, $(c_n)_{n\ge1}$; f, $(f_n)_{n\ge1}$ satisfying (64), (65) (but not necessarily uniformly). We denote by A^n_α the corresponding operators. Then we have

Corollary III.2 : *With the above notations, we assume that* φ_n *converges uniformly on* $K \times A$ *to* φ *for* $\varphi = \sigma,b,c,f$ *and for all compact sets* K *included in* $\mathcal{O}$. *Assume that* $u^n \in W^{1,\infty}_{loc}(\mathcal{O})$ *satisfies* (8'), (7") (*with* A^n_α *and* C^n_δ) *and :* $\sup_{\alpha\in A} \{A^n_\alpha u^n(x) - f^n(x,\alpha)\} \ge 0$ *in* $\mathcal{O}$ *; assume finally that* u^n *converges uniformly on compact sets to some function* u.

Then we have :

i) For any open set $\omega \subset \mathcal{O}_\delta$ *(for some* $\delta > 0$*) we have :*

$$u(x) \geq \sup_{\mathcal{A}} E \int_0^\sigma f(X_t,\alpha_t) \exp(-\int_0^t c) + u(X_\sigma) \exp(-\int_0^\sigma c)$$

where $\sigma = \inf(t \geq 0, X_t \notin \bar{\omega})$.

ii) In particular u *is a viscosity supersolution of the HJB equation :*

$$\sup_{\alpha \in A} \{A_\alpha u(x) - f(x,\alpha)\} \geq 0 \quad \text{in } \mathcal{O} .$$

iii) In addition if u *is locally semi-concave in* $\mathcal{O}$ *and if we have equalities instead of inequalities for* u^n *then* $u \in W^{1,\infty}_{loc}(\mathcal{O})$ *and* (8'), (4) *hold.*

<u>Proof of Corollary III.2</u> : Let ω be an open included in $\mathcal{O}_\delta$ (for some $\delta > 0$). Theorem III.1 and its proof implies :

$$u^n(x) \geq \inf_{\mathcal{A}} E \int_0^{\tau^h_{x_0}} f(X^n_t,\alpha_t) \exp(-\int_0^t c) + u^n(X^n_{\tau^h_{x_0}}) \exp(-\int_0^{\tau^h_{x_0}} c)$$

where $x_0 \in \omega$, $\tau^h_{x_0} = \inf(t \geq 0, X_t \notin \overline{B(x_0,h)})$, h is small enough. (A slight extension of the proof enables us to replace $\tau^h_{x_0}$ by $\sigma \wedge h$). And this implies exactly as in Part 2 [37] that u^n is a viscosity supersolution of :

$$\sup_{\alpha \in A} \{A^n_\alpha u^n - f^n(x,\alpha)\} \geq 0 \quad \text{in } \mathcal{O} .$$

In view of the stability results for viscosity solutions (or supersolutions) proved in Part 2 [37], we see that $u \in C(\mathcal{O})$ is a viscosity supersolution of the HJB equation :

$$\sup_{\alpha\in A} \{A_\alpha u(x) - f(x,\alpha)\} \geq 0 \quad \text{in } \mathcal{O} .$$

Then i) also follows from the results and proofs made in Part 2 [37]. Finally, if equalities hold for u^n instead of inequalities, u is a viscosity solution and iii) follows from the results of Part 2 [37].

We could give various other stability results : all of them use the idea of viscosity solutions which (cf. [37]) is stable with respect to the uniform convergence (on compact sets).

III.2. Proof of Theorem III.1

In a first step, we reduce the result to the situation when $\mathcal{O} = \mathbb{R}^N$ by a trick we already used in Parts 1 and 2 [36], [37]. Then we recall the proof we gave in [39] to show Theorem III.1.

Step 1 : Reduction to the case $\mathcal{O} = \mathbb{R}^N$.

Let $\xi \in \mathcal{D}_+(\mathcal{O})$, $0 \leq \xi \leq 1$ in $\bar{\mathcal{O}}$, $\xi \equiv 1$ on $\mathcal{O}_\delta$ neighborhood of $\bar{\mathcal{O}}_\delta$ (where $\delta > 0$ is fixed). We then take $\eta \in \mathcal{D}_+(\mathcal{O})$, $0 \leq \eta \leq 1$, $\eta \equiv 1$ on a neighborhood of Supp ξ. Extending the various functions considered below with compact support in $\mathcal{O}$ by 0 outside $\mathcal{O}$, we have clearly : $\tilde{u} = \eta u \in W^{1,\infty}(\mathbb{R}^N)$, $\tilde{A}_\alpha u \in L^\infty(\mathbb{R}^N)$ and

$$\sup_{\alpha\in A} \|\tilde{A}_\alpha u\|_{L^\infty(\mathbb{R}^N)} < \infty, \quad \sup_{\alpha\in A} \{\tilde{A}_\alpha \tilde{u} + \gamma\tilde{u} - g(x,\alpha)\} = 0 \quad \text{a.e. in } \mathbb{R}^N ;$$

where $\gamma > 0$ is fixed, $\tilde{A}_\alpha = \xi^2 A_\alpha$, $g = \xi^2 f + \gamma \tilde{u}$.

Clearly the above equation corresponds to the control problem with $\mathcal{O}$ replaced by $\mathbb{R}^N$, σ by $\xi\sigma$, b by $\xi^2 b$, c by $\tilde{c} = \xi^2 c + \gamma$, f by g. Finally we observe that since $u \in W^{1,\infty}_{loc}(\mathcal{O})$, we also have :

$$\Delta\tilde{u} \le C \quad \text{in } \mathcal{D}'(\mathbb{R}^N) \ .$$

Thus if we prove the Theorem in this special case ($\mathcal{O} = \mathbb{R}^N$), then we have : $\tilde{u}(x) = \inf_{\mathcal{A}} E \int_0^\infty g(\tilde{X}_t,\alpha_t) \exp(-\int_0^t \tilde{c})dt$ and thus, applying the dynamic programming principle, we find :

$$\forall x \in \bar{\mathcal{O}}_\delta,\ u(x) = \tilde{u}(x) = \inf_{\mathcal{A}} E\int_0^{\tau_\delta} \{f(X_t,\alpha_t) + \gamma u(X_t)\}\exp(-\int_0^t (c+\gamma))dt$$

And we conclude letting $\gamma \to 0_+$.

Step 2 : Proof in the case $\mathcal{O} = \mathbb{R}^N$.

This case was already treated in P.L. Lions [38], [39]. We briefly recall the proof. Let $\varepsilon, \delta > 0$; we consider $\varphi_\varepsilon = \varphi * p_\varepsilon$ where :

$$p_\varepsilon = \frac{1}{\varepsilon^N} p(\frac{\cdot}{\varepsilon}) \ , \ p \in \mathcal{D}_+(\mathbb{R}^N) \ , \ \mathrm{Supp}\ p \subset B(0,1), \quad \int_{\mathbb{R}^N} p \, d\xi = 1.$$

Then (denoting $\tilde{A}_\alpha = A_\alpha$, $u \equiv \tilde{u}$) we have :

$$-\frac{\delta^2}{2} \Delta u_\varepsilon + \{\sup_{\alpha \in A} \{(A_\alpha + \gamma)u - g(.,\alpha)\}\} * p_\varepsilon \ge - C\delta^2 \ ,$$

and we deduce from the following Lemma :

$$-\frac{\delta^2}{2}\Delta u_\varepsilon + \sup_{\alpha\in A}\{(A_\alpha+\gamma)u_\varepsilon - g(.,\alpha)\} \geq -C\delta^2 - h_\varepsilon$$

where $h_\varepsilon \in L^\infty_+(\mathbb{R}^N)$, $0 \leq h_\varepsilon \leq C$, $h_\varepsilon \to 0$ a.e. as $\varepsilon \to 0_+$. Indeed we have on one hand :

$$\sup_{\alpha\in A}\{[(A_\alpha+\gamma)u - g(.,\alpha)]*p_\varepsilon\} \leq [\sup_{\alpha\in A}\{A_\alpha u - g(.,\alpha)\}]*p_\varepsilon$$

$$\text{a.e. } \liminf_{\varepsilon\to 0_+}\{\sup_{\alpha\in A}[\{(A_\alpha+\gamma)u - g(.,\alpha)\}*p_\varepsilon]\} \geq \sup_{\alpha\in A}\{(A_\alpha+\gamma)u - g(.,\alpha)\}$$

and on the other hand, the following Lemma and its proof imply :

$$\bar{h}_\varepsilon = \sup_{\alpha\in A}|A_\alpha u_\varepsilon - (A_\alpha u)*p_\varepsilon| \to 0 \text{ a.e. as } \varepsilon \to 0_+,\ \bar{h}_\varepsilon \leq C.$$

<u>Lemma III.1</u> : *Let* $w \in W^{1,p}(\mathbb{R}^N)$ *with* $1 \leq p \leq \infty$. *Then we have :*

i) $\bar{h}_\varepsilon = \sup_{\alpha\in A}|A_\alpha w_\varepsilon - (A_\alpha w)*p_\varepsilon|$ *is bounded in* $L^p(\mathbb{R}^N)$; $\bar{h}_\varepsilon \xrightarrow[\varepsilon\to 0_+]{} 0$ *a.e.*

ii) $\bar{h}_\varepsilon \xrightarrow[\varepsilon\to 0_+]{} 0$ *in* $L^p(\mathbb{R}^N)$ *if* $p < \infty$; $\bar{h}_\varepsilon \xrightarrow[\varepsilon\to 0_+]{} 0$ *weakly in* $L^\infty(\mathbb{R}^N)*$ *if* $p = \infty$.

The proof of this Lemma will be given below. We next consider

$$u^\delta(x) = \inf_{\mathcal{A}} E\int_0^\infty g(X_t^\delta,\alpha_t)\exp(-\int_0^t c)dt$$

where X_t^δ is the solution of :

$$dX_t^\delta = \sigma(X_t,\alpha_t)dB_t + \delta\, dW_t + b(X_t,\alpha_t)\, dt$$

and W_t is N-dimensional Brownian motion independent of F^t. Of course, in view of the assumptions made upon σ,b,f,c (and thus g) we see that : $u^\delta \xrightarrow[\delta\to 0_+]{} \tilde{u}$ in $L^\infty(\mathbb{R}^N)$ where $\tilde{u} \in B \cup C(\mathbb{R}^N)$ is given by :

$$\tilde{u}(x) = \inf_{\mathcal{A}} E \int_0^\infty f(X_t,\alpha_t) \exp(-\int_0^t c)dt \ ;$$

recall that we have to prove : $u \equiv \tilde{u}$ in $\mathbb{R}^N$. Actually, since u satisfies : $u \in W^{1,\infty}(\mathbb{R}^N)$, $A_\alpha u \le f(.,\alpha)$ in $\mathcal{D}'(\mathbb{R}^N)$ for all $\alpha \in A$, we already proved : $u \le \tilde{u}$ in $\mathbb{R}^N$ in Part 1 [36]. Therefore we just need to prove ; $u \ge \tilde{u}$ in $\mathbb{R}^N$.

We denote by τ_R^δ the first exit time of X_t^δ from $\bar{B}_R$: since the matrix $a^\delta = a + \frac{\delta^2}{2} I_N$ is uniformly definite positive, it is easy to build near-optimal feedbacks using the results of N.V. Krylov [29] (cf. also [24]) and to deduce from the equality satisfied by u_ε the following :

$$u_\varepsilon(x) \ge \inf_{\mathcal{A}} E \int_0^{\tau_R^\delta} \{g(X_t^\delta,\alpha_t) - C\delta^2 - h_\varepsilon(X_t^\delta)\} \exp(-\int_0^t c)dt +$$

$$+ u_\varepsilon(X^\delta_{\tau_\delta^R}) \exp(-\int_0^{\tau_R^\delta} c)$$

- see for similar arguments M. Nisio [64], N.V. Krylov [25], A. Bensoussan and J.L. Lions [4] ... -. But we have also :

$$u^\delta(x) = \inf_{\mathcal{A}} \{E\int_0^{\tau_R^\delta} g(X_t^\delta,\alpha_t) \exp(-\int_0^t c)dt + u^\delta(X^\delta_{\tau_R^\delta}) \exp(-\int_0^{\tau_R^\delta} c)\}.$$

And we deduce :

$$\max_{\bar{B}_R}(u^\delta - u_\varepsilon)^+ \le \frac{C}{\gamma}\delta^2 + \sup_{\mathcal{A}} E \int_0^{\tau_R^\delta} h_\varepsilon(X_t^\delta)dt + C \sup_{\mathcal{A}} E\{e^{-\gamma\tau_\delta^R}\}.$$

Using N.V. Krylov inequalities [32], [33], [24] we find :

$$\max_{\bar{B}_R}(u^{\delta}-u_{\varepsilon})^{+} \le \frac{C}{\gamma}\,\delta^2 + C(R,\delta)\,\|h_{\varepsilon}\|_{L^N(B_R)} + C\sup_{\mathcal{A}} E\{e^{-\gamma\tau^R_{\delta}}\} .$$

We now let $\varepsilon \to 0_+$, $R \to \infty$, $\delta \to 0_+$ and we conclude :

$$\tilde{u} \le u \quad \text{in} \quad \mathbb{R}^N .$$

<u>Proof of Lemma III.1</u> : It is easy to check that Lemma III.1 is proved if we prove i), ii) for : $\bar{h}_{\varepsilon} = \sup_{\alpha\in A} |\beta^{\alpha}\frac{\partial w_{\varepsilon}}{\partial x_1} - (\beta^{\alpha}\frac{\partial w}{\partial x_1}) * p_{\varepsilon}|$, where $(\beta^{\alpha})_{\alpha\in A}$ is a bounded set in $W^{1,\infty}(\mathbb{R}^N)$ and where $w \in L^p(\mathbb{R}^N)$. To express the dependence on w, we will sometimes write : $\bar{h}_{\varepsilon} = \bar{h}_{\varepsilon}(w)$.

We first show that we have for some $C > 0$:

$$\|\bar{h}_{\varepsilon}(w)\|_{L^p(\mathbb{R}^N)} \le C\,\|w\|_{L^p(\mathbb{R}^N)} ,\ \forall\varepsilon > 0,\ \forall w \in L^p(\mathbb{R}^N) .$$

Remarking that : $|\bar{h}_{\varepsilon}(w_1) - \bar{h}_{\varepsilon}(w_2)| \le \bar{h}_{\varepsilon}(w_1-w_2)$, we see that is enough to prove the above inequality for w smooth. For such a w we have :

$$\beta^{\alpha}\partial_1 w_{\varepsilon} - (\beta^{\alpha}\partial_1 w) * p_{\varepsilon} = \int \beta^{\alpha}(x)\, w(y)\partial_1 p_{\varepsilon}(x-y)dy -$$

$$- \int \beta^{\alpha}(y)w(y)\partial_1 p_{\varepsilon}(x-y)dy + \int (\partial_1\beta^{\alpha}(y))w(y)p_{\varepsilon}(x-y)dy$$

and this yields :

$$\bar{h}_{\varepsilon}(w) \le C\ |w| * p_{\varepsilon} + C|w| * \tilde{p}_{\varepsilon} \quad \text{in} \quad \mathbb{R}^N$$

where $\tilde{p}_\varepsilon = \frac{1}{\varepsilon^N} \tilde{p}(\frac{\cdot}{\varepsilon})$, $\tilde{p}(\xi) = |\xi| \, |\partial_1 p_\varepsilon(\xi)|$. And this proves our claim.

To conclude we just have to prove that for w smooth in L^p we have $\bar{h}_\varepsilon \xrightarrow[\varepsilon \to 0_+]{} 0$ a.e. and in $L^p(\mathbb{R}^N)$ for $p < \infty$. Indeed the general case is then obtained remarking again that :

$$|\bar{h}_\varepsilon(w_1) - \bar{h}_\varepsilon(w_2)| \le \bar{h}_\varepsilon(w_1 - w_2) \quad \text{in} \quad \mathbb{R}^N .$$

Now for $w \in \mathcal{D}(\mathbb{R}^N)$, we have

$$\bar{h}_\varepsilon(w) = \sup_{\alpha \in A} \left| \int_{\mathbb{R}^N} (\beta^\alpha(x) - \beta^\alpha(y)) \partial_1 w(y) \; p_\varepsilon(x-y) dy \right|$$

$$\le C \varepsilon (|\partial_1 w| * p_\varepsilon)$$

and we conclude easily.

III.3. <u>Variants and extensions</u> :

We first give a result extending Theorem III.1 where we relax the assumptions made upon u : $\mathcal{O}$ is still a bounded open set in $\mathbb{R}^N$ (for example).

<u>Theorem III.2</u> : *We assume* (64) *and* (65). *Let* $u \in C(\mathcal{O}) \cap W^{1,N}_{loc}(\mathcal{O})$ *satisfy* :

$$A_\alpha u \le f(.,\alpha) \quad \textit{in} \quad \mathcal{D}'(\mathcal{O}) \quad \textit{for all} \quad \alpha \in A ; \tag{68}$$

$$\exists g \in L^N_{loc}(\mathcal{O}) , \; \Delta u \le g \quad \textit{in} \quad \mathcal{D}'(\mathcal{O}) . \tag{69}$$

In particular (68) *implies that* $A_\alpha u$ *is a measure on* $\mathcal{O}$ *and we assume that the following equation holds in the sense of measures :*

$$\sup_{\alpha\in A} \{A_\alpha u - f(.,\alpha)\} = 0 \quad in \ \mathcal{D}'(\mathcal{O}) \tag{4'}$$

Then we have for all $\delta > 0$ *and for all* $x \in \bar{\mathcal{O}}_\delta$:

$$u(x) = \inf_{\mathcal{A}} \{E \int_0^{\tau_\delta} f(X_t,\alpha_t) \exp(-\int_0^t c)dt + u(X_{\tau_\delta}) \exp(-\int_0^{\tau_\delta} c)\}.$$

Remark III.4. : This result has exactly the same applications and variants as Theorem III.1. In particular the analogues of Remarks III.1-3 and of Corollaries III.1-2 hold.

Proof of Theorem III.1 : We first prove the above result assuming in addition :

$$\exists\alpha_0 \in A \ , \ A_{\alpha_0} u \in L^N_{loc}(\mathcal{O}). \tag{70}$$

We just have to show how one modifies the argument made above in Step 2 : we keep the same notation with obvious adaptations. Clearly we have :

$$0 \geq \sup_{\alpha\in A} \{[(A_\alpha+\gamma)u - g(.,\alpha)]*p_\varepsilon\} \geq \{(A_{\alpha_0}+\gamma)u - g(.,\alpha_0)\}*p_\varepsilon = z_\varepsilon$$

and because of (68), z_ε converges in $L^N(\mathbb{R}^N)$ to $[(A_{\alpha_0}+\gamma)u - g(.,\alpha_0)]$. In addition recalling that, if $\mu_\alpha = (A_\alpha+\gamma)u - g(.,\alpha)$, we have :

$$\mu_\alpha^\varepsilon = [(A_\alpha+\gamma)u - g(.,\alpha)]*p_\varepsilon \xrightarrow[\varepsilon\to 0_+]{} \frac{d\mu_\alpha}{dx} \quad \text{a.e.}$$

we deduce, since $\mu_\alpha \leq 0$, that we have a.e. :

$$\liminf_{\varepsilon \to 0_+} \sup_{\alpha\in A}(\mu_\alpha * p_\varepsilon) \geq \sup_{\alpha\in A}(\frac{d\mu_\alpha}{dx})$$

and $0 \geq \sup_{\alpha\in A}(\frac{d\mu_\alpha}{dx}) \geq \sup_{\alpha\in A}(\mu_\alpha) = 0$ in $\mathcal{D}'(\mathbb{R}^N)$.

Thus in conclusion we deduce that $\sup_{\alpha\in A}(\mu_\alpha * p_\varepsilon)$ converges to 0 a.e. and in $L^N(\mathbb{R}^N)$ as ε goes to 0. Using Lemma III.1 (recall that $u \in W^{1,N}(\mathbb{R}^N)$) and (69) we finally obtain :

$$-\frac{\delta^2}{2}\Delta u_\varepsilon + \sup_{\alpha\in A}\{(A_\alpha+\gamma)u_\varepsilon - g(x,\alpha)\} \geq -\delta^2 g_\varepsilon - h_\varepsilon$$

where h_ε converges in $L^N(\mathbb{R}^N)$ (and a.e.) to 0, while $g_\varepsilon \geq 0$ converges in $L^N(\mathbb{R}^N)$ (and a.e.) to some g. Therefore for any $\mu > 0$, we may find $g_\varepsilon^1, g_\varepsilon^2 \geq 0$ such that $g_\varepsilon^1 \leq C$ in $\mathbb{R}^N$, $\|g_\varepsilon^2\|_{L^N(\mathbb{R}^N)} \leq \mu$, $g_\varepsilon = g_\varepsilon^1 + g_\varepsilon^2$. We thus have proved :

$$-\frac{\delta^2}{2}\Delta u_\varepsilon + \sup_{\alpha\in A}\{(A_\alpha+\gamma)u_\varepsilon - g(x,\alpha)\} \geq -C_\mu\delta^2 - \delta^2 g_\varepsilon^2 - h_\varepsilon .$$

Arguing as in Step 2, we find :

$$\max_{\bar{B}_R}(u^\delta - u_\varepsilon)^+ \leq \frac{C_\mu\delta^2}{\gamma} + \delta^2 \sup_{\mathcal{A}} E\int_0^{\tau_R^\delta} g_\varepsilon^2(X_t^\delta)e^{-\gamma t}\,dt +$$

$$+ \sup_{\mathcal{A}} E\int_0^{\tau_R^\delta} h_\varepsilon(X_t^\delta)e^{-\gamma t}\,dt + \sup_{\mathcal{A}} E[e^{-\gamma\tau_R^\delta}]$$

$$\leq \frac{C_\mu\delta^2}{\gamma} + C(R,\delta)\,\|h_\varepsilon\|_{L^N(B_R)} + \kappa(R) + \delta^2 \sup_{\mathcal{A}} E\int_0^{\tau_R^\delta} g_\varepsilon^2(X_t^\delta)e^{-\gamma t}\,dt.$$

where $\kappa(R) \to 0$ as $R \to \infty$.

Applying Theorem 2 in Chapter 2 of N.V. Krylov [24], we see that we have :

$$\sup_{\mathcal{A}} E \int_0^{\tau_R^\delta} g_\varepsilon^2(X_t^\delta) e^{-\gamma t}\, dt \le C_R\, \delta^{-2} \, \|g_\varepsilon^2\|_{L^N(B_R)} \le C_R\, \delta^{-2} \mu \quad .$$

We finally get :

$$\max_{\bar{B}_R} (u^\delta - u_\varepsilon)^+ \le \frac{C_\mu \delta^2}{\gamma} + C(R,\delta) \, \|h_\varepsilon\|_{L^N(B_R)} + C_R \mu + \kappa(R)$$

and we conclude letting $\varepsilon \to 0_+$, $\delta \to 0_+$, $\mu \to 0_+$ and $R \to \infty$.

We now explain how to prove Theorem III.1 without assumption (70). We then introduce $A' = A \cup \{\alpha'\}$ (where $\alpha' \notin A$) and we set

$$\sigma(x,\alpha') \equiv 0, \; b(x,\alpha') \equiv 0, \; c(x,\alpha') \equiv 1, \; f(x,\alpha') = C$$

where $C \ge 1 + \|u\|_{L^\infty(K)}$ with Supp $\eta \subset K \subset \mathcal{O}$.

Clearly we still have with the above notation :

$$\sup_{\alpha \in A'} \{(A_\alpha + \gamma)u - g(.,\alpha)\} = 0 \quad \text{in } \mathcal{D}'(\mathbb{R}^N)$$

$$A_{\alpha'} u = u \in L^N(\mathbb{R}^N)$$

and thus applying the first part of the proof we find :

$$u(x) = \inf_{\mathcal{A}'} E \int_0^\infty g(X_t, \alpha_t) \exp(-\int_0^t \tilde{c}) dt$$

where the infimum has to be taken over all admissible systems such that α_t takes now its values in A'.

Next, remark that we have :

$$(A_{\alpha'}+\gamma)u = (\xi^2+\gamma)u < \xi^2 C + \gamma u = g(x,\alpha') \quad \text{in} \quad \mathbb{R}^N$$

and this will imply that u is given by :

$$u(x) = \inf_{\mathcal{A}} E \int_0^\infty g(X_t,\alpha_t) \exp(-\int_0^t c)dt$$

and we will then be able to conclude.

Smoothing the coefficients σ,b,f,g if necessary we may assume that (9) holds and using Proposition IV.1 in P.L. Lions [39] (simple consequence of the regularity results) we find for all $\varepsilon > 0$ u^ε satisfying :

$$\begin{cases} \|u^\varepsilon - u\|_{L^\infty(\mathbb{R}^N)} \le \varepsilon, \; u^\varepsilon \in W^{1,\infty}(\mathbb{R}^N), \; \sup_{\alpha\in A'} \|A_\alpha u^\varepsilon\|_{L^\infty(\mathbb{R}^N)} \le C_\varepsilon \\ \exists C_\varepsilon > 0, \forall \xi \in \mathbb{R}^N : |\xi| = 1, \; \partial_\xi^2 u^\varepsilon \le C_\varepsilon \text{ in } \mathcal{D}'(\mathbb{R}^N) \\ \sup_{\alpha\in A'} \{(A_\alpha+\gamma)u^\varepsilon - g(x,\alpha)\} \ge \varepsilon \quad \text{a.e. in } \mathbb{R}^N . \end{cases}$$

But we have clearly :

$$(A_{\alpha'}+\gamma)u^\varepsilon \le (A_{\alpha'}+\gamma)u + \varepsilon(\xi^2+\gamma) \le \xi^2(C-1) + \gamma u + \varepsilon\xi^2 + \varepsilon\gamma$$

$$\le g + \xi^2(\varepsilon-1) + \varepsilon\gamma < g + \varepsilon \text{ if } \varepsilon \in (0,1), \; \gamma \in (0,1) .$$

Therefore we have also :

$$\sup_{\alpha\in A} \{(A_\alpha+\gamma)u^\varepsilon - g(x,\alpha)\} \geq \varepsilon \quad \text{a.e. in } \mathbb{R}^N .$$

Applying now Theorem III.1 (and its proof) we find :

$$u^\varepsilon(x) \geq \inf_{\mathcal{A}} \{E \int_0^\infty g(X_t,\alpha_t) \exp(-\int_0^t c)dt\} - C_\varepsilon$$

and thus letting $\varepsilon \to 0_+$: $u(x) \geq \inf_{\mathcal{A}} E \int_0^\infty g(X_t,\alpha_t) \exp(-\int_0^t \tilde{c})dt.$

Since we already proved that u was equal to the infimum over a larger class of admissible systems, we are able to conclude.

Remark III.5 : It is clear from the proof that it is enough to assume (for example) $\partial_i u \in L^N(\mathbb{R}^N)$ for $1 \leq i \leq p$ if for all $(x,\alpha) \in \mathbb{R}^N \times A$ $a_{k\ell}(x,\alpha) = 0$ when k or $\ell \geq p+1$.

We conclude this section by giving one example which shows that Theorem III.2 is "optimal" : indeed let $\mathcal{O} = \mathbb{R}^N$ (with $N \geq 2$), $\sigma \equiv 0$, $b \equiv \alpha$, $A = \{\xi \in \mathbb{R}^N / |\xi| \leq 1\}$, $c \equiv \lambda > 0$, $f \equiv 1$. Then it is obvious to see that $u_o \equiv 1/\lambda$ is the associated optimal cost function. Next, for $\beta > 0$, we define

$$u_\beta(x) = \frac{1}{\lambda} - \beta e^{-\lambda|x|}$$

Clearly $u_\beta \in W^{1,\infty}(\mathbb{R}^N)$, $\sup_{\alpha\in A} \|A_\alpha u_\beta\|_{L^\infty(\mathbb{R}^N)} < \infty$, u_β satisfies

$$\sup_\alpha \{A_\alpha u_\beta - f(x,\alpha)\} = |\nabla u_\beta| + \lambda u_\beta - 1 = 0 \quad \text{a.e. in } \mathbb{R}^N$$

(actually on $\mathbb{R}^N - \{0\}$) and

$$\Delta u_\beta = \frac{\beta\lambda(N-1)}{|x|} e^{-\lambda|x|} - \beta\lambda^2 e^{-\lambda|x|} \in L^p(\mathbb{R}^N) \quad \text{for all} \quad p < N$$

(and even in $M^N(\mathbb{R}^N)$). This shows that one cannot weaken the asumption on Δu in Theorem III.2.

III.4. Time-dependent problems

We will use the notations of section II.2 and we will make the following assumptions : let $T \in]0,+\infty[$ be fixed,

$$\left\{ \begin{array}{l} \sup_{\alpha\in A, t\in[0,T]} \|\varphi(.,t,\alpha)\|_{W^{1,\infty}(\mathbb{R}^N)} < \infty \text{ , for } \psi = \sigma_{ij} \ (1 \le i \le n) \text{ ,} \\ \qquad\qquad\qquad\qquad\qquad\qquad\qquad\qquad b_i \ (1 \le i \le n) \\ \psi \text{ is bounded, continuous on } \bar{B}_R \times [0,T] \text{ uniformly in } \alpha \in A, \\ \quad \text{for } \psi = \sigma,b,c,f \\ \psi \in C(\mathbb{R}^N \times [0,T] \times A) \text{ for } \psi = \sigma_{ij},b_i,c,f. \end{array} \right. \qquad (64')$$

Theorem III.3 : *We assume* (64'). *Let* $u \in C(\mathcal{O}\times(0,T))$ *be such that* $D_x u \in L^{N+1}_{loc}(\mathcal{O}\times(0,T))$. *We assume that* u *satisfies :*

$$-\frac{\partial u}{\partial t} + A_\alpha u - f(.,\alpha) \le 0 \text{ in } \mathcal{D}'(\mathcal{O}\times(0,T)) \text{ for all } \alpha \in A \qquad (68')$$

$$\left\{ \begin{array}{l} \exists g \in L^{N+1}_{loc}(\mathcal{O}\times(0,T)), \exists C_1,C_2 > 0, -\frac{\partial u}{\partial t} - C_1\Delta u \ge -C_2 \\ \qquad \text{in } \mathcal{D}'(\mathcal{O}\times(0,T)). \end{array} \right. \qquad (69')$$

We assume that the HJB equation holds in the sense of measures :

$$\sup_{\alpha\in A} \{- \frac{\partial u}{\partial t} + A_\alpha u - f(.,\alpha)\} = 0 \quad in \ \mathcal{D}'(\mathcal{O}\times(0,T)) .$$

Then for all $\varepsilon,\delta > 0$, *for all* $(x,t) \in \bar{\mathcal{O}}_\delta \times [\varepsilon,T-\varepsilon]$, *we have :*

$$u(x,t) = \inf_{\mathcal{A}} \{E\int_0^{\tau_\delta\wedge(T-\varepsilon-t)} f(X_s,s+t,\alpha_s) \exp(-\int_0^s c)ds +$$

$$+ u(X_{\tau_\delta},T-\varepsilon) \exp(-\int_0^{\tau_\delta\wedge(T-\varepsilon-t)} c)\}.$$

Remark III.6 : Exactly as before there are many possible extensions of this result replacing $C_1\Delta$ by some uniformly elliptic (in $\mathcal{O} \times (0,T)$) operator with coefficients Lipschitz in x, continuous in t and $-C_2$ by $g \in L^{N+1}_{loc}(\mathcal{O}\times(0,T))$. In addition we could state the analogues of Remarks III.1-6 and of Corollaries III.1-2 but we will not do so here.

Remark III.7 : Let us recall that the regularity results proved in the case of time-dependent problems imply :

$$\frac{\partial u}{\partial t} \geq - C,\ D_x u \in L^\infty(\mathcal{O}\times(0,T)) \ , \quad \Delta u \leq C \text{ in } \mathcal{O}\times (0,T)$$

and thus (69') holds.

We will not prove Theorem III.3 since its proof is a straightforward adaptation of the above proofs : let us only indicate the main lines of the proof.

One first reduces the problem to the case $\mathcal{O} = \mathbb{R}^N$ by the same trick as before. Then for $\gamma > 0$ small enough, we observe that :

$$\sup\{\sup_{\alpha\in A} (-\frac{\partial u}{\partial t} + A_\alpha u - f(.,\alpha)), -\frac{\partial u}{\partial t} - C_1\Delta u - \frac{1}{\gamma}\} \geq 0 \text{ in } \mathcal{D}'(\mathbb{R}^N)$$

and smoothing u first in x and then in t, we obtain u_ε smooth in $\mathbb{R}^N \times (\varepsilon, T-\varepsilon)$ satisfying

$$\sup\{\sup_{\alpha\in A} (-\frac{\partial u_\varepsilon}{\partial t} + A_\alpha u_\varepsilon - f(.,\alpha)), -\frac{\partial u_\varepsilon}{\partial t} - C_1\Delta u_\varepsilon - \frac{1}{\gamma}\} \geq - h_\varepsilon$$

where $h_\varepsilon \geq 0$, $h_\varepsilon \xrightarrow[\varepsilon\to 0_+]{} 0$ in $L^{N+1}(\mathbb{R}^N)$. Now for $\delta > 0$ we have :

$$\delta^2(-\frac{\partial u}{\partial t} - C_1\Delta u) + \sup\{\sup_{\alpha\in A} (-\frac{\partial u_\varepsilon}{\partial t} + A_\alpha u_\varepsilon - f(.,\alpha)) ,$$

$$-\frac{\partial u_\varepsilon}{\partial t} - C_1\Delta u_\varepsilon - \frac{1}{\gamma}\} \geq - h_\varepsilon - C_2\delta^2$$

and this yields taking $\varepsilon \to 0$, $\gamma \to 0$ and $\delta \to 0$:

$$u(x) \geq \inf_{\mathcal{A}} E \int_0^\infty f(X_t,\alpha_t) \exp(-\int_0^t c)dt.$$

On the other hand (68') implies the reversed inequality (cf. Part 1 [36]).

REFERENCES

[1] A.D. Alexandrov, Dirichlet's problem for the equation $\text{Det}\,\|z_{ij}\| = \Phi(z_1,\ldots,z_n,z,x_1,\ldots,x_n)$. I, Vestnik Leningrad Univ. Sem. Mat. Astr., 13 (1958), 5-24 (in Russian).

[2] S. Belbas and S. Lenhart, A system of nonlinear partial differential equations arising in the optimal control of stochastic systems with switching cost. Preprint.

[3] R. Bellman, Dynamic Programming, Princeton Univ. Press, Princeton, N.J., 1957.

[4] A. Bensoussan and J.L. Lions, Applications des inéquations variationnelles en contrôle stochastique. Dunod, Paris, 1978.

[5] J.M. Bony, Principe du maximum dans les espaces de Sobolev. Comptes Rendus Paris, 265 (1967), 333-336.

[6] H. Brézis and L.C. Evans, A variational inequality approach to the Bellman-Dirichlet equation for two elliptic operators. Arch. Rat. Mech. Anal., 71 (1979), 1-13.

[7] L. Caffarelli, J. Spruck and L. Nirenberg : To appear.

[8] S.Y. Cheng and S.T. Yau : On the regularity of the Monge-Ampère equation $\det(\partial^2 u/\partial x_i \partial x_j) = F(x,u)$. Comm. Pure Appl. Math., 30 (1977), 41-68.

[9] S.Y. Cheng and S.T. Yau, On the regularity of the Monge-Ampère equation and affine flat structures. Preprint.

[10] M.G. Crandall, L.C. Evans and P.L. Lions, Some properties of viscosity solutions of Hamilton-Jacobi equations. To appear in Trans. Amer. Math. Soc.

[11] M.G. Crandall and P.L. Lions, Condition d'unicité pour les solutions généralisées des équations de Hamilton-Jacobi du premier ordre Comptes-Rendus Paris, 252 (1981), 183-186.

[12] M.G. Crandall and P.L. Lions, Viscosity solutions of Hamilton-Jacobi equations. To appear in Trans. Amer. Math. Soc.

[13] L.C. Evans, Classical solutions of fully nonlinear, convex, second-order elliptic equations. Comm. Pure Appl. Math., 25 (1982), 333-363.

[14] L.C. Evans, Classical solutions of the Hamilton-Jacobi-Bellman equation for uniformly elliptic operators. Preprint.

[15] L.C. Evans and A. Friedman, Optimal stochastic switching and the Dirichlet problem for the Bellman equation. Trans. Amer. Math. Soc., 253 (1979), 365-389.

[16] L.C. Evans and P.L. Lions, Résolution des équations de Hamilton-Jacobi-Bellman pour des opérateurs uniformément elliptiques. Comptes-Rendus Paris, 290 (1980), 1049-1052.

[17] W.H. Pleming and R. Rishel, Deterministic and stochastic optimal control. Springer, Berlin, 1975.

[18] M.I. Freidlin, Smoothness of solutions of degenerate elliptic equations. Math. USSR Izv., 2 (1968), 1391-1413.

[19] B. Gaveau, Méthode de contrôle optimal en analyse complexe I. J. Funct. Anal., 25 (1977), 391-411.

[20] I.L. Genis and N.V. Krylov, An example of a one-dimensional controlled process. Th. Proba. Appl., 21 (1976), 148-152.

[21] I.I. Gihman and A.V. Skorohod, Stochastic differential equations. Springer, Berlin, 1972.

[22] N. Ikeda and S. Watanabe, Stochastic differential equations and diffusion process. North-Holland, Amsterdam, 1981.

[23] R. Jensen, Boundary regularity for variational inequalities. Preprint.

[24] N.V. Krylov, Controlled diffusion processes. Springer, Berlin, 1980.

[25] N.V. Krylov, Control of a solution of a stochastic integral equation. Th. Proba. Appl., 17 (1972), 114-131.

[26] N.V. Krylov, On control of the solution of a stochastic integral equation with degeneration. Math. USSR Izv., 6 (1972), 249-262.

[27] N.V. Krylov, Control of the diffusion type processes. Proceedings of the International Congress of Mathematicians, Helsinki.

[28] N.V. Krylov, Some new results in the theory of controlled diffusion processes. Math. USSR Sbornik, 37 (1980), 133-149.

[29] N.V. Krylov, On the selection of a Markov process from a system of processes and the construction of quasi-diffusion processes. Math. USSR Izv., 7 (1973), 691-708.

[30] N.V. Krylov, On passing to the limit in degenerate Bellman equations. I. Math. USSR Sbornik, 34 (1978), 765-783.

[31] N.V. Krylov, On passing to the limit in degenerate Bellman equations. II. Math. USSR Sbornik, 35 (1979), 351-362.

[32] N.V. Krylov, Some estimates in the theory of stochastic integrals. Th. Proba. Appl.,18 (1973), 54-63.

[33] N.V. Krylov, Some estimates of the probability density of a stochastic integral. Math. USSR Izv., 8 (1974), 233-254.

[34] J.M. Lasry and P.L. Lions, To appear.

[35] S. Lenhart, Partial Differential Equations from Dynamic Programming equations. Ph. D., Univ. of Kentucky-Lexington, 1981.

[36] P.L. Lions, Optimal control of diffusion processes and Hamilton-Jacobi-Bellman equations. Part 1. Dynamic Programming Principle and applications. To appear in Comm. P.D.E.

[37] P.L. Lions, Optimal control of diffusion processes and Hamilton-Jacobi-Bellman equations. Part 2 : Viscosity solutions and uniqueness. To appear in Comm. P.D.E.

[38] P.L. Lions, Contrôle de diffusions dans $\mathbb{R}^N$. Comptes-Rendus Paris, 283 (1979), 339-342.

[39] P.L. Lions, Control of diffusion processes in $\mathbb{R}^N$. Comm. Pure Appl. Math., 34 (1981), 121-147.

[40] P.L. Lions, Equations de Hamilton-Jacobi-Bellman dégénérées. Comptes-Rendus Paris, 289 (1979), 329-332.

[41] P.L. Lions, Equations de Hamilton-Jacobi-Bellman. In "Séminaire Goulaouic-Schwartz. 1979-1980, Ecole Polytechnique, Palaiseau.

[42] P.L. Lions, Optimal stochastic control and Hamilton-Jacobi-Bellman equations. In "Mathematical Control Theory", Banach Center publications, Warsaw, to appear.

[43] P.L. Lions, Optimal control of diffusion type processes and Hamilton-Jacobi-Bellman equations. In "Advances in Filtering and Optimal Stochastic Control", Eds W.H. Fleming and L. Gorostiza, Springer, Berlin.

[44] P.L. Lions, Fully nonlinear elliptic equations and applications. In "Nonlinear Analysis, Function Spaces and Applications". Teubner , Leipzig, 1982.

[45] P.L. Lions, Un problème de contrôle géométrique et les équations de Hamilton-Jacobi-Bellman. Ann. de Toulouse, Vol. II (1980), 67-78.

[46] P.L. Lions, Un problème de Cauchy pour les équations de Hamilton-Jacobi-Bellman. Ann. de Toulouse, Vol. III (1981), 59-69.

[47] P.L. Lions, Résolution de problèmes quasilinéaires. Arch. Rat. Mech. Anal., 74 (1980), 335-354.

[48] P.L. Lions, Existence results for first-order Hamilton-Jacobi equations. To appear in Ricerche Mat.

[49] P.L. Lions, Une méthode nouvelle pour l'existence de solution régulières de l'équation de Monge-Ampère. Comptes-Rendus Paris, 293 (1981), 589-592.

[50] P.L. Lions, Sur les équations de Monge-Ampère. I. To appear in Manuscripta Math.

[51] P.L. Lions, Sur les équations de Monge-Ampère. II. To appear in Arch. Rat. Mech. Anal.

[52] P.L. Lions, A remark on Bony maximum principle. To appear in Proc. Amer. Math. Soc.

[53] P.L. Lions, Résolution des problèmes généraux de Bellman-Dirichlet. Comptes-Rendus Paris, 287 (1978), 747-750.

[54] P.L. Lions, Some problems related to the Bellman-Dirichlet equation for two operators. Comm. in P.D.E., 5 (1980), 753-771.

[55] P.L. Lions, Résolution analytique des problèmes de Bellman-Dirichlet. Acta Math., 146 (1981), 151-166.

[56] P.L. Lions, Bifurcation and optimal stochastic control. To appear in Nonlinear Anal. T.M.A.

[57] P.L. Lions, To appear.

[58] P.L. Lions, Generalized solutions of Hamilton-Jacobi equations. Pitman, London, 1982.

[59] P.L. Lions and J.L. Menaldi, Optimal control of stochastic integrals and Hamilton-Jacobi-Bellman equations. Part 1. SIAM J. control. Optim., 20 (1982), 58-81.

[60] P.L. Lions and J.L. Menaldi, Optimal control of stochastic integrals and Hamilton-Jacobi-Bellman equations. Part 2. SIAM J. Control Optim., 20 (1982), p. 82-95.

[61] P.L. Lions and J.L. Menaldi, Problèmes de Bellman avec le contrôle dans les coefficients de plus haut degré. Comptes-Rendus Paris, 287 (1978), 503-506.

[62] P.L. Lions and A.S. Sznitman, To appear.

[63] L. Nirenberg, Monge-Ampère equations and some associated problems in geometry. Proceedings of the International Congress of Mathematicians, Vancouver.

[64] M. Nisio, Some remarks on stochastic optimal controls. Proc. Third USSR-Japan Sympos. Proba. Theory, Lecture Notes in Math. 550, Springer, Berlin, 1976.

[65] O. Oleinik : Alcuni risultati sulle equazioni lineari e quasilineari ellitico paraboliche a derivate parziali del secondo ordine. Rend. Acad. Naz. Lincei, 40 (1966), 775-784.

[66] A.N. Pogorelov, On the Minkowski multidimensional problem. J. Wiley, New-York, 1978.

[67] C. Pucci, Limitazioni per soluzioni di equazioni ellitiche. Ann. Mat. Pura Appl, 74 (1966), 15-30.

[68] M.V. Safonov, On the Dirichlet problem for Bellman's equation in a plane domain. Math. USSR Sbornik, 31 (1977), 231-248.

[69] M.V. Safonov, On the Dirichlet problem for Bellman's equation in a plane domain. Math. USSR Sbornik, 34 (1978), 521-526.

[70] D.W. Stroock and S.R.S. Varadhan, On degenerate elliptic-parabolic operators of second order and their associated diffusions. Comm. Pure Appl. Math., 25 (1972), 651-714.

[71] D.W. Stroock and S.R.S. Varadhan, Duffision processes with boundary conditions. Comm. Pure Appl. Math., 24 (1971), 147-225.

[72] D.W. Stroock and S.R.S. Varadhan, Multidimensionnal diffusion processes, Springer, Berlin, 1979.

[73] H. Tanaka, Stochastic differential equations with reflecting boundary boundary conditions in convex regions. Hiroshima Math. J., 9 (1979), 163-177.

[74] N.S. Trudinger, Elliptic equations in non-divergence form. Proc. Miniconf. P.D.E., Canberra, 1981.

Pierre-Louis LIONS

Université Paris IX - Dauphine
Ceremade
Place de Lattre de Tassigny
75775 - PARIS CEDEX 16
FRANCE

F MIGNOT & J P PUEL

Flambage d'une tige viscoélastique

I. INTRODUCTION

Position du problème. On considère une tige inextensible de longueur 1 en matériau viscoélastique, dont les extrêmités A et B sont pincées et astreintes à se déplacer sur un axe. La tige est soumise à une charge axiale dépendant du temps $P(t)$ en son extrêmité B. On cherche à étudier la déformation quasistatique de la tige, supposée être au repos à l'instant $t = 0$.

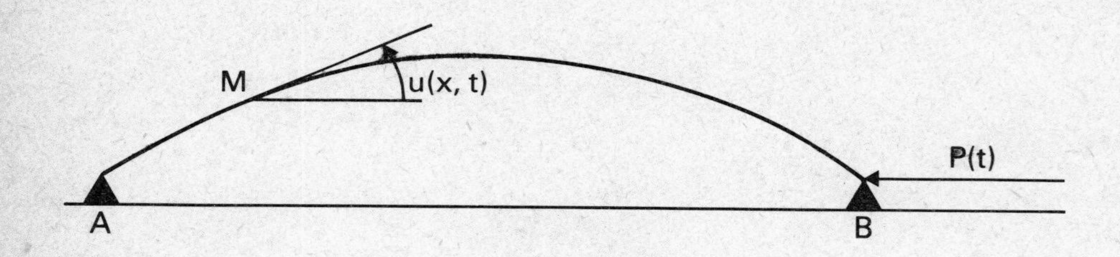

Un point M de la tige est repéré par son abscisse curviligne x ($x \in [0,1]$) et on définit par chaque x de t :

i) l'angle $u(x,t)$ que fait la tangente en M en l'axe AB

ii) le moment fléchissant $m(x,t)$ de la tige en M .

Equation d'équilibre. Le théorème des moments appliqué au problème quasi-statique conduit à l'équation d'équilibre :

$$m_x(x,t) + P(t) \sin u(x,t) = 0. \tag{1}$$

La tige est pincée en A et B, les conditions aux limites correspondantes sont :

$$m(0,t) = m(1,t) = 0 . \tag{2}$$

L'extrémité B de la tige est astreinte à se déplacer sur l'axe supportant la tige au repos, ce qui se traduit par :

$$\int_0^1 \sin u(x,t)dx = 0 . \tag{3}$$

Pour $t = 0$ la tige est au repos :

$$u(x,0) = 0 \quad \text{et} \quad P(0) = 0 .$$

Loi de comportement. On considère une loi de comportement viscoélastique linéaire (que nous avons normalisée pour simplifier) :

$$m(x,t) = u_x(x,t) - \beta_0 \int_0^t \beta(t-s)u_x(x,s)ds , \tag{4}$$

où β est une fonction définie sur $\mathbb{R}^+$, positive, continue, décroissante, telle que $\int_0^\infty \beta(s)ds = 1$ et où β_0 est un réel appartenant à $[0,1[$.

Exceptée la situation concernée par le Théorème 1 on se limitera au cas $\beta(t) = e^{-t}$.

En résumé, supposant connue l'évolution de la charge $P(t)$, nous cherchons les solutions u du système suivant :

$$\left\{\begin{array}{l} u_{xx}(x,t) - \beta_0 \int_0^t \beta(t-s)u_{xx}(x,s)ds + P(t)\sin u(x,t) = 0 , \\ u_x(0,t) = u_x(1,t) = 0 , \\ \int_0^1 \sin u(x,t)dx = 0 . \end{array}\right. \tag{5}$$

L'étude de ce problème nous a été proposée par M.E. Gurtin qui l'expose d'autre part dans [1] dont on pourra consulter la bibliographie.

Remarque 1. Le cas $\beta_o = 0$ correspond au cas purement élastique. On trouvera dans Love [2] une analyse détaillée de ce problème (Euler élastica) posé et résolu par Euler.

2. Dans l'établissement de ce modèle nous n'avons pas fait d'hypothèses restrictives sur u : ce modèle est notamment valable en grand déplacement.

3. Pour toute charge $P(.)$ si u est solution de (5) et si nous considérons une fonction quelconque $k : [0,T] \to \mathbb{Z}$ la fonction $u_1(x,t) = u(x,t) + k(t)\pi$ est encore solution de (5) sur $[0,T]$: pour éliminer ces solutions parasites nous ferons des hypothèses de régularité sur la dépendance en temps de la solution.

Par ailleurs notons que si u est solution, $-u$ est aussi solution.

Concluons cette première partie en soulignant que les résultats que nous allons énoncer mettent en évidence des phénomènes nouveaux de non unicité des branches bifurquées ainsi qu'un couple de paramètres critiques liés à la valeur et à la dérivée de P.

II. NOTATIONS ET RESULTATS

Nous noterons λ_o , $\lambda_1,\ldots,$ les valeurs propres rangées en ordre croissant du problème de Neumann.

$$\begin{cases} - u_{xx} = \lambda u \quad , \quad x \in]0,1[\quad , \\ u_x(0) = u_x(1) = 0 \ , \end{cases}$$

et ϕ_o, ϕ_1,... les vecteurs propres normalisés associés. ($\lambda_o = 0$, $\phi_o = 1$; $\lambda_n = n^2\pi^2$, $\phi_n = \sqrt{2}\cos n\pi x$ si $n \geq 1$).

Nous noterons π_n la projection dans $L^2(]0,1[)$ sur l'orthogonal de $(\phi_o,\ldots,\phi_n)$ et,

$$V = \{v, v \in H^1(]0,1[) \quad , \quad \int_0^1 v dx = 0\} ,$$

$$W = \{w, w \in V , \quad \int_0^1 w\, \phi_1\, dx = 0\} .$$

Si $u \in H^1(]0,1[)$ nous pouvons le mettre sous la forme

$$u = a + v , \qquad a \in \mathbb{R}, v \in V ,$$

ou

$$u = a + b\phi_1 + w , \qquad a \in \mathbb{R}, b \in \mathbb{R}, w \in W ,$$

$$(a = \int_0^1 u\, dx , \quad b = \int_0^1 u\, \phi_1\, dx) .$$

Enfin le problème

$$\begin{cases} - u_{xx} = f , & x \in]0,1[, \quad (f \in L^2(]0,1[) , \\ u_x(0) = u_x(1) = 0 , \end{cases}$$

admet une solution u unique dans V si et seulement si $\int_0^1 f(x)dx = 0$. Nous notons alors

$$u = Gf \quad \text{où} \quad G : \pi_o L^2(]0,1[) \to V .$$

Nous pouvons alors énoncer les principaux résultats. Le premier théorème concerne l'unicité de la branche triviale $u = 0$ en dessous de la valeur critique λ_1 .

Théorème 1. *S'il existe* ε , $\varepsilon > 0$ *tel que* $|P(t)| \leq \lambda_1 - \varepsilon$ *pour* t *appartenant à* $[0,T]$ *, alors toute solution* u *de* (5) *vérifie :* $\pi_o u(t) = 0$ *pour* $t \in [0,T]$ *et* $u(t) = k(t)\pi$ *où* k *est une application arbitraire de* $[0,T]$ *dans* $\mathbb{Z}$.

Corollaire 1. *Si* P(.) *est continue sur* $[0,T[$ *et si* $P(t) < \lambda_1$ *pour* $t \in [0,T[$, *la seule solution de* (5) *continue sur* $[0,T[$ *à valeurs dans* $H^1(]0,1[)$, *vérifiant* $u(0) = 0$ *est la solution nulle.*

Remarque 1. Ce résultat est valable pour toute fonction β vérifiant les conditions imposées.

Pour la suite de l'étude nous précisons ce que nous entendons ici par point de bifurcation.

Définition. Soit P(t) une charge continue définie sur $[0,T[$, $P(0) = 0$. Un point t_1 , $t_1 \in [0,T[$ est un point de bifurcation pour (5) s'il existe une solution u de (5) $u \in \mathcal{C}([0,t_1+\varepsilon(t_1)[, H^1(]0,1[))$, $\varepsilon(t_1) > 0$, avec $u \not\equiv 0$ dans tout voisinage de t_1 .

Remarque. 1) Ce temps t_1 dépend étroitement de P.

2) La définition permet de couvrir le cas où une branche u(.) non nulle sur $]t_1-\varepsilon(t_1),t_1[$ s'annule à partir de t_1 .

Le théorème suivant rattache le cas étudié ici à la situation classique.

Théorème 2. *Soit* P(t) *une charge continue définie sur* $[0,T[$. *Soit un point de bifurcation de* (5) , τ , $\tau \in [0,T[$ *, alors* $P(\tau)$ *est une valeur propre du problème* $-u'' = \lambda u, u'(0) = u'(1) = 0$, $u \in V$, *(c.à.d.* $P(t) = n^2\pi^2$, $n \geq 1$).

Dans les énoncés suivants nous donnons une description des branches bifurquées au voisinage de t_o , $P(t_o) = \lambda_1$. Plus précisément : on suppose que la fonction $P(t)$ est continue, atteint pour la première fois la valeur λ_1 en t_o et est régulière à droite de t_o , c'est-à-dire que :

$$\begin{cases} P(.) \in \mathcal{C}^o([0,T[) , & T > t_o , \\ P(.) \in \mathcal{C}^2([t_o,T[) , & \\ P(t) < \lambda_1 , \quad \forall\, t \in [0,t_o[, & P(t_o) = \lambda_1 , \\ P(t) = \lambda_1 + p_1(t-t_o) + Q(t)(t-t_o)^2 , & \forall\, t \in [t_o,T[. \end{cases} \tag{6}$$

Nous cherchons les solutions u de (5) avec $u \in \mathcal{C}^o([0,\tilde{t}[, H^1(]0,1[))$, $\tilde{t} \in]t_o,T[)$; d'après le Corollaire 1 $u(t) = 0$, $\forall\, t \in [0,t_o]$. Soit alors u une telle solution de (5) associée à $P(t)$ qui vérifie (6). La fonction $U(x,\tau) = u(x,t+\tau)$, $\tau \in [0,T-t_o[$ vérifie

$$\begin{cases} U_{xx}(x,\tau) - \beta_0 \int_0^\tau \beta(\tau-s)\, U_{xx}(x,s)ds + P(t_o+\tau)\sin U(x,\tau) = 0 , \\ U_x(0,\tau) = U_x(1,\tau) = 0 , \\ \int_0^1 \sin U(x,\tau)dx = 0 , \\ U(x,0) = 0 . \end{cases}$$

Autrement dit, en considérant la fonction $\tau \to P(t_o+\tau)$ on se ramène au cas $t_o = 0$, $P(0) = \lambda_1$, $u(0) = 0$, c'est ce que nous ferons dans la suite.

Rappelons que nous cherchons u sous la forme

$$\begin{cases} u(t) = a(t) + b(t)\,\phi_1 + w(x,t) = a(t) + v(t) \\ \\ (a(t),b(t) \in \mathbb{R}^2 , \quad w(t) \in W , \quad v(t) \in V) . \end{cases} \tag{7}$$

Nous montrerons qu'au voisinage du point $(\lambda_1,0)$ toutes les branches solutions de (5) sont repérées par leur composante b (a et w sont des fonctions de b) et que

$$|a(t)| \le C \|b\|^3_{\mathcal{C}([0,t])}$$

$$\|w(t)\|_W \le C \|b\|^3_{\mathcal{C}([0,t])} ,$$

ces propriétés justifient la forme des énoncés ci-dessous (qui, rappelons le, correspondent au cas $\beta(t) = e^{-t}$).

Théorème 3. (cas $p_1 \ge 0$) *Sous les hypthèses* (6) *sur* $P(t)$ $(t_o = 0)$, *si* $p_1 \ge 0$ *il y a bifurcation en* $t = 0$: *il existe une seule branche bifurquée, elle appartient à* $\mathcal{C}^1(]0,\tilde{T}[,H^1(]0,1[))$ *et sa composante* b *vérifie*

$$b(t) = \left(\frac{3p_1+2\lambda_1\beta_o}{3\mu\lambda_1}\right)^{1/2} t^{1/2}(1+0(t^{1/2})) \tag{9}$$

(on a posé $\mu = \frac{1}{6}\int_0^1 \phi_1^4\, dx$).

Remarque. Ce résultat est donc tout à fait analogue à celui de la bifurcation dans le cas purement élastique.

Théorème 4. *Sous les hypthèses* (6) *sur* $P(t)$ $(t_o = 0)$ *si* $p_1 \in]-\frac{2}{3}\lambda_1\beta_o,0[$ *il y a bifurcation en* $t = 0$: *il existe une infinité de branches bifurquées en* 0 *qui se répartissent en*

i) *une branche* $u_o : [0,T_o[\to H^1(]0,1[)$, $u_o \in \mathcal{C}^1(]0,T_o[\, H^1(]0,1[)$, *dont la composante* b_o *vérifie*

$$b_o(t) = \left(\frac{3p_1+2\lambda_1\beta_o}{3\mu\lambda_1}\right)^{1/2} t^{1/2}(1+0(t^{1/2})) \tag{10}$$

ii) *une famille de branches* u_γ *indexées par* γ, $\gamma > 0$ $u_\gamma : [0,T_\gamma[\mapsto H^1(]0,1[)$ *dont les composantes* b_γ *vérifient* :

$$b_\gamma(t) = \left(\frac{t}{\gamma}\right)^{-\frac{p_1+\lambda_1\beta_o}{p_1}} (1+\varepsilon_\gamma(t)) , \qquad (11)$$

si $\quad p_1 \in]-\frac{2}{3}\lambda_1\beta_o, -\frac{\lambda_1\beta_o}{2}[$, $\quad u_\gamma \in \mathcal{C}^1(]0,T_\gamma[, H^1(]0,1[))$

si $\quad p_1 \in [-\frac{\lambda_1\beta_o}{2}, 0[$, $\qquad u_\gamma \in {}^1([0,T_\gamma[, H^1(]0,1[))$

Enfin ce sont les seules branches bifurquées en 0 . □

Remarque. D'après le Théorème 3 il y a dans le cas $p_1 = 0$ une seule branche bifurquée en 0 ; grâce aux Théorèmes 3 et 4 nous pouvons préciser la situation dans un voisinage de $t = 0$:

Corollaire. *Si* $p_1 = 0$ *il existe* $\tilde{T}_o$, $\tilde{T}_o > 0$ *tel que pour tout* t , $t \in]0,\tilde{T}_o[$, *vérifiant* $Q(t) = 0$ *il y a bifurcation en* t . *On peut décrire l'allure des branches bifurquées au moyen des Théorèmes 3 et 4* , $P'(t)$ *jouant alors le rôle de* p_1 .

En particulier si $Q(t) \equiv 0$ *sur* $[0,T[$ *il y a bifurcation en tout point* τ *de* $[0,T[$ *et la branche bifurquée* u *vérifie* :

$$b_\tau(t) = \left(\frac{3\beta_o}{2\mu}\right)^{1/2} ((t-\tau)^+)^{1/2} (1+O(((t-\tau)^+)^{1/2})) \qquad (12)$$

Théorème 5. *Sous les hypothèses* (6) *sur* $P(t)$, *si* $p_1 < -\frac{2}{3}\beta_o\lambda_1$ *il n'y a pas de bifurcation dans un voisinage* $[0,T_{p_1}[$ de 0 : *la seule solution de* (5) *sur* $[0,T_{p_1}[$ *est* $u = 0$.

Ces différents théorèmes donnent ainsi une description complète de la situation au voisinage de $t = 0$.

<u>Remarques</u>. 1. Dans le cas purement élastique si $P(t) \leq \lambda_1$ il n'y a pas de bifurcation, ici il apparait un second paramètre critique p_1 : si $P(t)$ atteint la valeur λ_1 en t_o puis décroit, il y a une pente de descente critique $-\frac{2}{3}\lambda_1\beta_o$ telle que si $P'(t_o^+) < -\frac{2}{3}\lambda_1\beta_o$ il n'y a pas de bifurcation alors que si $P'(t_o^+) > -\frac{2}{3}\lambda_1\beta_o$ il y a bifurcation avec, pour les valeurs négatives de $P'(t_o^+)$, une infinité de solutions ce qui est nouveau.

2. La stabilité des branches bifurquées reste à définir et à étudier. On pourra consulter [1] pour l'introduction d'une fonctionnelle type énergie.

3. Le schéma suivant résume les résultats en fonction du graphe de $P(t)$.

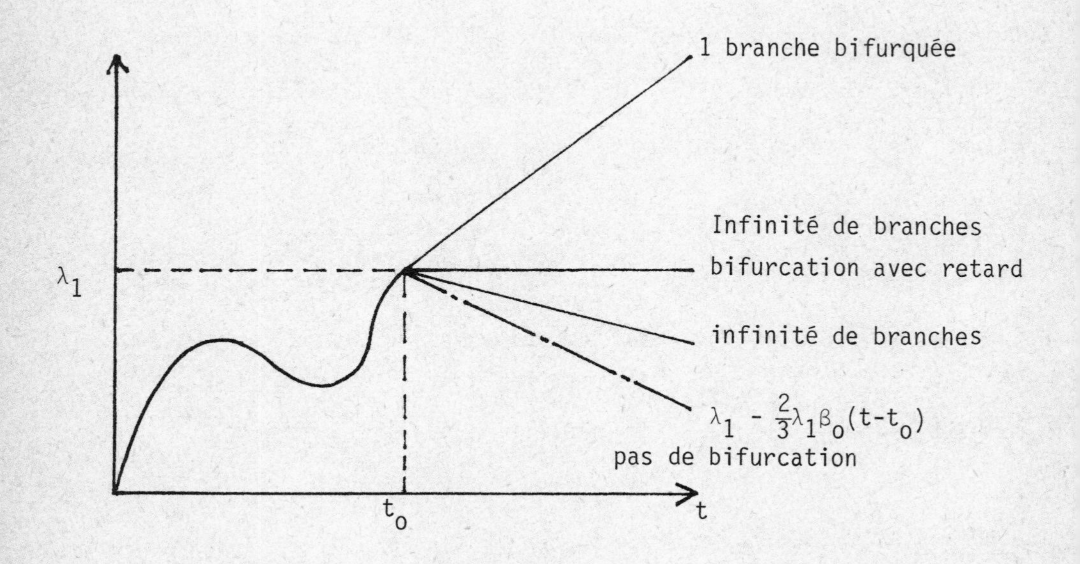

3. DEMONSTRATIONS

Démonstration du Théorème 1. (unicité de la solution nulle si $|P(t)| < \lambda_1 - \varepsilon$ sur $]0,T[$). La démonstration repose sur le lemme de Gronwall. Avec les notations, $u(t) = a(t) + v(t)$, $a(t) \in \mathbb{R}$, $v(t) \in V$, l'équation (5) s'écrit :

$$- v_{xx}(t) + \beta_0 \int_0^t \beta(t-s) v_{xx}(s) ds = P(t)\pi_0 \sin(a(t)+v(t)) . \tag{13}$$

multipliant cette équation par v et intégrant en x nous obtenons :

$$\left\{\begin{aligned} & |v_x(t)|^2_{L^2(0,1)} - \beta_0 \int_0^t \int_0^1 \beta(t-s) v_x(x,s) v_x(x,t)\, dxds \\ & \qquad = P(t) \int_0^1 \pi_0 \sin(a(t)+v(x,t)) v(x,t) dx . \end{aligned}\right. \tag{14}$$

Comme $v \in V$ on a $\int_0^1 \sin(a(t)).v(x,t)dx = 0$, donc

$$|v_x(t)|^2_{L^2} - \beta_0 \int_0^t \int_0^1 \beta(t-s) v_x(x,s) v_x(x,t) dx\, ds =$$

$$P(t) \int_0^1 \pi_0 [\sin(a(t)+v(x,t)) - \sin a(t)]\, v(x,t) dx ;$$

nous avons d'autre part,

$$|P(t) \int_0^1 \pi_0 [\sin(a(t)+v(x,t)) - \sin\, a(t)]\, v(x,t) dx| \leq$$

$$\leq |P(t)|\, |v(t)|^2_{L^2} \leq \frac{|P(t)|}{\lambda_1} |v_x(t)|^2_{L^2}$$ (d'après l'inégalité de Poincaré), et pour $\alpha > 0$,

$$|\beta_0 \int_0^t \int_0^1 \beta(t-s) v_x(x,s) v_x(x,t) dx\, ds| \leq \beta_0 \int_0^t \beta(t-s) |v_x(s)|_{L^2} |v_x(t)|_{L^2} ds$$

$$\leq \beta_0 \frac{\alpha}{2} |v_x(t)|^2_{L^2} \int_0^t \beta(t-s) ds + \frac{\beta_0}{2\alpha} \int_0^t \beta(t-s) |v_x(s)|^2_{L^2} ds$$

$$\leq \beta_0 \frac{\alpha}{2} |v_x(t)|^2_{L^2} + \frac{\beta_0}{2\alpha} \int_0^t \beta(t-s) |v_x(s)|^2_{L^2} ds .$$

Par suite (14) implique :

$$|v_x(t)|_{L^2}^2 \left(1 - \frac{\beta_o\alpha}{2} - \frac{|P(t)|}{\lambda_1}\right) \leq \frac{\beta_o}{2\alpha} \int_0^t \beta(t-s)|v_x(s)|_{L^2}^2 \, ds$$

$$\leq \frac{\beta_o}{2\alpha} \beta(0) \int_0^t |v_x(s)|_{L^2}^2 \, ds ,$$

si $1 - \frac{|P(t)|}{\lambda_1} \geq \frac{\varepsilon}{\lambda_1}$, en prenant α tel que $\frac{\beta_o\alpha}{2} \leq \frac{\varepsilon}{2\lambda_1}$, nous obtenons

$$|v_x(t)|_{L^2}^2 \leq \frac{\beta_o \, \beta(0)\lambda_1}{\alpha\varepsilon} \int_0^t |v_x(s)|_{L^2}^2 \, ds ,$$

ce qui, d'après le lemme de Gronwall, donne,

$$\forall \, t \in [0,T], \; v_x(t) = 0 ,$$

donc $v(t) = 0$, car $v \in V$;

par suite $u(t) = a(t)$, et $\int_0^1 \sin a(t)dx = \sin a(t) = 0$, donc

$\forall \, t \in [0,T]$, $a(t) = k(t)\pi$, où $k : [0,T] \to \mathbb{Z}$.

Le Théorème 1 est ainsi démontré et le Corollaire 1 s'en déduit aisément.

<u>Démonstration du Théorème 2</u>. Soit $u(t)$ la solution bifurquée en τ, elle est définie sur $[0,\tau+\varepsilon(\tau)[$, et $u(t) = a(t) + v(t)$, la fonction v vérifiant l'équation (13) à laquelle on applique l'opérateur G, on obtient ainsi :

$$\left\{ \begin{aligned} & v(t) - \beta_o \int_0^t \beta(t-s)v(s)ds - P(t) \; G\pi_o \sin(a(t)+v(t)) = 0 , \\ & \qquad\qquad\qquad\qquad \forall \, t \in [0,\tau+\varepsilon[, \end{aligned} \right. \tag{15}$$

Au point τ (15) permet d'écrire (car $\beta(t) = e^{-t}$)

$$- \beta_o \int_0^\tau \beta(t-s)v(s)ds = 0 ,$$

et par différence on a,

$$v(t) - P(t)G\pi_0 \quad \sin(a(t)+v(t)) = \beta_0 \int_\tau^t \beta(t-s)v(s)ds . \qquad (16)$$

Si $P(\tau) \neq n^2\pi^2$ $(n \geq 1)$ nous allons mettre cette équation, pour t voisin de τ, sous la forme,

$$v(t) = H(t,a(t),\beta_0 \int_\tau^t \beta(t-s)v(s)ds) , \qquad (17)$$

avec

$$\begin{cases} |H(t,a(t),\beta_0 \int_\tau^t \beta(t-s)v(s)ds)| \leq C \left| \int_\tau^t \beta(t-s)v(s)ds \right| , \\ \qquad\qquad\qquad\qquad \forall\, t \in [\tau-\varepsilon',\tau+\varepsilon'] , \end{cases} \qquad (18)$$

Comme $v(\tau) = 0$ le Lemme de Gronwall impliquera que $v(t) = 0$ sur $[\tau-\varepsilon',\tau+\varepsilon']$, l'équation (5) initiale se réduira alors à $P(t) \sin a(t) = 0$, donc $a(t) \equiv 0$ sur $[\tau-\varepsilon',\tau+\varepsilon']$ (le cas $P(\tau) = 0$ est exclus d'après le théorème 1) d'où la contradiction avec l'hypothèse que τ est un point de bifurcation.

Il reste à montrer (17) : l'opérateur non linéaire ϕ,

$$\phi : \mathbb{R} \times \mathbb{R} \times V \longrightarrow V$$

$$(t,a,v) \mapsto v - P(t)G\pi_0 \sin(a+v) ,$$

est continu, C^1 par rapport à (a,v), son opérateur dérivé $D_v\phi(t,a,v).\tilde{v}$ est inversible en $(\tau,0,0)$ car $D_v\phi(\tau,0,0)\tilde{v} = \tilde{v} - P(\tau)G\,\tilde{v}$ et $P(\tau) \neq n^2\pi^2$, $n \geq 1$. Le théorème des fonctions implicites avec paramètres permet alors d'écrire que pour t voisin de τ et (v_1,v_2) petits, l'équation $\phi(t,a(t),v_1) = v_2$ est équivalente à $v_1 = H(t,a(t),v_2)$ avec $|H(t,a(t),v_2)| \leq C|v_2|$ ce qui implique (17).
Le Théorème 2 est ainsi démontré.

Démonstration des Théorèmes 3.4.5. Nous montrerons successivement :

- l'existence des branches bifurquées décrites dans les énoncés des Théorèmes 3 et 4 : grâce à la méthode de Lyapunov Schmidt on ramènera la résolution de (5) à la résolution d'une équation intégrale non linéaire avec une singularité. On traitera cette équation par une méthode de perturbation à partir d'une équation différentielle dont on explicitera les solutions ;
- l'absence de branches bifurquées pour $p_1 < -\frac{2}{3}\lambda_1\beta_o$;
- qu'il n'y a pas pour $p_1 > -\frac{2}{3}\lambda_1\beta_o$ d'autres solutions que celles obtenues à la première étape, et que celles ci sont régulières : cette troisième étape, relativement technique, est mise en annexe.

Nous ramenons donc le problème (5) à la recherche de t_1 positif assez petit et de (a,b,w) appartenant à $(\mathcal{C}([0,t_1],\mathbb{R}))^2 \times \mathcal{C}([0,t_1],W)$, (avec $u(t) = a(t) + b(t)\phi_1+w(t))$, solution du système suivant, obtenu en projetant (5) sur $\mathbb{R}$, W, et $\mathbb{R}\phi_1$ et en utilisant l'opérateur G :

$$F_1(a,b,w)(t) = \int_0^1 \sin(a(t)+ b(t)\phi_1+w(t))\,dx = 0 \tag{19}$$

$$\left\{\begin{aligned} F_2(a,b,w)(t) &= w(t) - \beta_o \int_0^t e^{-(t-s)}w(s)ds \\ &\quad - P(t)\pi_1 G\pi_o \sin(a(t)+b(t)\phi_1 + w(t)) = 0 \end{aligned}\right. \tag{20}$$

$$\left\{\begin{aligned} F_3(a,b,w)(t) &= \lambda_1 b(t) - \lambda_1\beta_o \int_0^t e^{-(t-s)}b(s)ds \\ &\quad - P(t)\int_0^1 \sin(a(t)+b(t)\phi_1 + w(t))\phi_1 dx = 0 \end{aligned}\right. \tag{21}$$

Montrons que a et w sont fonctions de b.

Lemme 0. *L'application*

$$(a,b,w) \mapsto (F_1(a,b,w),F_2(a,b,w),F_3(a,b,w))$$

de $(\mathcal{C}([0,t_1]))^2 \times \mathcal{C}([0,t_1],W]$ *dans* $(\mathcal{C}([0,t_1]))^2 \times \mathcal{C}([0,t_1],W)$ *est analytique.*

En effet elle est composée de l'application $u \mapsto \sin u$ de $(\mathcal{C}([0,t_1]),H^1)$ dans lui-même qui est analytique car $H^1(]0,1[)$ est une algèbre de Banach et d'applications linéaires telles que $\pi_1\pi_o G$ ou $\int \cdot\phi_1 dx$.

Lemme 1. *Il existe* $t_1 > 0$, *il existe un voisinage* U_1 *de* 0 *dans* $\mathcal{C}([0,t_1])$, *un voisinage* U_2 *de* $(0,0)$ *dans* $\mathcal{C}([0,t_1]) \times \mathcal{C}([0,t_1],W)$ *tels que le système*

$$\begin{cases} F_1(a,b,w)(t) = 0 \ , \\ F_2(a,b,w)(t) = 0 \ , \qquad t \in [0,t_1] \ , \end{cases}$$

avec $b \in U_1$ *et* $(a,w) \in U_2$, *admette l'unique solution* $(a(b),b,w(b))$ *où* $b \mapsto (a(b),w(b))$ *est une application analytique de* U_1 *dans* U_2 *telle que*

$$\begin{cases} (a(0),w(0)) = (0,0) \ , \\ |a(b)(t)| \leq C(\|b\|_{\mathcal{C}[0,t]})^3 \ , \\ \|w(b)(t)\|_W \leq C(\|b\|_{\mathcal{C}[0,t]})^3 \ . \end{cases} \tag{22}$$

Démonstration. Elle repose, bien sûr, sur le théorème des fonctions implicites appliqué au système

$$\begin{cases} F_1(a,b,w) = 0 \ , \\ F_2(a,b,w) = 0 \ . \end{cases}$$

Nous avons,

$$\begin{cases} F_1(0,0,0) = 0 , \\ F_2(0,0,0) = 0 . \end{cases}$$

Montrons que la différentielle du système par rapport à (a,w) prise en (0,0,0) est un isomorphisme de $\mathcal{C}([0,t_1]) \times \mathcal{C}([0.t_1],W)$ sur lui-même. Soit donc à résoudre, pour (a_o,w_o) donné dans $\mathcal{C}([0,t_1])\times\mathcal{C}([0,t_1],W)$ le système,

$$\begin{cases} \frac{\partial F_1}{\partial a}(0,0,0)\ [a] + \frac{\partial F_1}{\partial w}(0,0,0)\ [w] = a_o , \\ \frac{\partial F_2}{\partial a}(0,0,0)\ [a] + \frac{\partial F_2}{\partial w}(0,0,0)\ [w] = w_o , \end{cases}$$

qui s'écrit,

$$\begin{cases} a(t) = a_o(t) , \\ w(t) - \beta_o \int_0^t e^{-(t-s)} w(s)ds - P(t)\pi_1 G\pi_o w(t) = w_o(t), \quad \forall\ t \in [0,t_1] \end{cases}$$

ou encore puisque $\pi_1 G\pi_o w(t) = Gw(t)$

$$\begin{cases} a(t) = a_o(t) \\ [I-P(t)G]w(t) - \beta_o \int_0^t e^{-(t-s)} w(s)ds = w_o(t) , \quad \forall\ t \in [0,t_1] . \end{cases}$$

Comme $P(0) = \lambda_1$ et que $(I-\lambda_1 G)$ est inversible sur W, pour t voisin de 0, $L(t) = (I-P(t)G)$ est inversible sur W, et la deuxième équation s'écrit

$$w(t) - \beta_o (L(t))^{-1} \int_0^t e^{-(t-s)} w(s)ds = (L(t))^{-1} w_o(t) ,$$

ou encore si,

$$y(t) = \int_0^t e^s w(s)ds ,$$

$$\begin{cases} y'(t) - \beta_o(L(t))^{-1}y(t) = (L(t))^{-1}(e^t w_o(t)), \\ y(0) = 0 \ ; \end{cases}$$

cette équation différentielle admet une solution unique dans $\mathcal{C}^1([0,t_1],W)$ et w est défini de manière unique.

Nous pouvons donc appliquer le théorème des fonctions implicites au voisinage de $(0,0,0)$ et la première partie du lemme est démontrée.

L'unicité dans le théorème des fonctions implicites appliquée sur $[0,t]$ avec $t \in]0,t_1[$ permet de dire que $a(b)$ et $w(b)$ sur $[0,t]$ ne dépendent que des valeurs de b sur $[0,t]$. De plus l'analyticité de F_1 et F_2 par rapport à leurs arguments implique que l'application $b \to (a(b),w(b))$ est analytique. En calculant les premiers termes du développement de $(a(b),w(b))$ au voisinage de $b = 0$, on voit que

$$a(b)(t) = 0\left(\|b\|^3_{\mathcal{C}[0,t]}\right),$$

$$w(b)(t) = 0\left(\|b\|^3_{\mathcal{C}[0,t]}\right),$$

de plus l'analyticité de $b \to (a(b),w(b))$ implique

$$|Da(b)(\delta b)(t)| \le C \quad \|b\|^2_{\mathcal{C}[0,t]} \quad \|\delta b\|_{\mathcal{C}[0,t]} ,$$

$$\|Dw(b)(\delta b)(t)\|_W \le C \quad \|b\|^2_{\mathcal{C}[0,t]} \cdot \|\delta b\|_{\mathcal{C}[0,t]}$$

ce qui complète la démonstration du Lemme 1.

D'après ce lemme, le système (19)(20)(21) se réduit donc à l'équation

$$\begin{cases} F(b)(t) = F_3(a(b),b,w(b))(t) = \\ \lambda_1 b(t) - \lambda_1 \beta_o \int_0^t e^{-(t-s)} b(s)ds - \\ \qquad - P(t) \int_0^1 \sin[a(b)(t) + b(t)\phi_1 + w(b)(t)]\phi_1 \, dx . \end{cases} \tag{23}$$

Maintenant nous allons chercher les solutions b de (23) astreintes à $b \in \mathcal{C}([0,t_1])$ (t_1 assez petit) et $b(0) = 0$.

Tout d'abord nous décomposons $F(b)$ en une partie principale $\tilde{F}(b)$ et une perturbation. L'analyticité de $a(b)$ et $w(b)$ permet d'écrire :

$$a(b)(t) = \sum_{n\geq 3} a_n(b)(t) \qquad w(b)(t) = \sum_{n\geq 3} w_n(b)(t) ,$$

$(a_n \in \mathcal{L}_s^n(\mathcal{C}([0,t]), \mathcal{C}([0,t]))$, $w_n \in \mathcal{L}_s^n(\mathcal{C}([0,t])$, $\mathcal{C}([0,t],W)))$,

($\mathcal{L}_s^n$: applications n linéaires symétriques), donc

$$\sin(b\,\phi_1 + a(b) + w(b)) = b\phi_1 + (a(b)+w(b)) - \frac{1}{6}(b\,\phi_1 + a(b)+w(b))^3 +$$

$$+ \sum_{\sigma\geq 2} \frac{1}{(2\sigma+1)!} (b\phi_1 + a(b)+w(b))^{2\sigma+1} = b\phi_1 - \frac{1}{6} b^3\phi_1^3 + a(b) + w(b) + \theta_5(b) ,$$

où $\theta_5(b)$ est une fonction analytique de b, d'ordre 5 :

$$\theta_5(b)(t) = O(\| b \|^5_{\mathcal{C}[0,t]}) ,$$

donc (θ_5 désignant toujours une fonction analytique d'ordre 5)

$$\int_0^1 \sin(b\,\phi_1 + a(b) + w(b))\phi_1 dx = b - \mu b^3 + \theta_5(b) ,$$

(on a posé $\mu = \frac{1}{6}\int_0^1 \phi_1^4 dx$) puis comme $P(t) = \lambda_1 + p_1 t + Q(t)t^2$

$$P(t) \int_0^1 \sin(b\phi_1 + a(b)+w(b))\phi_1 dx = \lambda_1 b + p_1 tb - \lambda_1 \mu b^3$$

$$+ Q(t)t^2 b - (p_1 + Q(t)t)tb^3 + \theta_5(b) .$$

Enfin on obtient,

$$F(b)(t) = F_3(a(b),b,w(b))(t) = -p_1 tb - \lambda_1\beta_o \int_0^t e^{-(t-s)}\, b(s)ds$$

$$+ \lambda_1\mu e^{2t}b^3 - Q(t)t^2 b - (\lambda_1\mu(e^{2t}-1) - t(p_1+Q(t)t)b^3 + \tilde{\theta}_5(t.b) ,$$

on peut poser $r(t) = -\lambda_1\mu \left(\frac{e^{2t}-1}{t}\right) - (p_1+Q(t)t)$, et résumer le résultat obtenu par l'énoncé suivant :

Lemme 2. *L'équation* $F(b)(t) = 0$ *se met sous la forme* :

$$F(b)(t) = \tilde{F}(b)(t) + \psi_1(t,b)(t) + \psi_2(t,b)(t) + \psi_3(t,b)(t) = 0 \qquad (24)$$

avec

$$\tilde{F}(b)(t) = -p_1 tb(t) - \lambda_1\beta_o \int_0^t e^{-(t-s)}b(s)ds + \lambda_1\mu e^{2t}b^3(t) , \qquad (24)\text{bis}$$

$$\begin{cases} \psi_1(t,b)(t) = -Q(t)t^2 b(t) = O(t^2 b(t)) , \\ \psi_2(t,b)(t) = -r(t)tb^3(t) = O(tb^3(t)) , \\ \psi_3(t,b)(t) = \tilde{\theta}_5(t,b)(t) = O(\|b\|^5_{\mathcal{C}[0,t]}) , \\ |D_b\psi_3(t,b)(\delta b)(t)| \le C\,\|b\|^4_{\mathcal{C}[0,t]} \cdot \|\delta b\|_{\mathcal{C}[0,t]} . \end{cases} \qquad (24)\text{ter}$$

Cette dernière inégalité résulte de la construction de $\tilde{\theta}_5(t,b)$ *qui est analytique par rapport à* b *et d'ordre* 5.

Nous allons maintenant exhiber des solutions de $\tilde{F}(b)(t) = 0$ par la méthode d'intégration des équations du type Lagrange. Nous montrerons dans l'annexe que nous avons obtenu ainsi *toutes* les solutions continues, s'annulant en 0, définies sur $[0,t_1[$, de $\tilde{F}(b)(t) = 0$.

Posons $y(t) = \int_0^t e^s b(s)ds$, $y \in \mathcal{C}^1([0,t[)$, alors $e^t\tilde{F}(b)(t) = 0$ se se met sous la forme d'une équation de Lagrange

$$\begin{cases} \mathcal{F}(t,y,y') = -p_1 ty' - \lambda_1\beta_o y + \lambda_1\mu y'^3 = 0 \\ y(0) = y'(0) = 0 \,. \end{cases} \tag{25}$$

Cette équation présente une singularité en 0 (terme ty') ce qui explique (un peu) la présence de deux conditions initiales et l'apparition de plusieurs solutions.

1) *Cas* $p_1 = 0$. Les solutions non nulles de :

$$\begin{cases} -\beta_o y + \mu y'^3 = 0 \\ y(0) = y'(0) = 0 \end{cases}$$

sont

$$y(t) = \left(\frac{2}{3}\ \frac{\beta_o}{\mu}\right)^{1/2} \left((t-t_o)^+\right)^{3/2} \qquad (t_o > 0)$$

2) *Cas* $p_1 \neq 0$. On cherche les solutions non nulles de (25), appartenant à $\mathcal{C}^1([0,+\infty[) \cap \mathcal{C}^2(]0,+\infty[)$ paramètrisables par $y' = p$, $p \in [0,+\infty[$. Par dérivation de (25) on obtient,

$$\begin{cases} (p_1+\lambda_1\beta_o)\, p\, \dfrac{dt}{dp} = -p_1 t + 3\lambda_1\mu p^2 \,, \\ t(0) = 0 \,, \\ \dfrac{dy}{dp} = p\,\dfrac{dt}{dp} \,, \end{cases} \tag{26}$$

et $t \in \mathcal{C}^1([0,+\infty[)$, $t(p) > 0$ si $p > 0$.

Résolution de (26) : C'est une équation linéaire que l'on résout par la méthode de variation des constantes.

$$\frac{dt}{t} = - \frac{p_1}{p_1+\lambda_1\beta_o} \quad \frac{dp}{p} ,$$

$$t = C\, p^{-\frac{p_1}{p_1+\lambda_1\beta_o}}$$

puis

$$C' = \frac{3\mu\lambda_1}{(p_1+\lambda_1\beta_o)} \, p^{\frac{2p_1+\lambda_1\beta_o}{p_1+\lambda_1\beta_o}}$$

2) i)

$$3p_1 + 2\lambda_1\beta_o = 0 ,$$

ceci correspond à $\frac{2p_1+\lambda_1\beta_o}{p_1+\lambda_1\beta_o} = -1$,

alors

$$C = \frac{9\mu}{\beta_o} \text{Log}|p| + \gamma \qquad (\gamma \in \mathbb{R})$$

$$t(p) = \frac{9\mu}{\beta_o} p^2 \text{Log}|p| + \gamma p^2 ,$$

cette fonction ne peut être retenue car pour p positif, voisin de 0 , $t(p) < 0$.

2) ii)

$$3p_1 + 2\lambda_1\beta_o \neq 0$$

alors

$$C = \frac{3\mu\lambda_1}{3p_1+2\lambda_1\beta_o} \, p^{\frac{3p_1+2\lambda_1\beta_o}{p_1+\lambda_1\beta_o}} + \gamma$$

$$t(p) = \frac{3\mu\lambda_1}{3p_1+2\lambda_1\beta_o} \, p^2 + \gamma \, p^{-\frac{p_1}{p_1+\lambda_1\beta_o}} . \qquad (27)$$

Il faut distinguer trois possibilités :

α) $\quad p_1 < -\frac{2}{3}\lambda_1\beta_o$, comme $t(p) \sim \frac{3\mu\lambda_1}{3p_1+2\lambda_1\beta_o}\, p^2$ dans $]0,\varepsilon[$ on a $t(p) < 0$, donc cette solution n'est pas admissible

β) $\quad -\frac{2}{3}\lambda_1\beta_o < p_1 < 0$,

il existe une infinité de solutions de (26) paramétrée par γ , $\gamma \in [0,\infty[$, et donnée par (27) .

γ) $\quad 0 < p_1$,

Comme $\quad -\frac{p_1}{p_1+\lambda_1\beta_o} < 0$ il faut dans (27) prendre $\gamma = 0$, et l'équation (26) admet la solution :

$$t(p) = \frac{3\mu\lambda_1}{3p_1+2\lambda_1\beta_o}\; p^2 \quad .$$

A partir des résultats précédents on obtient les solutions suivantes de $\tilde{F}(b)(t) = 0$.

$$\left\{\begin{array}{lll}
\cdot\; p_1 \in]-\infty, -\frac{2}{3}\lambda_1\beta_o] , & b(t) = 0 & \\
\cdot\; p_1 \in]-\frac{2}{3}\lambda_1\beta_o, 0[, & \left\{\begin{array}{l} b(t) \sim (\frac{t}{\gamma})^{-\frac{p_1+\lambda_1\beta_o}{p_1}}\; e^{-t} , \quad (\gamma \in \mathbb{R}^+) \\ b(t) = (\frac{3p_1+2\lambda_1\beta_o}{3\mu\lambda_1})^{1/2}\; t^{1/2}\; e^{-t} \end{array}\right. & \\
\cdot\; p_1 = 0, & b(t) = (\frac{2\beta_o}{3\mu})^{1/2}\; ((t-t_o)^+)^{1/2}\; e^{-t} \quad (t_o \geq 0) & \\
\cdot\; p_1 \in]0,+\infty[, & b(t) = (\frac{3p_1+2\lambda_1\beta_o}{3\mu\lambda_1})^{1/2}\; t^{1/2}\; e^{-t} &
\end{array}\right. \qquad (28)$$

Nous sommes alors en mesure de chercher les solutions de $F(b)(t) = 0$. Tout d'abord le lemme suivant nous fournit un encadrement utile pour toute solution de $F(b)(t) = 0$.

Lemme 3. *Soit* b *une solution quelconque de* $F(b)(t) = 0$, $(t \in [0,t_1[)$, *alors il existe une constante* M *dépendant de* b *telle que*

$$|b(t)| \le M \sqrt{t} \quad , \quad \forall t \in [0,t_1[\ . \tag{29}$$

Preuve. Il revient au même de montrer que $c(t) = b(t)e^t = O(t^{1/2})$. On a à partir de (18)

$$- p_1 tc(t) - \lambda_1 \beta_o \int_0^t c(s)ds + \lambda_1 \mu c^3(t) + \tilde{\psi}_1(c) + \tilde{\psi}_2(c) + \tilde{\psi}_3(c) = 0$$

donc

$$\lambda_1 \mu c^3(t) = p_1 tc(t) + \lambda_1 \beta_o \int_0^t c(s)ds - \tilde{\psi}_1(c) - \tilde{\psi}_2(c) - \tilde{\psi}_3(c) \ ,$$

soit $s_o \in [0,t]$ tel que $|c(s_o)| = \sup_{s \in [0,t]} |c(s)|$, au point s_o on a :

$$\lambda_1 \mu c^3(s_o) = p_1 s_o c(s_o) + \lambda_1 \beta_o \int_0^{s_o} c(s)ds - \tilde{\psi}_1(c)(s_o) - \tilde{\psi}_2(c)(s_o) - \tilde{\psi}_3(c)(s_o)$$

donc

$$\lambda_1 \mu |c^3(s_o)| \le |p_1| s_o |c(s_o)| + \lambda_1 \beta_o s_o |c(s_o)| + M_1 s_o^2 |c(s_o)| + M_2 s_o |c(s_o)|^3 + M_3 |c(s_o)|^4$$

soit

$$(\lambda_1 \mu - M_2 s_o - M_3 |c(s_o)|) \ |c(s_o)|^2 \le (|p_1| + \lambda_1 \beta_o + M_1 s_o) s_o \ ,$$

comme $c(s)$ est continue pour t assez petit on a

$$M_2 s_o + M_3 |c(s_o)| \leq \frac{\lambda_1 \mu}{2} ,$$

donc,

$$|c(s_o)| \leq M_4 |s_o|^{1/2}$$

et

$$\| c \|_{\mathcal{C}[0,t]} \leq M_4 \; t^{1/2} \quad .$$

Nous allons maintenant exhiber pour chaque solution b_o de $\tilde{F}(b)(t) = 0$ une solution b de $F(b)(t) = 0$ qui lui est tangente à l'origine. On pose donc,

$$b(t) = b_o(t) + d(t) \; , \quad d \in \mathcal{C}([0,t_1]) \; ,$$

avec

$$|d(t)| = o(\| b_o \|_{\mathcal{C}([0,t])}) = o(|b_o(t)|) \; .$$

La fonction d est donc solution de $F(b_o+d)(t) = 0$ que l'on développe par rapport à d :

$$0 = F(b_o+d)(t) = \tilde{F}(b_o+d)(t) + \sum_{i=1}^{3} \psi_i(t,b_o+d)(t)$$

$$= \tilde{F}(b_o)(t) + D_b\tilde{F}(b_o)(d)(t) + \int_0^1 D_b\tilde{F}(b_o+\tau d)(d)(t)d\tau \; +$$

$$+ \sum_{i=1}^{3} \psi_i(t,b_o+d)(t) \; .$$

Compte tenu de la forme de $\tilde{F}$ et de $\tilde{F}(b_o)(t_o) = 0$, l'égalité précédente devient

$$\left\{\begin{array}{l} - p_1 t d(t) - \lambda_1 \beta_o \int_0^t e^{-(t-s)} d(s)ds + 3\lambda_1 \mu e^{2t} b_o^2(t) d(t) \\ + 3\lambda_1 \mu e^{2t} b_o(t) d^2(t) + \lambda_1 \mu e^{2t} d^3(t) + \sum_1^3 \psi_i(t, b_o + d)(t) = 0 . \end{array}\right. \tag{30}$$

Posons,

$$\psi_4(t,b_o,d)(t) = 3\lambda_1 \mu e^{2t} b_o(t) d^2(t) + \lambda_1 \mu e^{2t} d^3(t) , \tag{31}$$

ce terme ψ_4 apparaitra comme une perturbation que nous traiterons de la même façon que ψ_1 , ψ_2 , ψ_3 . L'équation (30) s'écrit alors ,

$$\left\{\begin{array}{ll} \text{i/} & - p_1 t d(t) + 3\lambda_1 \mu e^{2t} b_o^2(t) d(t) - \lambda_1 \beta_o \int_0^t e^{-(t-s)} d(s)ds = \mathcal{F}(d)(t) , \\ \text{ii/} & \text{avec} \qquad \mathcal{F}(d)(t) = - \sum_{i=1}^4 \psi_i(t,b_o,d)(t) . \end{array}\right. \tag{32}$$

Ceci est une équation différentielle linéaire par rapport à $\int_0^t e^s d(s)ds$ avec un second membre $\mathcal{F}(d)$, on peut donc calculer explicitement d en fonction de $\mathcal{F}(d)$:

$$d = \phi(\mathcal{F}(d)) = \mathcal{C}(d) ;$$

nous cherchons donc les points fixes de $\mathcal{C}$ par une méthode adaptée à chaque b_o .

1er cas. $p_1 \geq 0$. Grâce à l'expression de b_o donnée par (28) l'équation (32) s'écrit :

$$2(p_1 + \lambda_1 \beta_o) t \, d(t) - \lambda_1 \beta_o \int_0^t e^{-(t-s)} d(s)ds = \mathcal{F}(d)(t) ,$$

après calcul et puisque $d(0) = 0$, nous obtenons pour $t > 0$,

$$d(t) = \mathcal{C}(d)(t) = \frac{\mathcal{F}(d)(t)}{2(p_1+\lambda_1\beta_o)t} +$$

$$+ \frac{\lambda_1\beta_o}{4(p_1+\lambda_1\beta_o)} t^{-\frac{2p_1+\lambda_1\beta_o}{2(p_1+\lambda_1\beta_o)}} \int_0^t \frac{\mathcal{F}(d)(s)e^{s-t}}{s^{(2p_1+3\lambda_1\beta_o)/2(p_1+\lambda_1\beta_o)}} ds$$

Montrons que $\mathcal{C}$ admet un point fixe dans $\mathcal{C}([0,t_1])$ muni de la norme uniforme, pour un t_1 assez petit.

Lemme 4. *Il existe* M *et* $\bar{t}_1 > 0$ *tel que pour tout* t_1 *dans* $]0,\bar{t}_1[$, $\mathcal{C}$ *envoie l'ensemble*

$$B_{t_1}(0,M) = \{d, d \in \mathcal{C}([0,t_1]) \ , \ \|d\|_{\mathcal{C}[0,t]} \leq M\, t^{1/2} \ , \ \forall\, t \in [0,t_1]\}$$

dans lui même.

Preuve. Nous savons que,

$$\|b_o\|_{\mathcal{C}[0,t]} \leq B_o\, t^{1/2} ,$$

soit $d \in B_{t_1}(0,M)$, d'après (24 ter) et (30), on a :

$$|\mathcal{F}(d)(t)| \leq A_oB_oM^2t^{3/2} + A_1M^3t^{3/2} + A_2(B_o+M)^4t^2$$
$$+ A_3(B_o+M)t^{5/2} + A_4(B_o+M)^3t^{5/2} ,$$

où les A_i sont des constantes indépendantes de M .

D'après (32)ii/ on a alors,

$$\|\mathcal{C}(d)\|_{\mathcal{C}[0,t]} \leq (A'_o+A'_1M)M^2t^{1/2} + h(M)t$$

où les A_i' sont indépendantes de M et $h(M)$ une fonction polynomiale en M. Alors pour M et t assez petits on aura :

$$(A_o'+A_1'M)M^2 + h(M)t^{1/2} \leq M ,$$

d'où le lemme.

Lemme 5. *On peut choisir* M *et* t_1 *assez petits* (>0) *pour que* $\mathcal{T}$ *soit contractante sur l'ensemble* $B_{t_1}(0,M)$.

Preuve. Soient d et $\hat{d}$ dans l'ensemble $B_{t_1}(0,M)$. Nous allons évaluer les quatre termes $(\psi_i(t,b_o+d)(t)-\psi_i(t,b_o+\hat{d})(t))$.

Tout d'abord d'après les définitions de ψ_1 et ψ_2

$$(\psi_1(t,b)(t) = - Q(t)t^2b(t) \quad , \quad \psi_2(t,b)(t) = r(t)\,tb^3(t))$$

nous avons immédiatement

$$|\psi_1(t,b_o+d)(t) - \psi_1(t,b_o+\hat{d})(t)| \leq A_1t^2 \,\|d-\hat{d}\|_{\mathcal{C}([0,t])}$$

$$|\psi_2(t,b_o+d)(t) - \psi_2(t,b_o+\hat{d})(t)| \leq A_2M^2t^2 \,\|d-\hat{d}\|_{\mathcal{C}([0,t])} \,.$$

La même technique s'applique à ψ_4

$$|\psi_4(t,b_o,d)(t) - \psi_4(t,b_o,\hat{d})(t)| \leq (A_o+A_1M+A_o'(1+M))Mt\,|(d-\hat{d})(t)|$$

$$\leq (A_o+A_1M+A_o'(1+M))Mt\,\|d-\hat{d}\|_{\mathcal{C}([0,t])} \,.$$

Quant à ψ_3 on utilise la dernière relation de (24)ter :

$$|\psi_3(t,b_o+d)(t) - \psi_3(t,b_o+\hat{d})(t)| = |\int_0^1 D_b\psi_3(t,b_o+\hat{d}+ \mathcal{T}(d-\hat{d}))(d-\hat{d})d\mathcal{T}|$$

ce qui implique

$$|\psi_3(t,b_o+d)(t) - \psi_3(t,b_o+\hat{d})(t)| < A_3M^3t^{3/2}\,\|d-\hat{d}\|_{\mathcal{C}([0,t])}$$

par suite

$$|\mathcal{F}(d)(t) - \mathcal{F}(\hat{d})(t)| \leq t\,\|d-\hat{d}\|_{\mathcal{C}[0,t]}\,[A'_o(M+M^2)+A_3M^2t^{1/2} + A_2M^2t + A_1t],$$

et en utilisant (32) on obtient :

$$|\mathcal{T}(d) - \mathcal{T}(\hat{d})|(t) \leq (A''_o(M+M^2) + A'_2(M\ )t^{1/2})\,\|d-\hat{d}\|_{\mathcal{C}([0,t])}$$

nous voyons que si M et t_1 sont choisis suffisamment petits, $\mathcal{T}$ envoie $B_{t_1}(0,M)$ dans lui même et est contractant dans cet ensemble d'où le Lemme 5.

Soient M_o et t_1 un tel choix et d_o le point fixe de $\mathcal{T}$ dans $B_{t_1}(0,M_o)$, alors (b_o+d_o) vérifie $F(b_o+d_o)(t) = 0$. Remarquons qu'en prenant $M_o < (\frac{3p_1+2\lambda_1\beta_o}{3\mu\lambda_1})^{1/2}$ on exhibe une solution non nulle !

Il reste à montrer que la solution b_o+d_o ainsi trouvée est tangente à b_o en 0 . Pour cela on montre, par une technique analogue à celle qui précède que $\mathcal{T}$ envoie dans lui même l'ensemble

$$\{d, d \in \mathcal{C}[0,t_1]\ ,\ \|d\|_{\mathcal{C}[0,t_1]} \leq N_o t\ ,\ \forall\, t \in \mathcal{C}[0,t_1]\}$$

et est contractante sur cet ensemble, pourvu que N_o soit choisi assez grand grand et t_1 suffisamment voisin de 0 . Par suite $\mathcal{T}$ a aussi un point fixe unique dans cet ensemble et d'après l'unicité ce point fixe ne peut être que d_o . Ceci entraine

$$|d_o(t)| = o(\|b_o\|_{\mathcal{C}[0,t]}) = o(b_o(t))\ ,$$

et donc la solution (b_o+d_o) est tangente à b_o en 0 .

Ceci montre l'existence de la branche bifurquée donnée par le Théorème 3.

2ème cas : $-\frac{2}{3}\lambda_1\beta_o < p_1 < 0$. Il y a deux possibilités :

i/ si $b_o(t) = (\frac{3p_1+2\lambda_1\beta_o}{3\lambda_1\mu})^{1/2} t^{1/2} e^{-t}$, on procède comme dans le premier cas et on trouve un unique d_o tel que (b_o+d_o) soit solution de (24) et soit tangente à b_o .

ii/ si $b_o(t) = (\frac{t}{k})^{-\frac{p_1+\lambda_1\beta_o}{p_1}} e^{-t}(1+\varepsilon(t))$,

l'équation (30) devient

$$\left\{\begin{aligned} & -p_1 t d(t) - \lambda_1\beta_o \int_0^t e^{-(t-s)} d(s)ds + \sum_{i=1}^{3} \psi_i(t,b_o+d)(t) \\ & \qquad + \psi_5(t,b_o,d)(t) = 0 \end{aligned}\right. \tag{34}$$

où

$$\left\{\begin{aligned} \psi_5(t,b_o,d)(t) &= 3\mu\lambda_1 b_o^2(t) e^{2t} d(t) + 3\mu\lambda_1 e^{2t} b_o(t) d^2(t) \\ & \qquad + \lambda_1\mu e^{2t} d^3(t) . \end{aligned}\right.$$

Compte tenu de $d(t) = \varepsilon(t) b_o(t) = \tilde{\varepsilon}(t) t^{-\frac{p_1+\lambda_1\beta_o}{p_1}}$, la fonction ψ_5 s'écrit encore :

$$\psi_5(t,b_o,d)(t) = 0(t^{-2(\frac{p_1+\lambda_1\beta_o}{p_1})}) d(t) ,$$

et l'équation (34) se transforme en :

$$-p_1 t d(t) - \lambda_1\beta_o \int_0^t e^{-(t-s)} d(s)ds = \mathcal{F}(d)(t)$$

ou $\quad \mathcal{F}(d)(t) = \sum_{i=1}^{3} \psi_i(t,b_o,d)(t) + \psi_5(t,b_o,d)(t) \; .$ (35)

Par suite $d(t)$ s'écrit :

$$\left\{ \begin{aligned} d(t) &= k_o \frac{\lambda_1\beta_o}{-p_1} t^{-\frac{p_1+\lambda_1\beta_o}{p_1}} + \frac{\mathcal{F}(d)(t)}{-p_1 t} \\ &+ \frac{\lambda_1\beta_o}{p_1^2} t^{-\frac{p_1+\lambda_1\beta_o}{p_1}} \int_0^t \frac{\mathcal{F}(d)(s).e^{-(t-s)}}{s^{(p_1-\lambda_1\beta_o)/p_1}} ds \; , \quad \text{où} \quad k_o \in \mathbb{R} \; . \end{aligned} \right. \tag{36}$$

Montrons que $k_o = 0$. En effet nous cherchons $d(t)$ tel que

$$d(t) = o(t^{-\frac{p_1+\lambda_1\beta_o}{p_1}}) \; , \tag{37}$$

Posons $\theta = - \dfrac{p_1+\lambda_1\beta_o}{p_1}$, la condition $p_1 \in]-\frac{2}{3}\lambda_1\beta_o,0[$ implique $\theta > \frac{1}{2}$.
Alors d'après (35) et le comportement des fonctions ψ_i nous voyons que

$$\mathcal{F}(d)(t) = O(t^{\nu})$$

où $\nu = \min(2+\theta,3\theta)$.

i/ Si $3\theta < 2+\theta$ c'est-à-dire si

$$p_1 \in]-\frac{2}{3}\lambda_1\beta_o \; , \; -\frac{\lambda_1\beta_o}{2}]$$

on a

$$\left|\frac{\mathcal{F}(d)(t)}{t}\right| = O(t^{3\theta-1}) \; ,$$

et

$$\frac{\lambda_1\beta_o}{p_1^2}\, t^\theta \int_0^t \frac{\mathcal{F}(d)(s)e^{-(t-s)}}{s^{2+\theta}}\, ds = O(t^{3\theta-1}) ,$$

comme $3\theta - 1 > \theta$, si $k_o \neq 0$ on a d'après (36)

$$d(t) \sim k_o \frac{\lambda_1\beta_o}{-p_1} t^\theta ,$$

ce qui est contradictoire avec (37), donc $k_o = 0$.

i/ Si $2+\theta < 3\theta$ c'est-à-dire si,

$p_1 \in]-\frac{\lambda_1\beta_o}{2},0[$, on a,

$$\frac{\mathcal{F}(d)(t)}{t} = O(t^{1+\theta}) ,$$

et

$$\frac{\lambda_1\beta_o}{p_1^2}\, t^\theta \int_0^t \frac{\mathcal{F}(d)(s)e^{-(t-s)}}{s^{2+\theta}}\, ds = O(t^{1+\theta}) ,$$

par suite si $k_o \neq 0$ d'après (36) on a encore

$$d(t) \sim k_o \frac{\lambda_1\beta_o}{-p_1} t^\theta ,$$

ce qui est contradictoire avec (37) donc $k_o = 0$.

Par conséquent (36) se réduit à :

$$d(t) = \frac{\mathcal{F}(d)(t)}{-p_1 t} + \frac{\lambda_1\beta_o}{p_1^2}\, t^\theta \int_0^t \frac{\mathcal{F}(d)(s).e^{-(t-s)}}{s^{2+\theta}}\, ds = \mathcal{C}(d)(t)$$

Nous allons montrer que $\mathcal{C}(d)$ est une contraction dans un espace de Banach convenable. Pour $t_1 > 0$ définissons

$$E_{t_1} = \{d, d\in \mathcal{C}([0,t_1]) , |||d||| = \sup_{t\in[0,t_1]} (\frac{\|d\|_{\mathcal{C}[0,t]}}{t^\theta}) < +\infty \}.$$

Alors E_{t_1} muni de $|||d|||$ est un espace de Banach et nous noterons $B_{t_1}(0,\rho)$

la boule de centre 0 et de rayon ρ dans E_{t_1} .

Lemme 6. *Soit* $M_o > 0$. *Il existe* $\overline{t}_1 > 0$ *tel que pour tout* $t_1 \in]0,\overline{t}_1[$ $\mathcal{T}$ *envoie* $B_{t_1}(0,M_o)$ *dans elle même.*

Démonstration. Si $d \in B_{t_1}(0,M_o)$, nous avons $|d(t)| \leq M_o t^\theta$, $\forall\ t \in [0,t_1]$, par ailleurs $|b_o(t)| \leq B_o\ t^\theta$, donc, grâce à (31), nous obtenons, (on désigne par B_o, A_i, A'_i des constantes indépendantes de M_o) :

$$|\mathcal{F}(d)(t)| \leq A_1(B_o+M_o)t^{2+\theta} + A_2 t^{1+3\theta}(B_o+M_o)^3$$
$$+ A_3(B_o+M_o)^4\ t^{4\theta} + A_4 M_o t^{3\theta}$$

puis grâce à (38) :

$$|\mathcal{T}(d)(t)| \leq A'_o(B_o+M_o)t^{1+\theta} + A'_1(B_o+M_o)^3 t^{3\theta}$$
$$+ A'_2(B_o+M_o)^4 t^{4\theta-1} + A'_3(M_o+M_o^2+M_o^3)\ t^{3\theta-1} \quad ,$$

donc

$$\frac{\|\mathcal{T}(d)\|_{\mathcal{C}([0,t])}}{t^\theta} \leq A'_o(B_o+M_o)t + A''_1(B_o+M_o)^3 t^{2\theta} +$$
$$+ A'_2(B_o+M_o)^4\ t^{3\theta-1} + A'_3(M_o+M_o^2+M_o^3)t^{2\theta-1} \quad ,$$

puisque $\theta > \frac{1}{2}$ nous voyons que pour M_o fixé nous pouvons avoir pour tout t dans $[0.t_1]$ (pourvu que t_1 soit assez petit) :

$$\frac{\|\mathcal{T}(d)\|\ \mathcal{C}([0,t])}{t^\theta} \leq M_o$$

Lemme 7. *Soit* $M_1 > 0$. *Il existe* t_1 , $t_1 > 0$ *tel que* $\mathcal{T}$ *soit une contrac-*

tion stricte sur la boule $B_{t_1}(0,M_1)$ *de* E_{t_1} .

Démonstration. Soient d et $\hat{d}$ dans la boule $B_{t_1}(0,M_1)$, nous allons évaluer l'ordre de $|\mathcal{F}(d)(t) - \mathcal{F}(\hat{d})(t)|$.

$$\psi_1(t,b_o+d)(t) - \psi_1(t,b_o+\hat{d})(t) = \int_0^1 D_b\psi_1(t,b_o+d+\mathcal{C}(\hat{d}-d))(\hat{d}-d)d\tau$$

et comme $D_b\psi_1(b)(t) = O(t^2)$, on a

$$|\psi_1(t,b_o+d)(t) - \psi_1(t,b_o+\hat{d})(t)| \leq A_1t^2 \|d-\hat{d}\|_{\mathcal{C}([0,t])}.$$

Nous procédons de même pour ψ_2 , ψ_3 et ψ_5 , et nous obtenons ainsi :

$$|\mathcal{F}(d)(t) - \mathcal{F}(\hat{d})(t)| \leq \|d-\hat{d}\|_{\mathcal{C}([0,t])} [A_1t^2 + A_2(B_o+M_1)^2t^{1+2\theta}$$

$$+ A_3(B_o+M_1)^3t^{3\theta} + A_4B_oM_1t^{2\theta} + A_5M_1t^{2\theta} + A_6t^{2\theta}] .$$

(Cette estimation ne nous permet pas d'obtenir dans (38) une majoration du terme intégral pour la norme uniforme, c'est ce qui a nécessité l'introduction de la norme $|||d|||$ et de l'espace E_{t_1}) .
De (39) nous tirons immédiatement :

$$\|\mathcal{F}(d)(t) - \mathcal{F}(\hat{d})(t)\| \leq |||d-\hat{d}||| [A_1t^{2+\theta}+(A_2(B_o+M_1)^2t$$

$$+ A_3(B_o+M_1)^3t^{\theta} + A_4B_oM_1 + A_5M_1^2 + A_6)t^{3\theta}] \leq A_7\phi(M_1)\ t^{\delta}\ |||d-\hat{d}||| ,$$

où $\delta = \inf(2+\theta,3\theta)$ et où $\phi(M_1)$ est un polynôme en M_1 .

1er cas. Si $3\theta \leq 2+\theta$ alors $\delta = 3\theta$ et d'après (40) on a

$$|\mathcal{F}(d)(t) - \mathcal{F}(\hat{d})(t)| \leq A_7\phi(M_1)t^{3\theta}\ |||d-\hat{d}|||$$

puis

$$|\mathcal{T}(d)(t) - \mathcal{T}(\hat{d})(t)| \leq A_8\phi(M_1)\, t^{3\theta-1} |||d-\hat{d}|||$$

et

$$|||\mathcal{T}(d) - \mathcal{T}(\hat{d})||| \leq A_8\phi(M_1)t^{2\theta-1} |||d-\hat{d}||| ,$$

comme $\theta > \frac{1}{2}$, pour M_1 fixé et t_1 suffisamment voisin de 0 , $\mathcal{T}$ est une contraction stricte dans $B_{t_1}(0,M_1)$.

2ème cas. Si $2 + \theta < 3\theta$ alors $\delta = 2 + \theta$ et

$$|\mathcal{F}(d)(t) - \mathcal{F}(\hat{d})(t)| \leq A_7\, \phi(M_1)t^{2+\theta} \quad |||d-\hat{d}|||$$

$$|\mathcal{T}(d)(t) - \mathcal{T}(\hat{d})(t)| \leq A_7\, \phi(M_1)t^{1+\theta} \quad |||d-\hat{d}|||$$

$$|||\mathcal{T}(d) - \mathcal{T}(\hat{d})||| \leq A_8\, \phi(M_1)t \quad |||d-\hat{d}||| ,$$

donc pour M_1 fixé et t_1 suffisamment petit, $\mathcal{T}$ est une contraction stricte dans $B_{t_1}(0,M_1)$. Ceci termine la démonstration du Lemme 7.

D'après les Lemmes 6 et 7 , pour M_1 fixé et t_1 suffisamment petit, $\mathcal{T}$ laisse invariant $B_{t_1}(0,M_1)$ et est une contraction dans cette boule pour la norme de E_{t_1} donc admet un point fixe d unique qui est donc solution de (38).

En fait d'après l'unicité du point fixe de $\mathcal{T}$ dans l'ensemble précédent, il est possible de préciser l'ordre de d_o au voisinage de 0 en montrant que $\mathcal{T}$ laisse invariant un ensemble du type

$$\{d,\, d \in \mathcal{C}([0,t_1]) \ ,\ |d(t)| \leq Mt^{\nu} \ ,\ \forall\, t \in [0,t_1]\ \},$$

où $\nu = \inf[1+\theta, 4\theta-1]$, M suffisamment grand et t_1 petit. Donc la solution d_o vérifie au voisinage de 0 ,

$$d_o(t) = O(t^{\nu})$$

Ceci termine les preuves d'existence des branches bifurquées indiquées dans les Théorèmes 3 et 4.

Preuve du Théorème 5. D'après (24) (24)bis (24)ter nous cherchons $b \in \mathcal{C}([0,t_1])$ tel que

$$\begin{cases} - p_1 t b(t) - \lambda_1 \beta_o \int_0^t e^{-(t-s)} b(s)ds + \lambda_1 \mu e^{2t} b^3(t) + \sum_{i=1}^{3} \psi_i(t,b)(t) = 0 \ , \\ b(0) = 0 \ , \end{cases} \tag{41}$$

avec

$$\psi_1(b)(t) = O(t^2 |b(t)|) \ ,$$

$$\psi_2(b)(t) = O(t \, \|b\|^3_{\mathcal{C}([0,t])}),$$

$$\psi_3(b)(t) = O(\|b\|^4_{\mathcal{C}[0,t]}) \ ,$$

et nous voulons montrer que la seule solution est $b \equiv 0$.

Tout d'abord d'après le Lemme 3 toute solution b de (41) vérifie :

$$b(t) = O(t^{1/2}) \ .$$

Nous posons alors $c(t) = e^t b(t)$ et nous modifions la "partie principale" de (41) e^t en considérant le terme $e^{3t} b^3(t)$ comme une perturbation :

$$- p_1 t c(t) - \lambda_1 \beta_o \int_0^t c(s)ds = F_o(c)(t) \ ,$$

avec

$$\begin{aligned} F_o(c)(t) &= - \lambda_1 \mu c^3(t) - \tilde{\psi}_1(c)(t) - \tilde{\psi}_2(c)(t) - \tilde{\psi}_3(c)(t) \\ &= - \lambda_1 \mu c^3(t) + F_1(c)(t) \ . \end{aligned}$$

En posant $y(t) = \int_0^t c(s)ds$ nous avons

$$\begin{cases} - p_1 t y'(t) - \lambda_1 \beta_o y(t) = F_o(c)(t) \\ y(0) = y'(0) = 0 \ , \end{cases}$$

donc

$$y(t) = k_o t^{-\frac{\lambda_1\beta_o}{p_1}} + \frac{t^{-\frac{\lambda_1\beta_o}{p_1}}}{-p_1} \int_0^t \frac{F_o(c)ds}{s^{(p_1-\lambda_1\beta_o)/p_1}} ds , \qquad (42)$$

(où $k_o \in \mathbb{R}$)

Comme $y(t) = 0(t^{3/2})$ et $p_1 \in]-\infty, -2/3\ \lambda_1\beta_o[$, on voit que nécessairement $k_o = 0$ (le terme contenant k_o ne pouvant pas être compensé par le deuxième terme).

Donc

$$y(t) = \frac{t^{-\frac{\lambda_1\beta_o}{p_1}}}{-p_1} \int_0^t \frac{F_o(c)(s)}{s^{(p_1-\lambda_1\beta_o)/p_1}} ds ,$$

et (43)

$$c(t) = \frac{F_o(c)(t)}{-p_1 t} + \frac{\lambda_1\beta_o}{p_1^2} t^{-\frac{\lambda_1\beta_o+p_1}{p_1}} \int_0^t \frac{F_o(c)(s)}{s^{(p_1-\lambda_1\beta_o)/p_1}} ds,$$

en remplaçant F_o par sa valeur on obtient :

$$\left\{ \begin{array}{l} c(t)\left[1 - \dfrac{\lambda_1\mu}{p_1 t} c^2(t)\right] + \dfrac{\lambda_1\beta_o}{p_1^2} t^{-\frac{p_1+\lambda_1\beta_o}{p_1}} \displaystyle\int_0^t \frac{\lambda_1\mu c^3(s)}{s^{(p_1-\lambda_1\beta_o)/p_1}} ds = \\[2ex] \dfrac{F_1(c)(t)}{-p_1 t} + \dfrac{\lambda_1\beta_o}{p_1^2} t^{-\frac{p_1+\lambda_1\beta_o}{p_1}} \displaystyle\int_0^t \frac{F_1(c)(s)ds}{s^{(p_1-\lambda_1\beta_o)/p_1}} . \end{array} \right. \qquad (44)$$

Posons $\theta = -\frac{\lambda_1\beta_o+p_1}{p_1}$ $(\theta < 1/2)$ et $\gamma(t) = c(t).t^{-1/2}$, $t \in]0,t_1]$, d'après le Lemme 3 pour t_1 assez petit

$$\sup_{t\in]0,t_1]} |\gamma(t)| \leq M < +\infty \quad ,$$

avec (42) nous obtenons :

$$\gamma(t)\,[1 - \frac{\lambda_1\mu}{p_1}\gamma^2(t)] + \frac{\lambda_1\beta_o}{p_1^2}\, t^{\,\theta-1/2} \int_0^t \frac{\lambda_1\mu\gamma^3(s)}{s^{\theta+1/2}}\, ds = F_2(\gamma,t), \quad (45)$$

où (les M_i sont des constantes indépendantes de t et de γ)

$$|F_2(\gamma,t)| \leq M_1 t\|\gamma\|_{L^\infty(0,t)} + M_2 t\,\|\gamma\|^3_{L^\infty(0,t)} + M_3 t^{1/2}\|\gamma\|^4_{L^\infty(0,t)} \quad ,$$

remarquons que

$$1 - \frac{\lambda_1\mu}{p_1}\,\gamma^2(t) \geq 1 \qquad (\text{car } p_1 < 0) \;;$$

en multipliant (45) par $\dfrac{\gamma^2(t)}{1 - \frac{\lambda_1\mu}{p_1}\gamma^2(t)}$, nous obtenons

$$\gamma^3(t) + h(t)t^{\theta-1/2} \int_0^t \frac{\gamma^3(s)}{s^{\theta+1/2}}\, ds = F_3(\gamma,t) \;, \qquad (46)$$

où

i/ $h(t)$ est une fonction positive et bornée

ii/ $F_3(\gamma,t)$ vérifie les (M_i' étant des constantes indépendantes de t et γ).

$$|F_3(\gamma,t)| \leq M_1' t\|\gamma\|^3_{L^\infty(0,t)} + M_2' t\,\|\gamma\|^5_{L^\infty(0\ t)} + M_3'{}^{1/2}\|\gamma\|^6_{L^\infty(0\ t)} \;.$$

En posant $F_3(t,\gamma) = g(t)$ et

$$y(t) = \int_0^t \frac{\gamma^3(s)}{s^{\theta+1/2}}\, ds \;,$$

l'équation (46) devient,

$$y'(t) + \frac{h(t)}{t}\, y(t) = \frac{g(t)}{t^{\theta+1/2}} \;. \qquad (47)$$

Posons encore

$$H(t) = \int_{t_1}^{t} - \frac{h(s)}{s} ds ,$$

alors H est définie sur $]0,t_1]$ et est décroissante, avec $H(t_1) = 0$, donc est positive sur $]0,t_1]$ et nous avons en résolvant (47)

$$y(t) = k_o e^{H(t)} + \int_0^t \frac{g(s)\, e^{H(t)-H(s)}}{s^{\theta+1/2}} ds ,$$

comme H est décroissante $e^{H(t)-H(s)} < 1$ et comme $\theta < 1/2$

$\frac{g(s)\, e^{H(t)-H(s)}}{s^{\theta+1/2}}$ est intégrable au voisinage de 0 .

Comme nous devons avoir $y(0) = 0$, nécessairement $k_o = 0$

donc ,

$$y(t) = \int_0^t \frac{g(s)\, e^{H(t)-H(s)}}{s^{\theta+1/2}} ds$$

et

$$\gamma^3(t) = g(t) - h(t)t^{\theta-1/2} \int_0^t \frac{g(s)\, e^{(H(t)-H(s)}}{s^{\theta+1/2}} ds ,$$

ceci entraîne

$\| \gamma \|^3_{L^\infty(0,t)} \leq M \| g \|_{L^\infty(0,t)}$, où M est constante;or $g(t) = F_3(\gamma)(t)$, compte tenu des majorations sur F_3 , nous obtenons (les constantes M''_i sont sont indépendantes de t et γ)

$$\|\gamma\|^3_{L^\infty(0,t)} \leq M'' t^{1/2} \|\gamma\|^3_{L^\infty(0,t)}$$

donc $\|\gamma\|_{L^\infty(0,t)} = 0$ pour t suffisamment voisin de 0 .

Ceci termine la démonstration du Théorème 5.

□

Démonstration du Corollaire 1 (cas $p_1 = 0$)

Soit $\hat{T}_o$, $\hat{T}_o > 0$ tel que sur $[0,\hat{T}_o[$,

$$Q(t)t^2 < \lambda_2 - \lambda_1 \quad \text{et} \quad Q'(t)t > -\frac{2}{3}\lambda_1\beta_o \quad .$$

Tout d'abord, d'après le Théorème 2 il n'y a pas de bifurcation en tout point t de $]0,\hat{T}_o[$ vérifiant $Q(t) \neq 0$.

Par contre soit $\tau \in]0,T_o[$ vérifiant $Q(\tau) = 0$, pour $t > \tau$ on peut écrire :

$$P(t) = \lambda_1 + P'(\tau)(t-\tau) + \tilde{Q}(t)\,(t-\tau)^2 \text{ , (avec } P'(\tau) = Q'(\tau)\tau)$$

il est alors évident que les résultats des Théorèmes 3 et 4 s'appliquent si on remplace le temps $t = 0$ par le temps $t = \tau$ il y a bifurcation simple ou multiple suivant le signe de $P'(\tau)$.

Le cas $Q(t) \equiv 0$ se traite de façon évidente.

ANNEXE

Dans cette annexe nous montrons que dans les Théorèmes 3 et 4 nous avons obtenu toutes les branches bifurquées continues et que celles-ci sont continuement différentiables sauf en 0 .

La démarche suivie comporte plusieurs étapes :

1. Si la fonction $b(t)$ (qui repère une branche bifurquée) est dérivable au point t alors $a(t)$ et $w(t)$ le sont aussi.

2. En tout point t où $b(t)$ est dérivable, $c(t) = e^t b(t)$ vérifie une équation différentielle (E) qui est une bonne perturbation de l'équation de Lagrange introduite dans la démonstration des Théorèmes 3 et 4.

3. Soit une branche bifurquée quelconque b $(p_1 > -\frac{2}{3}\lambda_1\beta_o)$ alors :

3i/ la fonction b ne s'annule pas sur un intervalle $]0,t_1]$,

3ii/ la fonction b est croissante, continûment dérivable

3iii/ la fonction b est solution de (E) sur $]0,t_1]$.

4. Toutes les solutions de (E) sont tangentes aux fonctions b données par les Théorèmes 3 et 4 donc leur sont égales sur un voisinage de 0 grâce aux théorèmes de point fixe utilisés.

1ère étape.

Lemme A1. *Si une branche bifurquée* $u(t) = a(t) + b(t)\phi_1 + w(t)$ *, définie sur* $[0,t_1[$ *, est telle que la fonction* b *est dérivable au point* t *alors les fonctions* a *et* w *le sont aussi et on a :*

$$\begin{cases} a'(t) = \varepsilon_1(t)b'(t) + \varepsilon_2(t)[w(t)] + \varepsilon_3(t)\left[\int_0^t e^s w(s)ds\right], \\ w'(t) = \varepsilon_4(t)b'(t) + \beta_o\alpha(t)\left[-e^{-t}\int_0^t e^s w(s)ds + w(t)\right] \end{cases} \tag{48}$$

avec

$$\varepsilon_i(t) = O(t) \quad , \qquad i = 1,2,3,4$$

$$\alpha(t) = (Id - \lambda_1\pi_1 G\pi_o)^{-1} + O(t).$$

Démonstration. On commence par étudier les quotients différentiels. Avec les notations du lemme A1 on a le :

Lemme A2. *On pose pour* $t \in [0,t_1[$, $h \in [-t,t_1-t[$, $\Delta a(t) = \frac{a(t+h)-a(t)}{h}$ *de même pour* Δw *et* Δb. *Alors on a* :

$$\begin{cases} i/ \quad \Delta a(t) = \mathcal{G}_1(b,t,h) \ \Delta b(t) + \mathcal{G}_2(b,t,h) , \\ \\ ii/ \quad \Delta w(t) = \mathcal{G}_3(b,t,h) \ \Delta b(t) + \mathcal{G}_4(b,t,h) , \end{cases} \qquad (49)$$

où les $_i$ *sont des fonctions continues de* (b,t,h) ,

$$(b,t,h) \in \mathcal{C}^0([0,t_1[) \times \{(t,h) \mid 0 \leq t < t_1 , \ 0 \leq t+h \leq t_1\} ,$$

et vérifiant :

pour $i - 1,2$

$$| \mathcal{G}_i(b,t,h)| \leq C|b^2|_{\mathcal{C}([0,t])}$$

$$| \mathcal{G}_3(b,t,h)|_W \leq C|b^2|_{\mathcal{C}([0,t])} ,$$

et

$$| \mathcal{G}_4(b,t,h)|_W \leq C|b|_{\mathcal{C}([0,t])} .$$

Remarque. Ce lemme donne une estimation des quotients différentiels de a et w au point t en fonction du quotient différentiel de b *au point* t et des valeurs de b sur $[0,t]$ alors que dans le Lemme 1 (application du théorème des fonctions implicites à $F_1(a,b,w) = 0$, $F_2(a,b,w) = 0$ pour déterminer $a(b)(t),w(b)(t))$ on obtient une évaluation de $a(t)$ et $w(t)$ en fonction des valeurs de b *sur* $[0,t]$).

2/ Si a,b,w sont dérivables au point t les estimations précédentes s'obtiennent immédiatement par dérivation de $F_i(a(t),b(t),w(t)) = 0$ $i = 1.2$.

3/ Nous avons introduit les quotients différentiels car les solutions b de $F_3(b)(t) = 0$ ne sont pas en général dérivables en $t = 0$ (par exemple pour $p_1 \geq 0$, $b(t) \sim At^{1/2}$) ce qui ne permet pas d'appliquer le théorème des fonctions implicites à $F_1 = F_2 = 0$ dans
$\mathcal{C}^1([0,t_1[) \times \mathcal{C}^1([0,t_1[) \times \mathcal{C}^1([0,t_1[,W)$. En revanche cette technique

s'applique avec succès si b est dérivable sur $[0,t_1[$ ce qui est le cas pour la famille

$$b(t) = \left(\frac{t}{\gamma}\right)^{-\frac{p_1+\lambda_1\beta_o}{p_1}} (1+\varepsilon(t)) \ , \quad \gamma \in \mathbb{R}^+, \ 0 > p_1 > -\frac{\lambda_1\beta_o}{2}$$

Démonstration du Lemme A2.

Nous avons $F_i(a(t),b(t),w(t)) = 0$, $\forall\, t \in [0,t_1[$,

donc

$$\frac{1}{h}(F_1(a,b,w)\ (t+h) - F_1(a,b,w)(t)) = 0$$

soit $\quad (\theta \in]0,1[)$

$$\int_0^1 \cos(a+b\phi_1+w)(t+\theta_x h)\ (\Delta a+\Delta b\phi_1+\Delta w)dx = 0 \ ,$$

comme a,b,w sont voisins de 0 , ceci s'écrit encore :

$$\int_0^1 (1+\phi(x,t,h))(a+b\phi_1+w)(t+\theta_x h)^2\ (\Delta a+\Delta b\phi_1+\Delta w)dx = 0 \ ,$$

donc

$$\Delta a = \mathcal{K}_1(b,t,h)\Delta b + \int_0^1 \mathcal{K}_2(b,t,h,x)\ (\Delta w)dx \ , \tag{50}$$

où les fonctions $\mathcal{K}_i$ sont des fonctions continues de leurs arguments et

$$|\mathcal{K}_1(b,t,h)| \le C|b(t)|^2$$

$$|\mathcal{K}_2(b,t,h,x)| \le C|b(t)|^2 \ .$$

La même technique appliquée à

$$\frac{1}{h}(F_2(a,b,w)(t+h) - F_2(a,b,w)(t)) = 0 \ ,$$

donne

$$\left\{\begin{array}{l}\Delta w = \beta_o(\frac{e^{-h}-1}{h}) \displaystyle\int_0^{t+h} e^{-(t-s)} w(s)ds \\ + \beta_o(\frac{1}{h}\displaystyle\int_t^{t+h} e^{-(t-s)} w(s)ds) + \frac{P(t+h)-P(t)}{h} \pi_1 G\pi_o \sin(a+b\phi_1+w)(t+h) \\ + P(t)\pi_1 G\pi_o\ (1+\phi(t,x,h))((a+b\phi_1+w)(t+\theta_x h))^2(\Delta a+\Delta b\phi_1+\Delta w)\ ,\end{array}\right.$$

en tenant compte de (49) on obtient

$$\left\{\begin{array}{l}\Delta w - \mathcal{K}_3(b,t,h)(\Delta w) = \beta_o(\frac{e^{-h}-1}{h}) \displaystyle\int_0^{t+h} e^{-(t-s)} w(s)ds \\ + \beta_o(\frac{1}{h} \displaystyle\int_t^{t+h} e^{-(t-s)} w(s)ds + \frac{P(t+h)-P(t)}{h} \pi_1 G\pi_o \sin(a+b\phi_1+w) \\ + \mathcal{K}_4(b,t,h)(\Delta a) + \mathcal{K}_5(b,t,h)\ (\Delta b)\ ,\end{array}\right.$$

avec,

$$|\mathcal{K}_3(b,t,h)|_{\mathcal{L}(W,W)} \leq C|b|^2_{L^\infty(0,t)}\ ,$$

$$i = 4,5\quad |\mathcal{K}_i(b,t,h)| \leq C|b|^2_{L^\infty(0,t)}\ .$$

Comme $\lim_{t\to 0} b(t) = 0$, l'opérateur $(Id-\mathcal{K}_3(b,t,h))$ est inversible pour t assez petit, ce qui permet à partir de (51) et des estimations fournies par le Lemme 1 d'obtenir 49i) puis 49ii).

Démonstration du Lemme A1

La dérivabilité de a et w au point t est conséquence immédiate du Lemme A2 si b est dérivable au point t . On peut alors dériver $F_1 = F_2 = 0$ au point t et obtenir un système résoluble en $(a'(t),w'(t))$:

$$\left\{\begin{array}{l}
\left(\int_0^1 \cos u(t)dx\right)a'(t) + \int_0^1 (\cos u(t))w'(t)dx = \\
\qquad\qquad - \left(\int_0^1 \cos u(t)\phi_1 dx\right)b'(t) \\
(-P(t)\pi_1 G\pi_o \cos u(t))a'(t) + w'(t) - P(t)\pi_1 G\pi_o \cos u(t)w'(t) = \\
\qquad (P(t)\pi_1 G\pi_o \cos u(t)\phi_1)b'(t) - \beta_o e^{-t}\int_0^t e^s w(s)ds + \beta_o w(t)
\end{array}\right.$$

Le premier membre peut s'écrire sous forme matricielle

$$\mathcal{G}(t)\begin{pmatrix} a' \\ w' \end{pmatrix} = \left\{\begin{pmatrix} 1 & 0 \\ 0 & Id-\lambda_1\pi_1 G\pi_o \end{pmatrix} + \right.$$

$$\left.\begin{pmatrix} -1 + \int_0^1 \cos u(t)dx & \int_0^1 \cos u(t)\, dx \\ - P(t)\pi_1 G\pi_o \cos u & - P(t)\pi_1 G\pi_o \cos u + \lambda_1\pi_1 G\pi_o \end{pmatrix}\right\}\begin{pmatrix} a' \\ w' \end{pmatrix}$$

La première matrice $\begin{pmatrix} 1 & 0 \\ 0 & Id - \lambda_1\pi_1 G\pi_o \end{pmatrix}$

est inversible dans $\mathcal{L}(\mathbb{R}\times W, \mathbb{R}\times W)$ et la seconde est un $O(t)$ dans le même espace, l'opérateur $\mathcal{G}(t)$ est donc inversible et $(\mathcal{G}(t))^{-1}$ dépend continuement de t. Le système précédent détermine donc de façon unique $a'(t)$ et $w'(t)$ en fonction de $b'(t)$ et de $(a,w) \in \mathcal{C}([0,t], \mathbb{R}) \times \mathcal{C}([0,t],W)$. Dans les formules (50) on a justement précisé les dépendances respectives de $(a',w')(t)$ par rapport d'une part aux valeurs des fonctions a,w,b,b' au point t et d'autre part à l'ensemble de leurs valeurs sur $[0,t]$.

□

2ème étape.

Lemme A3. *Soit* b *une solution continue de* $F_3(b)(t) = 0$ *sur* $[0,t_1[$ *(avec* $b(0) = 0$*). Alors en tout point* t *de* $[0,\tilde{t}_1[$ *(*$\tilde{t}_1$ *assez petit) où* b *est dérivable, la fonction* $c(t) = e^t b(t)$ *vérifie l'équation différentielle* (E) ,

$$(E) \begin{cases} p_1 t(1+\phi_1(t)t)c'(t) + (p_1+\lambda_1\beta_o)(1+\phi_2(t)t)c(t) \\ - 3\lambda_1\mu c^2(t)c'(t) + \xi_1(c)(t) + \xi_2(c,t)(t) = 0 , \\ c(0) = 0 , \end{cases}$$

où les fonctions ϕ_i *et* ξ_i *sont continues et vérifient*

$$\begin{cases} |\phi_i(t)| \leq M , & i = 1,2 , \\ |\xi_1(c)(t)| \leq M\, |c|^3_{\mathcal{C}([0,t])} \\ |\xi_2(c,t)(t)| \leq M\, t^2 |c|_{\mathcal{C}([0,t])} \end{cases}$$

Démonstration. D'après le Lemme A1 les fonctions a et w sont dérivables au point t , nous pouvons alors dériver l'équation de bifurcation (on a posé $\tilde{P}(t) = e^t P(t)$).

$$0 = e^t F_3(a,e^{-t}c,w)(t) = \lambda_1 c - \lambda_1\beta_o \int_0^t c(s)ds - \tilde{P}(t) \int_0^1 \sin(a+ce^{-t}\phi_1+w)\phi_1 dx ,$$

ce qui donne,

$$\begin{cases} 0 = \lambda_1 c' - \lambda_1\beta_o c - \tilde{P}'(t) \int_0^1 \sin(a+ce^{-t}\phi_1+w)\phi_1\, dx \\ - \tilde{P}(t) \int_0^1 \cos(a+ce^{-t}\phi_1+w)(a'+c'e^{-t}\phi_1-ce^{-t}\phi_1+w')\phi_1\, dx . \end{cases} \qquad (53)$$

Dans cette équation nous remplaçons a, w, a', w' par leur expression en fonction de b,b' donnée par (22) et (48), on obtient (on a posé $Q(0) = p_2$) successivement :

i/

$$\left\{\begin{aligned} &\tilde{P}'(t) \int_0^1 \sin(a+ce^{-t}\phi_1+w)\phi_1\, dx = (\lambda_1+p_1)c + (p_1+p_2)tc \\ &\qquad\qquad + \theta_1(c,t) + \theta_2(c,t) \end{aligned}\right. \tag{54i}$$

avec

$$|\theta_1(c,t)| \leq Mt^2|c|_{\mathcal{C}([0,t])}$$

$$|\theta_2(c,t)| \leq M|c|^3_{\mathcal{C}([0,t])}$$

ii/ pour le second terme, compte tenu de

$$0 = \int\phi_1\, dx = \int\varepsilon_2(t)w(t)\phi_1\, dx = \int\varepsilon_3(t)\left(\int_0^t e^s w(s)ds\right)\phi_1\, dx =$$

$$\int\alpha(t)\left(\int_0^t e^s w(s)ds\right)\phi_1\, dx\ ,$$

on obtient :

$$\left\{\begin{aligned} &\tilde{P}(t) \int_0^1 \cos(a+ce^{-t}\phi_1+w)(a'+e^{-t}c'\phi_1-e^{-t}c\phi_1+w')\phi_1\, dx \\ &= (\lambda_1+p_1t+\eta_1(t^2))c' - (\lambda_1+p_1t+\eta_2(t))c + 3\mu\lambda_1c^2c' + \theta_3(c)(t)\ , \end{aligned}\right. \tag{54ii}$$

où

$$|\eta_1(t^2)| \leq M\, t^2\ ,$$

$$|\eta_2(t)| \leq M\, t\ ,$$

$$|\theta_3(c)(t)| \leq M|c|^3_{\mathcal{C}([0,t])}$$

L'équation (E) résulte alors de (53) 54(i), (ii).

Le même calcul, où les dérivées $a'(t)$, $w'(t)$, $c'(t)$ sont remplacées par les quotients différentiels $\Delta a, \Delta w, \Delta c$ et où $\Delta a, \Delta w$ sont exprimés en fonction de c et Δc grace aux formules 49i) et ii) conduit au

Lemme A3bis. *Soit* b *une solution continue de* $F_3(b)(t) = 0$ *sur* $[0,t_1[$ *(avec* $b(0) = 0$). *Alors en tout point* t *de* $[0,\tilde{t}_1[$ *le quotient différentiel* $\Delta c(t) = \frac{c(t+h) - c(t)}{h}$ *vérifie* (E_h)

$$\left\{\begin{array}{l} (p_1 t(1+\phi_1(t)t) - 3\lambda_1\mu c^2(t) + \varepsilon_1(t,h))\,\Delta c = \\ -\,(p_1+\lambda_1\beta_o)(1+\phi_2(t)t)c(t) + \xi_1(c)(t) + \xi_2(c,t)(t) \\ +\,\varepsilon_2(t,h)\ , \end{array}\right. \qquad (E_h)$$

avec $\lim_{h\to 0} \varepsilon_i(t,h) = 0$.

Corollaire 1. *Soit* b *une solution continue de* $F_3(b)(t) = 0$ *sur* $[0,t_1[$ $(b(0) = 0)$. *En tout point* t *de* $[0,\tilde{t}_1[$ *où* $p_1 t(1+\phi_1(t)t) - 3\mu\lambda_1 c^2(t) \neq 0$ *la fonction* $b(t)$ *est dérivable.*

La démonstration est immédiate : il suffit de considérer (E_h) .

Corollaire 2. *Dans le cas* $p_1 < 0$ *toutes les branches bifurquées sont dérivables sur* $]0,\tilde{t}_1[$.

Dans le cas $p_1 = 0$ *, les branches bifurquées sont dérivables en tout point où elles sont non nulles.*

3ème étape. (Propriétés de régularité de b).

Lemme A4. *Soit une branche* $u(t) = a(t) + b(t)\phi_1 + w(t)$ *bifurquée en* 0 *continue sur* $[0,t_1[$ *, alors la composante* b *ne s'annule pas sur un intervalle* $]0,\tilde{t}_1[$ $(\tilde{t}_1 \leq t_1)$.

Démonstration (on travaille avec $c(t)$).

Si la conclusion du lemme est inexacte, il existe une suite t_n telle que $(n \geq 2)$.

i/ $t_{2n} > t_{2n+1} \geq t_{2n+2} > 0$,

ii/ $c(t_n) = 0$,

iii/ $0 < \sup_{t \in [0,t_{2n}]} |c(t)| = \sup_{t \in [t_{2n+1}, t_{2n}]} |c(t)| = c(\tau_n), \tau_n \in]t_{2n+1}, t_{2n}[$,

iv/ $c(t) > 0$ sur $]t_{2n+1}, t_{2n}[$.

Il y a alors deux possibilités pour le comportement de c au point τ_n :

1 : la fonction c est dérivable au point τ_n , donc $c'(\tau_n) = 0$ et l'équation (E) en τ_n s'écrit :

$$0 = (p_1+\lambda_1\beta_o)(1+\phi_2(\tau_n)\tau_n)c(\tau_n) + \xi_1(c)(\tau_n) + \xi_2(c,\tau_n)(\tau_n) \qquad (55)$$

mais d'après (52) et iii).

$$\xi_1(c)(\tau_n) = \theta_1(\tau_n)(c(\tau_n))^3 \quad , \quad |\theta_1(\tau_n)| \leq M$$

$$\xi_2(c,\tau_n)(\tau_n) = \theta_2(\tau_n)\, c(\tau_n) \quad , \quad |\theta_2(\tau_n)| \leq M ,$$

(55) devient alors, $(c(\tau_n) > 0)$,

$$c^2(\tau_n) = - \frac{(p_1+\lambda_1\beta_o)(1+\phi_2(\tau_n)\tau_n) + \theta_2(\tau_n)\tau_n^2}{\theta_1(\ _n)}$$

mais ceci est contradictoire avec $\lim_{t\to 0} c(t) = 0$, donc la possibilité 1 est à exclure. On étudie la deuxième possibilité :

2. La fonction c n'est pas dérivable en τ_n , donc d'après le Lemme A3bis : (d'après le corollaire 2 nécessairement $p_1 > 0$)

$$p_1\tau_n(1+\phi_1(\tau_n)\tau_n) - 3\lambda_1\mu c^2(\tau_n) = 0$$

soit

$$c({}_n) = \left(\frac{p_1}{3\lambda_1\mu}\right)^{1/2} (1+\phi_1(\tau_n)\tau_n)^{1/2}\tau_n^{1/2} \tag{56}$$

D'autre part,

$$0 = e^{\tau_n} F_3(e^{-\tau_n} c)(\tau_n) - e^{t_{2n+1}} F_3(e^{-t_{2n+1}} c)(t_{2n+1}) \ ,$$

soit,

$$\left| \begin{array}{l} 0 = - p_1\tau_n c(\tau_n) + \lambda_1\mu c^3(\tau_n) - \lambda_1\beta_o \displaystyle\int_{t_{2n+1}}^{\tau_n} c(s)ds \\ \quad + \displaystyle\sum_{i=1}^{3} (e^{\tau_n}\psi_i(e^{-\tau_n}c)({}_n) - e^{t_{2n+1}}\psi_i(e^{-t_{2n+1}}c)(t_{2n+1})) \ , \end{array} \right. \tag{57}$$

Mais comme

$$|c|_{L^\infty(0,t_{2n+1})} \le |c|_{L^\infty(0,\tau_n)} = c(\tau_n) \ ,$$

et que,

$$|c(\tau_n)| \le M\,\tau_n^{1/2} \ , \quad \text{(Lemme 4)}$$

on a,

$$\sum_{i=1}^{3} (e^{\tau_n}\psi_i - e^{t_{2n+1}}\psi_i) = \theta_n\,\tau_n^2 \ ,$$

avec $|\theta_n| \le M$, $\forall n$.

Avec (56) ceci permet d'écrire (57) sous la forme

$$-\theta_n\,\tau_n^{1/2} + \frac{2}{3}\,\frac{p_1^{3/2}}{(3\lambda_1\mu)^{1/2}}(1+\varepsilon(\tau_n)) = -\frac{\lambda_1\beta_o}{\tau_n^{3/2}}\int_{t_{2n+1}}^{\tau_n} c(s)ds \tag{58}$$

ce qui est impossible car d'après iv) $c(s) > 0$ sur $]t_{2n+1},\tau_n]$ et les deux membres de (58) sont de signes opposés.

Le lemme A4 est ainsi démontré.

<u>Lemme A5</u>. *Soit une branche bifurquée en* 0 : $u(t) = a(t) + b(t)\phi_1 + w(t)$, *la fonction* b (*ou* -b) *est continue strictement positive sur* $]0,\tilde{t}_1[$. *De plus la fonction* $c(t) = e^t b(t)$ (*ou* -c) *est croissante strictement, et continument dérivable sur* $]0,\tilde{t}_1[$.

<u>Démonstration</u>. D'après le Lemme A4 on sait que la fonction b (ou -b) est continue, strictement positive sur un intervalle $]0,\tilde{t}_1[$. On suppose que $b(t) > 0$. Si $c(t)$ n'est pas strictement croissante, soit t et τ, $\tau \in]0,t[$ tel que

$$c(\tau) = \sup_{s \in [0,t]} c(s) \geq c(t) .$$

Si la fonction c est dérivable au point τ , $c'(\tau) = 0$ et le Lemme A3 implique

$$0 = (\lambda_1\beta_o+p_1)(1+\phi_2(\tau)\tau)c(\tau) + \xi_1(c)(\tau) + \xi_2(c,\tau)(\tau) = 0 , \qquad (59)$$

or

$$\begin{cases} \xi_1(c)(\tau) = \theta_1(\tau)(c(\tau))^3 , & |\theta_1(\tau)| \leq M , \\ \xi_2(c,\tau)(\tau) = \theta_2(\tau)\tau^2 c(\tau) , & |\theta_2(\tau)| \leq M , \end{cases} \qquad (60)$$

et (59) s'écrit encore,

$$(c(\ ))^2 = \frac{-(\lambda_1\beta_o+p_1)(1+\phi_2(\tau)\tau) - \theta_2(\tau)\tau^2}{\theta_1(\tau)} ,$$

ce qui est contradictoire avec $\lim_{t\to 0} c(t) = 0$.

Si la fonction c n'est pas dérivable au point τ , d'après le Lemme A3bis on a

$$c(\tau) = \left(\frac{p_1}{3\lambda_1\mu}\right)^{1/2} (1+\phi_1(\tau)\tau)^{1/2} \ \tau^{1/2} , \qquad (61)$$

(nécessairement $p_1 > 0$ car pour $p_1 \leq 0$ les branches bifurquées sont dérivables (Corollaire 1))

donc

$$0 = e^{\tau} F_3(e^{-\tau}c)(\tau) = -\frac{2}{3}\frac{p_1^{3/2}}{(3\lambda_1\mu)^{1/2}}\tau^{3/2}(1+\varepsilon(\tau))$$

$$-\lambda_1\beta_o\int_0^{\tau} c(s)ds + \theta(\tau)\tau^2 \quad , \qquad |\theta(\tau)| \leq M \ ,$$

donc, comme $c(s) > 0$ sur $[0,\tau]$,

$$\frac{2}{3}\frac{p_1^{3/2}}{(3\lambda_1\mu)^{1/2}}(1+\varepsilon(\tau)) \leq \theta(\tau)\,\tau^{1/2} \qquad 61(\text{bis})$$

ce qui est impossible.

Il en résulte donc que la fonction c est strictement croissante.

Il reste à montrer que la fonction c est dérivable sur $]0,\tilde{t}_1[$. Tout d'abord cette propriété est vraie si $p_1 \leq 0$ d'après le Corollaire 2. Si donc c n'est pas dérivable en un point τ on a $p_1 > 0$ et comme $c(\tau) = \sup_{s\,[0,\tau]} c(s)$ (c est croissant) la valeur $c(\tau)$ vérifie (61) . Le raisonnemment précédent s'applique à nouveau et conduit à la contradiction (61)bis . La fonction c est donc dérivable et vérifie (E) . Comme

$c(t) \neq \left(\frac{p_1}{3\lambda_1\mu}\right)^{1/2}(1+\phi_2(t)t)^{1/2}\,t^{1/2}$, $c'(t)$ est continue et strictement positive sur $]0,\tilde{t}_1[$.

Le Lemme A.5 permet d'améliorer l'équation (E) : les relations (60) sont valables pour τ quelconque dans $]0,\tilde{t}_1[$, donc (E) s'écrit en remplaçant ξ_1 et ξ_2 par (60) :

$$0 = p_1 t(1+\phi_1(t)t)c'(t) + (p_1+\lambda_1\beta_o)\,[1+\phi_2(t)t) + \frac{1}{(p_1+\lambda_1\beta_o)}(\theta_1(t)c^2(t) + \theta_2(t)t)]c(t) - 3\lambda_1\mu c^2(t)c'(t)$$

et comme $|c(t)| \leq M\sqrt{t}$ on obtient

$$(\hat{E}) \begin{cases} 0 = p_1 t(1+\phi_1(t)t)c'(t) + (p_1+\lambda_1\beta_o)(1+\tilde{\phi}_2(t)t)c(t) \\ \\ - 3\lambda_1\mu c^2(t)c'(t) \ , \\ \\ c(0) = 0 \ . \end{cases}$$

4ème étape. Résolution de $(\hat{E})$

Lemme A6. *Toutes les solutions de* $\hat{E}$ *(et donc de* E *) sont tangentes en* 0 *aux fonctions données par les Théorèmes* 3 *et* 4 .

Démonstration. Comme $(t \to c(t))$ appartient à $\mathcal{C}^1(]0,\tilde{t}_1[)$, avec $c'(t) > 0$ on peut dans $\hat{E}$ prendre c comme variable et t comme fonction (de c) :

$$\begin{cases} 0 = p_1 t(1+\phi_1(t)t) + (p_1+\lambda_1\beta_o)(1+\tilde{\phi}_2(t)t)ct' \\ - 3\mu\lambda_1 c^2 \ , \\ t(0) = 0 \ . \end{cases} \tag{62}$$

Résolution de (62). Cas $p_1 \neq 0$. On applique une méthode de variation des constantes :

$$\frac{t'}{t}(1+\phi_3(t)t) = -\frac{p_1}{\lambda_1\beta_o+p_1}\ \frac{1}{c} \quad , \quad |\phi_3(t)| \leq M \ , \tag{63}$$

pour intégrer (63) on utilise le ,

Lemme : *Soit* $f(t) = \frac{1}{t} + \phi(t)$, ϕ *bornée continue sur* $]0,\tilde{t}_1[$, *alors il existe une primitive de* f, F *qui est de la forme*

$$F(t) = \text{Log}\ t(1+\theta(t)t) \ ,$$

où θ *est une fonction continue bornée sur* $]0,\tilde{t}_1[$. (Il suffit de prendre $\theta(t) = \frac{1}{t} \int_0^t \phi(\tau)e^{t-\tau}d\tau$). Alors (63) implique

$$t(1+\theta(t)t) = K\, c^{-\frac{p_1}{p_1+\lambda_1\beta_o}} , \tag{64}$$

comme $\lim_{c\to 0} t(c) = 0$, on peut écrire

$$\begin{cases} \theta(t)t = \varepsilon(c) \\ \lim_{c\to 0} \varepsilon(c) = 0 \end{cases}$$

si bien que (64) devient :

$$t = K\, c^{-\frac{p_1}{\lambda_1\beta_o+p_1}} (1+\tilde{\varepsilon}(c))$$

On revient alors à (62) en prenant K fonction de c et on remplace les $\varepsilon(t)$ par $\varepsilon(c)$

$$0 = K'(\lambda_1\beta_o+p_1)c^{\frac{\lambda_1\beta_o}{\lambda_1\beta_o+p_1}} (1+\varepsilon(c)) - 3\lambda_1\mu c^2 ,$$

par intégration :

$$K = K_o + \frac{3\mu\lambda_1}{3p_1+2\lambda_1\beta_o}\, c^{-\frac{3p_1+2\lambda_1\beta_o}{p_1+\lambda_1\beta_o}} (1+\varepsilon(c))$$

et

$$\begin{cases} t(c) = K_o\, c^{-\frac{p_1}{\lambda_1\beta_o+p_1}} (1+\varepsilon(c)) \\ \qquad + \frac{3\mu\lambda_1}{3p_1+2\lambda_1\beta_o} c^2 (1+\varepsilon(c)). \end{cases} \tag{65}$$

Si on se limite aux fonctions $t(c)$ telles que $t(c) > 0$ si $c > 0$ et

$\lim_{c \to 0} t(c) = 0$ on obtient pour $p_1 \neq 0$ des fonctions tangentes aux fonctions données par (27) $(p = c)$ qui sont les solutions de l'équation de Lagrange (28). Ceci montre que dans le cas $p_1 \neq 0$ les solutions de $F_3(b)(t) = 0$ sont de la forme imposée dans les démonstrations des Théorèmes 3 et 4 , et donc que l'on a obtenu ainsi *toutes* les branches bifurquées.

Cas $p_1 = 0$.

L'équation (62) s'écrit :

$$(\beta_o(1+\tilde{\phi}_2(t)t)t' - 3\mu c)c = 0 , \quad t(0) = 0 \tag{66}$$

on cherche les branches bifurquées en 0 , d'après le Lemme A5, $c(t) > 0$, donc (66) se réduit

$$\begin{cases} \beta_o(1+\tilde{\phi}_2(t)t)t' - 3\mu c = 0 \\ t(0) = 0 \end{cases}$$

dont la solution est

$$t = \left(\frac{3\mu}{2\beta_o}\right) c^2(1+\varepsilon(c))$$

donc

$$c(t) = \left(\frac{2\beta_o}{3\mu}\right)^{1/2} t^{1/2}(1+\varepsilon(t)) ,$$

nous avons donc la même conclusion que pour le cas $p_1 \neq 0$: il n'y a qu'une qu'une seule branche bifurquée en 0 et c'est celle qui est donnée par le Théorème 3.

REFERENCES

[1] M.E. Gurtin, Some Questions and Open Problems in Continuum Mechanics and Populations Dynamics (Prepint) (Depart. Math, Carnegie Mellon University, Pittsburg 1981. P.A. 15 213).

[2] M.E. Gurtin, V.J. Mizel, D.W. Reynolds. On the buckling of linear Viscoelastic Rods. Rapport MRC n°2222 (1981).

[3] A.E.H. Love, A treatise on the Mathematical Theory of Elasticity, Dover. New York 4ème édition 1944.

Fulbert MIGNOT

Centre de mathématiques
Ecole polytechnique
Route de Saclay

91128 PALAISEAU CEDEX
FRANCE

Jean-Pierre PUEL

Laboratoire d'Analyse Numérique
Tour 65-55, 5e étage
4, place Jussieu

75230 PARIS CEDEX 05
FRANCE

J C SAUT & B SCHEURER

Unique combination and uniqueness of the Cauchy problem for elliptic operators with unbounded coefficients

1. INTRODUCTION.

Let Ω be a connected open set in $\mathbb{R}^n$, $n \geq 2$ and P a uniformly elliptic operator of order $m \geq 2$. P is said to have the property of (weak) unique continuation if the following holds.

$$\left\{\begin{array}{l} \text{Let } u \in H^m_{loc}(\Omega) \quad (\text{more generally, } u \in W^{m,p}_{loc}(\Omega),\ p \geq 1) \\ \text{be a solution of } P\,u = 0 \text{ such that } u = 0 \text{ a.e. on a non-} \\ \text{empty open set in } \Omega. \\ \quad \text{Then } u \text{ vanishes identically in } \Omega. \end{array}\right. \qquad (U.C)$$

This property is equivalent to uniqueness in the Cauchy problem for any (smooth) hypersurface.

The aim of this lecture is to review some recent results on property (U.C) for elliptic operators having coefficients which are as irregular as possible. A part from their intrinsic interest, such results do have a large field of applications :

a) *Generic properties of nonlinear elliptic equations* (see [28], [29]) :

Let $F(.,\phi)$ be a family of nonlinear elliptic operators, depending on a parameter ϕ, which belongs to some Banach space Y. In order to study generic properties of the solution set of $F(u,\phi) = 0$ with respect to ϕ, using techniques of transversality, one is led to show (U.C) for the adjoint of the derivative $F'_u(u_o,\phi_o)$ where $F(u_o,\phi_o) = 0$. If u_o,ϕ_o belong to some Sobolev spaces of low order, the coefficients of $F'_u(u_o,\phi_o)$ are not necessarily locally bounded. The same property is also used in similar contexts, for instance generic properties of eigenvalues of elliptic equations.

b) *Asymptotic behavior of some equations of evolution.*

In [7], Dafermos shows asymptotic stability of solutions of linear thermoelasticity equations in generic domains using uniqueness in the Cauchy problem.

c) Control of systems having multiple states (cf. J.L. Lions [14]).

In this context property (U.C) or uniqueness in the Cauchy problem is used as a technical tool to characterize the optimal control. As in a) one deals with a linearization of a nonlinear operator around a non-smooth function, and the linear equation which results has in general unbounded coefficients.

d) Spectral properties of Schrödinger operator.

One consider the Schrödinger operator

$$Lu = - \Delta u + V(x)u \text{ in } \mathbb{R}^n$$

A classical method to show that L has no positive eigenvalue uses (U.C) for $L - \lambda$ (see [25]). Of course the case of an irregular potential V is of much interest for physical reasons.

The first work on unique continuation property seems to be due to Carleman [5] ; he proves (U.C) for second-order elliptic operators in $\mathbb{R}^2$. Then Müller [18] and Heinz [10] treated the case of operator with Δ as principal part. The first result for general second-order elliptic equations is due to Aronszajn [2] ; then Calderon [35] and Hörmander [12][13] considered more general operators, not necessarily elliptic. The book [12] contains the proof of (U.C) for elliptic operators with simple complex characteristics and a good bibliography.

All these results are established (in the optimal versions) for operators whose principal coefficients are bounded and lipschitzian, the others being locally bounded.

For elliptic operators with multiple complex characteristics, classical counter-examples show that (U.C) may be false : Plis [23] constructs $f,g \in C^\infty(\mathbb{R}^3)$ such that

$$Lu = (\partial_t^2+\partial_x^2+\partial_y^2)^2 u + t^2[(\partial_x^2+\partial_y^2)^2 - \frac{1}{2}\partial_x^4]u = (f\partial_x+g)u$$

doesn't satisfy (U.C).

In the same spirit, Cohen [6] exhibits a counter-example of the form $\Delta^3 + Q$, order of $Q \leq 5$.

However (U.C) subsists in some cases, but all terms of lower order are not allowed in general (cf. [9], [13], [16], [21], [22], [31], [33]).

As was mentioned above, we are interested in weakening the smoothness hypothesis on the coefficients of the operator involved.

This problem is hopeless for *principal* coefficients since Plis [24] built $a(t,x,y)$, $l(t)$, $f(t,x,y)$, $g(t,x,y)$, $h(t,x,y)$ with

$$Lu = \frac{\partial^2 u}{\partial t^2} + \frac{\partial^2 u}{\partial x^2} + l(t)\frac{\partial^2 u}{\partial y^2} = f(t,x,y)\frac{\partial u}{\partial x} + g(t,x,y)\frac{\partial u}{\partial y} = 0 \text{ in } \mathbb{R}^3,$$

L elliptic, $u,f,g,h \in C^\infty(\mathbb{R}^3)$, l is C^∞ for $t \neq 0$, $l \in C^\alpha(\mathbb{R})$, $\forall\alpha$, $0 < \alpha < 1$, and $u \equiv 0$ in the half space $t \geq 0$, but doesn't vanish identically on every neighborhood of 0.

Consequently we only focus our attention on non-principal coefficients : indeed (U.C) holds if these coefficients belongs to some L^p_{loc} (p depending on the order of the operator, on the multiplicity of the complex characteristics, on the order of the coefficients and on the space dimension).

We shall first deal with the case of general elliptic operators, then treat the case of Schrödinger type operators where more precise results are known, and we end up with some comments and open questions. Only brief sketchs of the proofs will be given ; we refer the interested reader to the original papers.

2. UNIQUE CONTINUATION FOR GENERAL ELLIPTIC EQUATIONS.

a) Operators with simple complex characteristics.

Let $P(x,D) = \sum_{|\alpha|\leq m} a^{\alpha}(x)D^{\alpha}$ be a uniformly elliptic operator of order $m \geq 2$, with real coefficients, in a connected open set Ω in $\mathbb{R}^n$.

Let $P_m(x,D) = \sum_{|\alpha|=m} a^{\alpha}(x)D^{\alpha}$ be the principal part of P. We suppose that $a_{\alpha} \in C^1(\bar{\Omega})$ for $|\alpha| = m$ and that P_m has simple complex characteristics. This means, by definition, that the polynomial in ξ_1 (for instance), $P_m(x,\xi_1,\xi_2,\ldots, \xi_n)$ has distinct (complex) zeroes for every $x \in \Omega$ and every $(\xi_2,\ldots, \xi_n)$ reals $\neq (0,0,\ldots, 0)$.

Theorem 1 : ([26][27]). *There exist real numbers* q_k, $2 \leq q_k < \infty$, $k = 0,\ldots, m-1$ *such that if* $a_{\alpha} \in L^{q_{|\alpha|}}_{loc}(\Omega)$, $|\alpha| = 0,1,\ldots, m-1$, *then* P(x,D) *satisfies* (U.C).

Remarks : 1 - One has explicit formulas for the q_k's but they are rather complicated, at least for large m. For instance,

$$q_o = \mathrm{Max}(2, \frac{2n}{2m-1}),\ q_1 = \mathrm{Max}(2, \frac{2n}{2m-3}),\ q_2 = \mathrm{Max}(2, \frac{2n(2m-1)}{(2m-3)(2m-5)}).$$

2 - One can replace $L^{q_{|\alpha|}}_{loc}(\Omega)$ by $L^{q_{|\alpha|}}_{loc}(\Omega\setminus S)$ where S is closed, of measure 0 and $\Omega \setminus S$ connected.

3 - For m = 2, Theorem 1 was obtained independently by Georgescu [8].

Idea of the proof : We rely heavily on Carleman inequalities, which are a usual tool in this context. Roughly speaking, they are continuity estimates for P in some weighted Sobolev spaces.

More precisely, let $\phi : \bar{\Omega} \to \mathbb{R}$ be a C^∞ function, with grad $\phi(x) \neq 0$, $x \in \Omega$ and which is strongly pseudo-convex (this is a convexity property relatively to the principal symbol of P). For instance, one can take quadratic ϕ's in the case of elliptic operators. Then, the following estimate holds (see [12]) : There exists K > 0 such that

$$\begin{cases} \sum_{|\alpha|\leq m} \tau^{2(m-|\alpha|)} \int_\Omega |D^\alpha u|^2 e^{2\tau\phi} dx \leq K\tau \int_\Omega |P_m(x,D)u|^2 e^{2\tau\phi} dx \\ \forall\, u \in C_o^\infty(\Omega),\quad \tau > 0 \text{ large enough} \end{cases} \tag{2.1}$$

If the non-principal coefficients of P(x,D) are locally bounded, one obtains the inequality (2.1) for P (it suffices to write $P_m(x,D) = P(x,D) - \sum_{|\alpha|\leq m-1} a^\alpha(x) D^\alpha$). Then, by a classical trick one first obtains uniqueness in the Cauchy problem (see [12], [32]) and property (U.C).

In the case where the non-principal coefficients belong only to some local L^p spaces, one can obtain a Carleman estimate for P from a Carleman estimate for P_m, but in a suitable L^r norm, $r > 2$. This motivates

Proposition 1 : *Let* $p_k = \frac{2q_k}{q_k - 2}$, $k = 0,\ldots, m-1$. *There exists* $K > 0$ *such that*

$$\left\{\begin{array}{l} \sum_{|\alpha|\leq m-1} \|D^\alpha u.e^{\tau\phi}\|_{L^{p_{|\alpha|}}} + \sum_{|\alpha|\leq m} \tau^{m-|\alpha|-\frac{1}{2}} \|D^\alpha u.e^{\tau\phi}\|_{L^2} \\ \\ \leq K\|P_m(x,D)\, ue^{\tau\phi}\|_{L^2} \qquad \forall u \in C_o^\infty(\Omega), \forall\tau > 0 \text{ large enough} \end{array}\right. \tag{2.2}$$

Proposition 1 is proved from (2.1) essentially by interpolation arguments (cf. [27]). The powers of τ in (2.1) are crucial here (see §3).

The final step is to prove (2.2) for P. This is carried out by Hölder inequality, for $u \in C_o^\infty(\Omega)$ with sufficiently small support. Then one gets uniqueness in the Cauchy problem by a classical argument (see [20]). □

Due to the fact that one has the L^2 norm of $P_m(x,D)ue^{\tau\phi}$ in (2.2), one cannot obtain $q_k < 2$ in Theorem 1. If the coefficients of P_m are *constant*, we can prove a Carleman inequality where the right-hand side of (2.2) is estimated in L , $1 < r < 2$. This allow us to treat the case of a perturbation of order 0 and to obtain

Theorem 2 :([27]). *Let* $a_o \in L^{2n/(2m-1)}_{loc}(\Omega)$ $2n > 2m-1$

$$a_o \in L^{1+\varepsilon}_{loc}(\Omega) \textit{ if } 2n \leq 2m-1 \; (\varepsilon > 0 \textit{ arbitrary})$$

Let $u \in W^{2,q}_{loc}(\Omega)$, $q \geq \text{Max}(1+\varepsilon, \frac{2n}{n+2m-1})$ *be a solution of* $P_m(D)u + a_o u = 0$.

If u *vanishes on a non-empty open set, then* u *vanishes identically in* Ω.

b) Operators with multiple complex characteristics.

As we mentioned in the Introduction, unique continuation is not necessarily true when the principal symbol of P has multiple complex characteristics. However, some positive results can be proved (see e.g. [9], [13], [16], [21], [22], [31], [33]), but the "sub-principal" coefficients have to be locally Lipschitz. One can then develop the technics of a) to derive (U.C) in the non-smooth case.

We shall just give 2 examples of operator having complex characteristics of multiplicity ≥ 3 (see [27] for other examples).

Let $A(x,D)$ be a uniformly homogeneous elliptic operator of order $m/3$, with coefficients in $C^{(2m/3)+1}(\bar{\Omega})$, and having simple complex characteristics. Let $B(x,D)$ be a homogeneous differential operator of order $m-1$, having locally Lipschitz coefficients.

We set $P(x,D) = A(x,D)^3 + B(x,D) + \sum_{|\alpha|\leq m-2} a^{\alpha}(x)D^{\alpha}$.

Theorem 3 : *There exist* q_k, $2 \leq q_k < \infty$, $k = 0,\ldots, m-2$ *such that if* $a^{\alpha} \in L^{q_{|\alpha|}}_{loc}(\Omega)$, $|\alpha| = 0,1,\ldots, m-2$, *then* $P(x,D)$ *satisfies* (U.C).

Remark : As in Theorem 1, the real q_k are defined by complicated formulas. For instance,

$$q_o = \mathrm{Max}(2, \frac{2n}{2m-3}),\ q_1 = \mathrm{Max}(2, \frac{2n}{2m-7}),\ q_2 = \mathrm{Max}(2, \frac{2n(2m-3)}{(2m-7)(2m-9)}).$$

The proof of Theorem 3 uses the Carleman estimates given by Watanabe [33]. □

Let now $P_m(D)$ be a homogeneous uniformly elliptic operator of order m, with constant coefficients, having complex characteristics of multiplicity

$r > 3$ (e.g. $P_m = (\Delta)^r$ with $m = 2r$).

Let $Q(x,D)$ be a homogeneous differential operator of order $m - [(r+1)/2] + 1$, having locally Lipschitz coefficients.

We set $P(x,D) = P_m(D) + Q(x,D) + \sum_{|\alpha| \leq m - [\frac{r+1}{2}]} a^{\alpha}(x) D^{\alpha}$.

Theorem 4 : *There exist* q_k, $2 \leq q_k < \infty$, $k = 0, \ldots, m - [\frac{r+1}{2}]$ *such that if* $a^{\alpha} \in L^{q_{|\alpha|}}_{loc}(\Omega)$, $|\alpha| = 0, 1, \ldots, m - [\frac{r+1}{2}]$, *then* $P(x,D)$ *satisfies* (U.C).

Remark. $q_o = \text{Max}(2, \frac{2n}{2m-r})$, $q_1 = \text{Max}(2, \frac{2n}{2m-2r-1})$, $q_2 = \text{Max}(2, \frac{2m(2m-r)}{(2m-2r-1)(2m-2r-3)})$

Here again, we use estimates of Watanabe [33].

3. SCHRÖDINGER OPERATORS.

Here $Pu = -\Delta u + V(x)u$ in $\mathbb{R}^n$.

Theorem 5 :

1°) $n = 2,3,4$. *Let* $V \in L^q_{loc}(\mathbb{R}^n)$, $q > \frac{n}{2}$. *Then* (U.C) *holds for* P *if* $u \in W^{2,1}_{loc}(\mathbb{R}^n)$, $q \leq 2$ *or* $u \in W^{2,2q/(q+2)}_{loc}(\mathbb{R}^n)$, $q \geq 2$.

2°) $n \geq 5$. *Let* $V \in L^{\frac{2n-1}{3}}_{loc}(\mathbb{R}^n)$. *Then* (U.C) *holds for* $u \in H^2_{loc}(\mathbb{R}^n)$. *Here one can have first-order terms with coefficients in* $L^{2n-1}_{loc}(\mathbb{R}^n)$.

Remark 1. Part 1 is due to Amrein-Berthier-Georgescu [1]. Part 2 is due to Georgescu [8]. Previous weaker results were due to Berthier [3] who obtained 1°) for $n = 3$, and to Schechter-Simon [30] who proved (U.C) for $V \in L^q_{loc}(\mathbb{R}^n)$, $q > \text{Max}(n-2, \frac{2n-1}{3})$.

Idea of the proofs :

1°) It is an improvement of the method of Schechter-Simon [30], which itself was based on the ideas of Heinz [10] : one shows an inequality of the type

$$\| |x|^k f \|_{L^p} \le c \| |x|^k \Delta f \|_{L^q} \qquad k = 0, \pm 1, \pm 2, \ldots, \tag{3.1}$$

By expanding f in spherical harmonics it suffices to prove the correspon-ding inequality in dimension 1. Heinz proved such an inequality for $p = q = 2$, so he could only add a perturbation with locally bounded coefficients (see part 2 a)). Amrein-Berthier-Georgescu, by a refinement of Heinz - Simon-Schechter arguments, prove the following :

Proposition 2 : *Let* $1 \le q \le 2 \le p < \infty$, $\frac{1}{\omega} = \frac{1}{q} - \frac{1}{p}$, $\mu = 2 - \frac{n}{\omega}$. *Let us suppose* $\omega > \frac{n}{2}$. *Then,* $\forall \tau \in \mathbb{R}$ *and* $\forall f \in W^{2,q}(\mathbb{R}^n \setminus \{0\})$ *with a compact support, one has*

$$\| |x| f \|_{L^p} \le c(\tau) \| |x|^{\tau+\mu} \Delta f \|_{L^q} \tag{3.2}$$

$c(\tau)$ *is a complicated constant, defined by a series.* □

The proof lies heavily on spherical harmonics.

2°) The proof of Georgescu is based on the following observation : the Carleman estimates of Hörmander are optimal in view of the powers of τ. (This is essentially in Hörmander's book [12]).

But one can improve them in *one and only one* direction. Let us recall a special case of (2.1) :

$$\tau^{1/2}\|e^{\tau\phi}\frac{\partial u}{\partial x_i}\|_{L^2} \le C\|e^{\tau\phi}\Delta u\|_{L^2} \quad \forall u \in C_o^\infty(\Omega) \quad \forall\tau >> 1. \qquad (3.3)$$

The following lemma says that one can improve the power of τ in (3.3) only for derivatives in a determined direction :

Lemma 1 ([8]) : *Let* Q *be a first-order differential operator in* Ω *with continous and bounded coefficients.*

If there exists $\mu > \frac{1}{2}$ *such that for every* $u \in C_o^\infty(\Omega)$,

$$\tau^{\mu}\|Q(e^{\tau\phi}u)\|_{L^2} \le c\|e^{\tau\phi}\Delta u\|_{L^2} \qquad \forall\tau >> 1,$$

then there exists $\lambda : \Omega \to \mathbb{R}$ *such that*

$$Q = \sum_{j=1}^{n} \lambda(x)\frac{\partial\phi}{\partial x_j} D_j \quad + \text{operator of order } 0$$

Moreover, if $\mu > 1$, *then* $\lambda = 0$. □

A similar lemma holds for Q a second-order differential operator.

Then Georgescu constructs such a first-order operator :

$$Q = p = \frac{1}{2}\left(\frac{\text{grad }\phi}{|\text{grad }\phi|} D + D\frac{\text{grad }\phi}{|\text{grad }\phi|}\right) = \frac{1}{2}\sum_{i=1}^{n}\left(\frac{\partial_i\phi}{|\text{grad }\phi|} D_i + D_i\frac{\partial_i\phi}{|\text{grad }\phi|}\right)$$

where $D_i = -i\frac{\partial}{\partial x_i}$.

Under suitable conditions on ϕ, he obtains the improved Carleman inequality :

$$\begin{cases} \tau^{3/2-s} \|e^{\tau\phi}u\|_{H^s} + \|p^2(e^{\tau\phi}u)\|_{L^2} + \tau\|p(e^{\tau\phi}u)\|_{L^2} \\ \qquad + \sum_{i=1}^{n} \|p\, D_j(e^{\tau\phi}u)\|_{L^2} \le c\, \|e^{\tau\phi}\Delta u\|_{L^2} \qquad (3.4) \\ \forall s \in [0,2] , \quad \forall\tau >> 1, \quad \forall u \in H^s(\Omega) \text{ with compact support.} \end{cases}$$

(Note that the left-hand side of (3.4) contains a second-order and a first term with τ to the power 0 and 1 respectively ; in (2.1) these powers are $-\frac{1}{2}$ and $\frac{1}{2}$).

In fact, Georgescu uses radial ϕ : $\phi = \phi(|x|)$; then p becomes the radial momentum $p = \frac{1}{2}(\frac{x}{|x|} D + D \frac{x}{|x|})$.

The final step uses (3.4) and an anisotropic Sobolev inequality involving p.

In all this proof, the fact that the laplacian is rotation-invariant is crucial, so it doesn't seem to generalize to arbitrary second-order equations.

4. COMMENTS AND OPEN QUESTIONS.

1. It seems to be a difficult problem to find the minimal assumptions on the coefficients in order that (U.C) hold.

For Schrödinger operator one expects (U.C) to be true if the potential V belongs to $L^q_{loc}(\mathbb{R}^n)$, $q > \frac{n}{2}$. In fact, this is true for $n \le 4$ (see §3). In any case no counter examples are known.

2. The problem of unique continuation for elliptic *systems* is a very challenging one. Very few are known - even when the coefficients are smooth - apart from some special cases we discuss now.

i) *Systems with diagonal principal part :* If the "diagonal" operators satisfy (U.C) and the lower terms have the right order (cf. §2), one is immediately reduced to the scalar case. The linearized reaction-diffusion systems belong to this category.

ii) *Dirac operator :* Let Ω be a connected open set in $\mathbb{R}^3$; $D_j = -\frac{\partial}{\partial x_j}$, $\vec{D} = (D_1, D_2, D_3)$, $\vec{\alpha} = (\alpha_1, \alpha_2, \alpha_3)$ where α_j are complex 4×4 self-adjoint matrices such that $\alpha_i \alpha_j + \alpha_j \alpha_i = 2\delta_{ij}$.

For $\psi : \Omega \to \mathbb{C}^4$, one set $\mathfrak{D}\psi = \sum_{i=1}^{3} \alpha_i D_i \psi$.

Let $V : \Omega \to \mathbb{R}_+$, $V \in L^5_{loc}(\Omega)$. Then $\mathfrak{D} + V$ satisfies (U.C). This result is due to Berthier-Georgescu [4]. They improve an earlier work of Hile and Protter [11] where V was supposed to be locally bounded. The idea is to use the identity $\mathfrak{D}^2 = \Delta$.

Unique continuation for some generalizations of Dirac system are given in [34].

iii) *Systems whose determinant of principal symbols has simple complex characteristics :* For these systems one can get Carleman estimates following an idea of Hörmander ([12]) : to use the cofactor matrix and to reduce to a Carleman estimate for the determinant of principal parts. Unfortunately, this method doesn't work for classical systems such as Stokes or Lamé.

iv) Derivation of Carleman estimates for systems of mixed order (e.g. Stokes) is an open question. However, one can obtain uniqueness of the (reduced) Cauchy problem for perturbations of Stokes system by some first-order operator with smooth coefficients (cf. [28]).

v) We have recently obtained (U.C) for some elliptic systems of the Lamé type. This will appear in a forthcoming paper.

3. It would be desirable to weaken the smoothness hypothesis in other continuation results, e.g. unique continuation for parabolic equations (Mizohata [17]) or coupled elliptic-parabolic systems (Matsumoto [15]). This could be of interest in control theory (cf. [14]).

REFERENCES.

[1] W.O. Amrein, A.M. Berthier, V. Georgescu, An L^p inequality for the laplacian and unique continuation, Ann. Inst. Fourier 31, 3 (1981), 153-168.

[2] N. Aronszajn, A unique continuation theorem for solutions of elliptic partial differential or inequalities of second order. J. Math. Pures Appl. 36 (1957), 235-249.

[3] A.M. Berthier, Sur le spectre ponctuel de l'opérateur de Schrödinger. C.R. Acad. Sci. Paris 290 (1980), 393-395.

[4] A.M. Berthier, V. Georgescu, Sur la propriété de prolongement unique pour l'opérateur de Dirac. C.R. Acad. Sci. Paris 291 (1980), 603-606.

[5] T. Carleman, Sur un problème d'unicité pour les systèmes d'équations aux dérivées partielles à deux variables indépendantes. Ark. Mat. Astr. Fys. 26 B, 17 (1939), 1-9.

[6] P. Cohen, The non-uniqueness of the Cauchy problem. Office of Naval Research Technical Report 93, Stanford, 1960.

[7] C. Dafermos, On the existence and the asymptotic stability of solutions to the equations of linear thermoelasticity. Arch. Rat. Mech. Analysis 29 (1968), 241-271.

[8] V. Georgescu, On the unique continuation property for Schrödinger hamiltonians. Helv.Phys. Acta (1980).

[9] P.M. Goorjian : The uniqueness of the Cauchy problem for partial differential equations which may have multiple characteristics. Trans. Amer. Math. Soc. 146 (1969), 493-509.

[10] E. Heinz, Über die Eindeutigkeit beim Cauchysches. Anfangswertproblem einer elliptischen Differentialgleichung zweiter Ordnung. Nachr. Akad. Wiss. Göttingen Math. Phys, Kl II a, n°1 (1955), 1-12.

[11] G.N. Hile, M.H. Protter, Unique continuation and the Cauchy problem for first order systems of partial differential equations. Comm. in P.D.E., 1 (5), (1976), 437-465.

[12] L. Hörmander, Linear Partial Differential Equations. Springer Verlag. Berlin, Heidelberg, New-York (1969).

[13] L. Hörmander, On the uniqueness of the Cauchy problem II, Math. Scand. 7 (1955), 177-190.

[14] J.L. Lions, Cours au Collège de France, Automne 1981.

[15] W. Matsumoto, Une remarque sur l'unicité du prolongement des solutions pour les systèmes mixtes de type parabolique et de type elliptique dégénérés. C.R. Acad. Sc. Paris, 292, (1981), 665-668.

[16] S. Mizohata, Unicité du prolongement des solutions des équations elliptiques du quatrième ordre. Proc. Japan Acad. 34 (1958), 687-692.

[17] S. Mizohata, Unicité du prolongement des solutions pour quelques opérateurs différentiels paraboliques. Memoires of the College of Science, University of Kyoto, Series A, Vol XXXI, Mathematics, 3 (1958), 219-239.

[18] C. Müller, On the behavior of the solutions of the differential equation $\Delta u = F(x,u)$ in the neighborhood of a point, Comm. Pure Appl. Math. 7 (1954), 505-515.

[19] L. Nirenberg, Uniqueness in Cauchy problem for differential equations with constant leading coefficients. Comm. Pure Appl. Math. 10 (1957), 89-105.

[20] L. Nirenberg, Lectures on linear partial differential equations. C.B.M.S. Regional Conference, (1973).

[21] R. Pederson, On the unique continuation theorem for certain second and fourth order elliptic equations. Comm. Pure Appl. Math. 11 (1958) 67-80.

[22] R. Pederson, Uniqueness in Cauchy's problem for elliptic equations with double characteristics. Ark. Mat. 6 (1966), 535-549.

[23] A. Plis, A smooth linear elliptic differential equation without any solution in a sphere. Comm. Pure Appl. Math. 14 (1968), 599-617.

[24] A. Plis, On non-uniqueness in Cauchy problem for an elliptic second order differential equation. Bull. Acad. Pol. Sci. Sér. Sci. Math. Astr. Phys. II, 3 (1963), 95-100.

[25] M. Reed, B. Simon, Methods of Modern Mathematical Physics. IV. Analysis of Operators. Academic Press. New-York. (1978).

[26] J.C. Saut, B. Scheurer, Un théorème de prolongement unique pour des operateurs elliptiques dont les coefficients ne sont pas localement bornés. C.R. Acad. Sci. Paris, 290 (1980), 595-598.

[27] J.C. Saut, B. Scheurer, Sur l'unicité du problème de Cauchy et le prolongement unique pour des équations elliptiques à coefficients non localement bornés, J. Diff. Eq. 43, 1 (1982), 28-43.

[28] J.C. Saut, R. Temam, Generic properties of Navier-Stokes equations : genericity with respect to the boundary values. Indiana J. Math. 29, (1980), 427-446.

[29] J.C. Saut, R. Temam, Generic properties of nonlinear boundary value problems. Comm. P.D.E., 4, 3 (1979), 293-319.

[30] M. Schechter, B. Simon, Unique continuation for Schrödinger operators with unbounded potentials. J. Math. Anal. Appl. 77 (1980), 482-492.

[31] M. Sussman, On uniqueness in Cauchy problem for elliptic partial differential operators with characteristics of multiplicity greater than two. Tôhoku Math. J. 29 (1977), 165-188.

[32] M. Taylor, Pseudo differential operators. Princeton University Press, Princeton (1981).

[33] K. Watanabe, On the uniqueness of the Cauchy problem for certain elliptic equations with triple characteristics. Tôhoku Math. J. 23 (1971), 473-490.

[34] N. Weck, Unique continuation for some systems of partial differential equations, Appl. Anal., 13, (1982), 53-63.

[35] A. Calderon, Uniqueness in the Cauchy problem of P.D.E., Amer. J. Math. 80, (1958), 16-36.

Jean Claude SAUT

Université Paris VII
Département de Mathématiques
4, Place Jussieu

75230 PARIS CEDEX 05
FRANCE

B. SCHEURER

CEA - Service MA
BP. 27

94190 VILLENEUVE ST-GEORGES
FRANCE

M SCHATZMAN

Spatial structuring in a model in neurophysiology

1. INTRODUCTION

The model considered in this paper arises in the neurophysiology of vision, and the phenomenon which it attempts to explain is interesting for its own sake, and needs a short description.

The cortex is the external envelope of the brain; if one could spread out its folds, it would appear as a large and thin sheet of tissue. One distinguishes on the basis of anatomical criteria several successive layers in the cortex, numbered from I to VI, from the outside to the inside of the brain.

The cells of the retina project (indirectly) onto the primary visual cortex, also called striate cortex, and mainly on layer IVc. By indirect projection we mean that axons coming from the retina establish contacts called synapses with cells in a first nucleus which in turn send axons to the cortical cells, and that signals are transferred through this pathway.

It has been observed, both by electrophysiological recording and by anatomical tracing methods, that the cortex of the adult animal is organized in ocular dominance regions, i.e. regions where neurones are activated by stimuli through either eye. This phenomenon is particularly clear in layer IVc and corresponds indeed to a segregation between the projections of the two eyes.

An ocular dominance region occupies all the depth of the layer; the planar pattern of these regions is made out of alternating stripes of approximately constant width; these stripes run generally locally parallel, with some occurrence of dead ends and branching.

Segregation into ocular dominance regions take place in some early phase of the development, which can be post-natal or embryonic, according to the species considered, starting from an initial state where projections are uniformly distributed, with, of course, some small fluctuations.

In a paper on the formation of ocular dominance stripes [11] , N.V. Swindale introduces an integro-differential system which is built on the following assumption : the growth of synapses of one kind, at a given point in cortex, is governed by the density of the neighboring synapses of both kinds in the following fashion : the growth is activated at short distances, and inhibited at larger distances by the presence of synapses of the same kind; it is inhibited at short distances, and activated at larger distances by the presence of synapses of the opposite kind.

This yields a linear convolution system, where the unknowns are the densities of synapses of either kind, u^L and u^R , which would not be very interesting because the asymptotic behavior of such a system would be given purely by spectral theory; according to the initial data, the solution would remain constant, go to zero or to infinity.

Therefore, N.V. Swindale introduces a non-linearity, which is devised to maintain the densities between 0 and some bounds N^N, N^R .

Last, it is sensible to consider only functions which vanish outside some set Ω; if 1_Ω is equal to 1 in Ω, and to zero outside of Ω, and if $*$ denotes the spatial convolution, Swindale's system can be written as

$$u_t^L = (w^{LL} * u^L + w^{LR} * u^R)(N^L - u^L)u^L . 1_\Omega$$

$$u^R_t = (w^{RL} * u^L + w^{RR} * u^R)(N^R - u^R)u^R.1_\Omega \quad \text{for} \quad x \in \Omega \quad \text{and} \quad t \geq 0 \quad (1.2)$$

with initial conditions

$$u^L(x,0) = u^L_o(x), \quad x \in \Omega \ ; \quad u^L(x,0) = 0 \ , \quad x \notin \Omega \ ; \qquad (1.3)$$

$$u^R(x,0) = u^R_o(x), \quad x \in \Omega \ ; \quad u^R(x,0) = 0 \ , \quad x \notin \Omega \ . \qquad (1.4)$$

As the nonlinearity of (1.1)-(1.2) is only meant to keep u^L, u^R in some convex set, it is possible as well to replace (1.1)-(1.2) by a system with unilateral constraints, which can be written in the formalism of maximal monotone operators [1] as

$$u^L_t - (w^{LL} * u^L + w^{LR} * u^R) + \beta^L(u^L) \ni 0 \qquad (1.5)$$

$$u^R_t - (w^{RL} * u^L + w^{RR} * u^R) + \beta^R(u^R) \ni 0 \qquad (1.6)$$

Here β^L is a multivalued maximal monotone graph in $\mathbb{R} \times \mathbb{R}$ defined by

$$\beta^L(r) = \begin{cases} \emptyset & \text{if} \quad r < 0 \quad \text{or} \quad r > N^L \\ \mathbb{R}^- & \text{if} \quad r = 0 \\ \{0\} & \text{if} \quad r \in (0, N^L) \\ \mathbb{R}^+ & \text{if} \quad r = N^L \end{cases} \qquad (1.7)$$

with a similar definition for β^R, where N^L is replaced by N^R.

Notice that (1.5) can be interpreted as follows : set

$$f^L = (w^{LL} * u^L + w^{LR} * u^R)1_\Omega$$

Then the right derivative in time of u^L is given by

$$\frac{\partial u^L}{\partial t}(x,t+0) = \begin{cases} f^L & \text{if} \quad 0 < u^L(x,t) < N^L \\ f^L & \text{if} \quad u^L(x,t) = 0 \text{ and } f^L \geq 0 \\ 0 & \text{if} \quad u^L(x,t) = 0 \text{ and } f^L < 0 \\ f^L & \text{if} \quad u^L(x,t) = N_L \text{ and } f^L \leq 0 \\ 0 & \text{if} \quad u^L(x,t) = N_L \text{ and } f^L > 0 \end{cases}$$

a.e. on $\Omega \times [0,\infty)$; and we have of course a similar interpretation for (1.6) This interpretation is proved in section 2. N.V. Swindale has performed a number of numerical experiments on (1.1) - (1.4), which he reports in [11], and he obtains the following behavior after a large number of iterations : stripes of approximately constant width appear; on each of these stripes, the value of u^L is either 0 or N^L , and the value of u^R is either 0 or N^R ; if some symmetry is assumed, the value of (u^L,u^R) is either $(N^L,0)$ or $(0,N^R)$; the stripes run normally into the boundary of Ω and narrow at branch points.

With an appropriate symmetry hypothesis, to be explained later, system (1.1) - (1.2) reduces to

$$\frac{\partial u}{\partial t} = (w * u)\,(1-u^2)1_\Omega \tag{1.9}$$

and (1.5)-(1.6) analogously to

$$\frac{\partial u}{\partial t} - w * u + \beta(x) \ni 0 \tag{1.10}$$

with $\beta(u) = 0$ if $u \in (-1,1)$, $\beta(-1) = \mathbb{R}^-$, $\beta(+1) = \mathbb{R}^+$.

The purpose of this paper is to investigate mathematically systems (1.1) - (1.2) and (1.5) - (1.6) and to present some numerical experiments*

---

* essentially due to E. Bienenstock, Neurobiologie du développement, Bât. 440, Université Paris-Sud, 91405 Orsay.

on one-dimensional configurations, which display some striking features of the asymptotic behavior of these systems.

From the mere neurophysiological point of view, the assumptions of Swindale might be thought of as over-simplifying; even if this were the case, this system retains two interesting features; the first one is that it can be used as an elementary trick for more sophisticated systems; the second one is that ideas pertaining to the same general frame appear in different contexts, for instance.

- a series of 3 papers by Gates and Penrose [2,3,4] , relative to the Van der Waals limit for classical systems, involves a variational principle with a convolution term plus a local term;
- a work [9] by G. Ruget extends some of the aspects of [2,3,4];
- an experimental work by J. Herault and C. Jutten* which consists in performing pattern recognition by observing the asymptotic state of a system which is slightly different from (1.10); the constraint is one-sided, and is modulated by the pattern to be recognized.
- models possessing the same self-organizing features have been studied by Manneville, Pommeau and Zaleski, with the object of understanding the formation of Taylor rolls in hydrodynamics [6,7,8].

2. EXISTENCE, UNIQUENESS, REGULARITY, GLOBAL BOUNDS

Let Ω be a measurable subset of $X = \mathbb{R}^N$ or $X = \mathbb{T}^N = (\mathbb{R}/Z)^N$, the N-dimensional torus.

---

* Laboratoire "Traitement d'images et Reconnaissance de formes", Enserg, 23, rue des Martyrs, 38031 Grenoble Cedex

The existence theory of the Cauchy Problem for (1.1) - (1.2) demands very mild hypotheses, as follows

$$w^{LL}, w^{LR}, w^{RL} \text{ and } w^{RR} \text{ belong to } L^1(X) \tag{2.1}$$

$$\begin{cases} u_o^L \text{ and } u_o^R \text{ are measurable, vanish a.e. on } \Omega^c \\ \text{and} \\ 0 \le u_o^L \le N^L \ , \ 0 \le u_o^R \le N^R \ , \text{ a.e. on } \Omega \end{cases} \tag{2.2}$$

Denote

$$\mathcal{U} = \{U = (u^L, u^R) \in L^\infty(X)^2 / U = 0 \ \text{ a.e. on } \Omega^c\} \tag{2.3}$$

where Ω^c is the complement of Ω, and let

$$\begin{cases} f^L(U) = (w^{LL} * u^L + w^{LR} * u^R) 1_\Omega \\ \\ f^R(U) = (w^{RL} * u^L + w^{RR} * u^R) 1_\Omega \ . \end{cases} \tag{2.4}$$

We define an operator T on $\mathcal{U}$ by

$$T(U) = \begin{pmatrix} f^L(U)(N^L - u^L)u^L) \\ \\ f^R(U)(N^R - u^R)u^R) \end{pmatrix} \tag{2.5}$$

Under assumption (2.1), T is continuously differentiable and locally Lipschitz continuous from $\mathcal{U}$ equipped with the norm topology to itself.

Theorem 2.1. *Assume* (2.1) *and* (2.2). *Then, there exists a unique* U *in* $C^1([0,\infty);\mathcal{U})$ *such that*

$$\frac{\partial U}{\partial t} = T(U), \qquad U(0) = \begin{pmatrix} u_o^L \\ u_o^R \end{pmatrix}$$

and, for all t

$$0 \leq u^L(.,t) \leq N^L \ , \quad 0 \leq u^R(.,t) \leq N^R \ \text{ a.e. on } \Omega \tag{2.7}$$

Moreover, there exists a null set M *such that*

$$\begin{cases} t \mapsto U(x,t) \ \textit{is continuously differentiable} \\ \textit{with respect to } t, \textit{ uniformly on } M^c \times [0,\infty) \end{cases} \tag{2.8}$$

Proof. According to the properties of T pointed out previously, there exists obviously a solution on an interval $[0,\tau(\|U_o\|))$. To be able to apply a Gronwall's lemma, in order to prove (2.7), we apply the regularity properties of processes described in [5, Proposition III. 5.2]. Then, there exists a null set M_o in Ω such that for x in M_o^c the trajectories $t \mapsto U(x,t)$ and $t \mapsto \frac{\partial U}{\partial t}(x,t)$ are continuous. If we denote by M_1 a null set such that $0 \leq u^L(x) \leq N^L$ for $x \in M_1^c$, then, a Gronwall inequality on $\min(u^L(x,.),0)$ and on $\max(u^L(x,.),N^L)$ shows that
$0 \leq u^L(x,t) \leq N^L$ for $x \in M_o^c \cap M_1^c$ and $t \in [0,\tau(\|U_o\|))$.
Similarly, there exists a null set M_2 such that
$0 \leq u^R(x,t) \leq N^L$ for $x \in M_o^c \cap M_2^c$ and $t \in [0,\tau(\|U_o\|))$.
These two estimates show that $\tau(\|U_o\|)$ is arbitrarily large, and that (2.7) holds for all time. Moreover, (2.8) holds on the semi-infinite interval. The uniqueness is clear. Details are given in [10, Theorem 3] .

A similar result holds for (1.5)-(1.6). For a correct formulation, let

$$K = [0,N^L] \times [0,N^R] \tag{2.9}$$

and, for U in $\mathbb{R}^2$

$$j(U) = \begin{cases} 0 & \text{if } U \in K \\ +\infty & \text{if } U \notin K \end{cases} \tag{2.10}$$

According to [1, Example 2.1.4] we define the subdifferential of the lower semi-continuous function j by

$$\partial j(U) = \{Z \in \mathbb{R}^2 / j(U+Y) - j(U) \geq Z.Y \ , \ \forall\, Y \in \mathbb{R}^2\} \tag{2.11}$$

Then, we know that if $V = (v^L, v^R) \in K$ and $F = (f^L, f^R)$ are given, the projection of 0 on the convex set $-\partial j(V) + F$ is given by

$$(-\partial j(V) + F)^o = (y^L, y^R) \tag{2.12}$$

with

$$\begin{cases} y^L = \begin{cases} f^L & \text{if } u^L \in (0, N^L) \text{ or if } u^L = 0, f^L > 0 \\ & \text{or if } u^L = N^L \ , \ f^L < 0 \\ 0 & \text{in the other cases} \end{cases} \\ y^R = \begin{cases} f^R & \text{if } u^R \in (0, N^R) \text{ or if } u^R = 0, f^R > 0 \ , \\ & \text{or if } u^R = N^R, f^R < 0 \\ 0 & \text{in the other cases} \end{cases} \end{cases} \tag{2.13}$$

Finally, we let $\mathcal{K} = \{U \in \mathcal{V} / U(x) \in K \text{ a.e. on } \Omega\}$.

We have the following theorem :

Theorem 2.2. *Assume* (2.1) *and* (2.2) *; then, there exists a unique* U *, which is Lipschitz continuous from* $\mathbb{R}^+$ *to* $\mathcal{V}$ *and satisfies*

$$\left(\frac{\partial U}{\partial t} - F(U), W - U\right) + \int_\Omega (j(W) - j(U))dx \geq 0 \quad \forall\, W \in \mathcal{V} \tag{2.14}$$

$$U(.t) \in \mathcal{K} \quad \text{a.e. on } \Omega \ . \tag{2.15}$$

Moreover, there exists a null set M *such that on* $M^c \times [0,\infty)$

$$t \mapsto U(x,t) \text{ is uniformly Lipschitz continuous} \tag{2.16}$$

$$U(x,t) \in K \tag{2.17}$$

$$\frac{\partial^+ U}{\partial t}(x,t) = [F(U(x,t)) - \partial j(U(x,t))]^o \tag{2.18}$$

(as defined in (2.12)-(2.13)).

Proof. The existence of an U satisfying (2.14)-(2.15) is completely trivial in a Hilbert space setting, as (2.14) can be considered as a Lipschitz perturbation of a maximal monotone operator. But the statements (2.16)-(2.18) are more precise that the ones which can be obtained by this method. Therefore, we devise a different one, based on the study of the multivalued ordinary differential equation

$$\begin{cases} \frac{dU}{dt}(t) + \partial j(U(t)) \ni F(t) \\ U(0) = U_o \end{cases} \tag{2.19}$$

with U_o given in K, and F given in $L^2_{loc}([0,\infty);\mathbb{R})$.
We know from [1,thm 3.6] that (2.19) has a unique solution U such that

$$\begin{cases} \frac{dU}{dt} \in L^2_{loc}([0,\infty);\mathbb{R}^2) \\ U(t) \in K, \quad \forall\, t \\ \frac{d^+U}{dt}(t) = [F(t) - \partial j(U(t))]^o, \quad \text{a.e. in } t. \end{cases} \tag{2.20}$$

This last relation implies $\left|\frac{d^+U}{dt}(t)\right| \le |F(t)|$, a.e. in t.

We can prove by energy estimates that, if U is the solution of (2.19) with U_o replaced by $\hat{U}_o$, then

$$|U(t) - \hat{U}(t)| \leq \sqrt{2t}\left(\int_0^t |F(s)-\hat{F}(s)|^2 ds\right)^{1/2} . \qquad (2.21)$$

Denote the solution of (2.19) by $\Phi(U_o,F)$. Then, solving (2.14) is equivalent to solving

$$U(x,.) = \Phi(U_o(x),F(U)(x,.)) \qquad (2.22)$$

Thanks to (2.21), we can show that (2.22) has a unique solution : let $\Psi : U \to \Phi(U_o,F(U))$. Then, for T given and large enough n, $\Psi^n\big|_{[0,T]}$ is a strict contraction in $L^2(0;T;\mathcal{V})$.

The remainder of the theorem uses [5, Proposition III.5.2] and (2.20) □

3. ASYMPTOTIC BEHAVIOR

We shall restrict ourselves to a special case which nevertheless displays some very interesting features. Assume

$$N^R = N^L = N \qquad (3.1)$$

$$w^{LL} = w^{RR} = - w^{RL} = - w^{LR} = w \qquad (3.2)$$

Then, if

$$u_o^L + u_o^R = N \text{ a.e. on } \Omega , \qquad (3.3)$$

the solution of (1.1) - (1.2) satisfies for all $t \geq 0$

$$u^L(.,t) + u^R(.,t) = N \text{ a.e. on } \Omega . \qquad (3.4)$$

This assertion can be proved by letting

$$z = u^L + u^R$$
$$v = u^L - u^R ,$$

Then $z - N$ satisfies

$$\frac{\partial}{\partial t}\{z-N\} = (w * v)(N-z)v.1_{\Omega}$$

and by a Gronwall inequality, (3.3) implies (3.4).

After scaling u and w, we are reduced to the study of

$$u_t = (w * u)(1-u^2)1_{\Omega} \tag{3.5}$$

Analogously, under assumptions (3.1)-(3.3), the system (1.5)-(1.6) becomes

$$u_t - (w * u)1_{\Omega} + \beta(u) \ni 0 \tag{3.6}$$

where β is defined by

$$\beta(r) = \begin{cases} \emptyset & \text{if } |r| > 1 \\ \{0\} & \text{if } |r| < 1 \\ \mathbb{R}^- & \text{if } r = -1 \\ \mathbb{R}^+ & \text{if } r = +1 \end{cases}$$

To obtain some results on the asymptotic behavior of (3.5) and (3.6), we shall make the following assumptions on w and Ω .

$$\Omega \text{ is bounded} \tag{3.8}$$

$$w \text{ is even} \tag{3.9}$$

$$\left\{ \begin{array}{l} \text{if, for } u \text{ in } L^{\infty}(\Omega),\ w * u \text{ vanishes on a} \\ \text{set of positive measure, then } u = 0 \text{ a.e. on } \Omega . \end{array} \right. \tag{3.10}$$

Notice that assumption (3.10) is very strong, but is satisfed on a dense subset of $L^1(\Omega)$: in the case $X = \mathbb{R}^N$, it is enough to approximate w by the sequence w_n defined as follows :

$$\rho_n(x) = n^{+N} e^{-rn^2|x|^2}$$

$$w_n(x) = [w.1_{\{|x| \leq n\}}]*\rho_n ,$$

and it is quite easy to show that w_n satisfies (3.10), as $w_n * u$ is analytic. In the case $X = \mathbb{T}^N$, the approximation process is very close : first approximate w by a sequence $\tilde{w}_n$ such that no Fourier coefficient of $\tilde{w}_n$ vanishes, and then convolve $\tilde{w}_n$ with ρ_n .

We may now state several results on the asymptotic behavior of the solutions of (3.5). Let γu_o be the set $\{u(t)/t \in \mathbb{R}^+\}$, where $u(t)$ is the solution of (3.5) starting at u_o ; we recall that the ω-limit set of γu_o is defined, for a given topology $\mathcal{T}$ as

$$\omega^+ \gamma u_o = \bigcap_t \overline{\{\gamma u(t)\}}^{\mathcal{T}} ;$$

in other words, $\omega^+ \gamma u_o$ is the set of limit points of $\{u(t)\}_{t \geq 0}$, as t tends to infinity. We shall take here for $\mathcal{T}$ the weak $*$ topology of $L^\infty(\Omega)$, and we observe that the unit ball of $L^\infty(\Omega)$, equipped with this topology, and denoted by E is metrizable, and compact.

Finally, we let Φ be the functional defined by

$$\Phi(u) = \int_{\Omega\times\Omega} w(x-y)u(x)u(y)dxdy \equiv \int_\Omega (w * u)(x)u(x)dx. \qquad (3.11)$$

Clearly, Φ is continuous from E to $\mathbb{R}$.

Theorem 3.1. *Assume* (3.8)-(3.10), *and let* u_o *take its values in* $(-1,1)$, *for almost every* x *in* Ω . *Then either*

$$\omega^+ \gamma u_o = \{0\} \qquad (3.12)$$

$$\begin{cases} 0 \notin \omega^+ \gamma u_0 \text{ , and moreover } \Phi \text{ is constant positive on } \omega^+ \gamma u_0 \text{ , and} \\ \lim_{t\to\infty} \int_\Omega (1-|u^2(x,t)|)dx = 1. \end{cases} \tag{3.13}$$

Proof. As E is metrizable and compact, $\omega^+ \gamma u_0$ is not empty. Assume that (3.12) is not true; then, there exists a sequence t_n tending to infinity, such that

$$u_n \equiv u(.,t_n) \longrightarrow u_\infty \quad \text{in } E$$

and u_∞ is not null. Assumption (3.10) implies that

$$(w * u_\infty)(x) \neq 0 \quad \text{a.e. on } \Omega \tag{3.14}$$

and

$$w * u_n \rightarrow w * u_\infty \quad \text{in } C^0(\Omega) \text{ (with the uniform topology)} \tag{3.15}$$

On the other hand, we observe that

$$\begin{aligned} \frac{d}{dt} \Phi(u(t)) &= 2 \int_\Omega (w * u)(x,t)\, u_t(x,t)\, dx = \\ &= 2 \int_\Omega (1-u^2(x,t))(w * u)^2(x,t)dx. \end{aligned}$$

As we have seen that $u(x,t)$ stays always between -1 and 1, $-\Phi$ is a Liapunov function for (3.5), and if we let

$$\Psi(u) = \int_\Omega (1-u^2(x))\,(w * u)^2(x)dx,$$

we can see that

$$\int_0^t \Psi(u(s))ds \leq \frac{1}{2} \{ \Phi(u(t)) - \Phi(u(0))\} \leq \int_{\Omega\times\Omega} |w(x-y)|\, dxdy .$$

Therefore, $s \mapsto \Psi(u(s))$ is integrable on $[0,\infty)$; as one readily checks that $\Psi(u(s))$ is Lipschitz continuous with respect to s, uniformly on $[0,\infty)$, it follows that

$$\lim_{t\to\infty} \Psi(u(t)) = 0 . \tag{3.16}$$

In particular,

$$\lim_{n\to\infty} \int (1-u_n^2(x))(w * u_n)^2(x)\, dx = 0 ,$$

and, thanks to (3.14)-(3.15),

$$\lim_{n\to\infty} \int (1-u_n^2(x))\, dx = 0 . \tag{3.17}$$

Let us show now that for all t, and almost everywhere on Ω, $|u(x,t)| < 1$:

$$\frac{d}{dt} (1- |u(x,t)|) = - (u_t \operatorname{sgn} u)(x,t) =$$

$$= - [(w * u)(1+|u|) \operatorname{sgn} u] (x,t)(1- |u(x,t)|) \geq - K(1-|u|),$$

where $K \geq 2 \sup\{|w * u|_\infty / u \in E\}$. Then, by a Gronwall's inequality,

$$1 - |u(x,t)| \geq (1 - |u_o(x,t)|)e^{-Kt} ,$$

which proves our claim. If we multiply both sides of (3.5) by $u/(1-u^2)$, we obtain

$$- \frac{\partial}{\partial t} \{\frac{1}{2} \operatorname{Log}(1 - |u(x,t)|^2)\} = [(w * u)u] (x,t) ;$$

if we integrate this relation on $\Omega \times [0,t_n]$, we have

$$\int_\Omega \operatorname{Log}(1 - u_o^2(x))dx - \int_\Omega \operatorname{Log}(1 - u_n^2(x))dx = 2 \int_0^{t_n} \Phi(u(s))ds. \tag{3.18}$$

Let ℓ be the limit of $\Phi(u(s))$ as s tends to infinity; from (3.17), it follows that the left-hand side of (3.18) goes to $+\infty$ as n goes to infinity. As $\Phi(u(s)) \leq \ell$, for all s, this is possible if and only if ℓ is positive.

Therefore, as Φ is constant on $\omega^+ \gamma u_o$, 0 does not belong to $\omega^+ \gamma u_o$.

Finally, if there were a subsequence $t_m \to +\infty$, such that

$$\int_\Omega (1-|u(x,t_m)|^2)dx \geq \alpha > 0 ,$$

then, possibly after a new extraction,

$$u(t_m) \rightharpoonup \tilde{u}_\infty \qquad \text{as} \quad m \to \infty ,$$

and $\tilde{u}_\infty$ is not null. This yields a contradiction, by the argument used at the beginning of the proof of the present theorem. □

Any u such that $|u| = 1$ a.e. is an equilibrium point of (3.5). But not all of them are stable, even in the mildest sense !

Lemma 3.2. *Assume that* $|u_\infty| = 1$ a.e. *Then* u_∞ *is linearly stable for* (3.5) *only if*

$$u_\infty(w * u_\infty) \geq 0 \quad \text{a.e. } \textit{on } \Omega .$$

Proof. Let $u = u_\infty + v$; if we consider only the first-order terms in v in (3.5), we obtain

$$\frac{\partial v}{\partial t} = -2(w * u_\infty)u_\infty v ,$$

and the conclusion of the lemma is immediate. □

Proposition 3.3. *Under the hypotheses of Theorem* 3.1 , *assume that* $\omega^+ \gamma u_o$ *contains only one element* u_∞ , *which is not null. Then*

$$|u_\infty| = 1 \text{ a.e. } \textit{on } \Omega \textit{ and } u_\infty(w * u_\infty) \geq 0 \text{ a.e. } \textit{on } \Omega .$$

Proof. The assumption of Proposition 3.3. imply that $w * u(t) \to w * u_\infty$ in $C^0(\Omega)$ (with the uniform topology)

and

$$w * u_\infty \neq 0 \text{ a.e.}$$

If we divide both sides of (3.5) by $(1-u^2)$, integrate from 0 to t, we obtain

$$\tanh^{-1} u(x,t) - \tanh^{-1} u_0(x) = \frac{1}{2}\int_0^t \frac{u_t(x,s)}{1-u^2(x,s)}\, ds = \frac{1}{2}\int_0^t (w * u)(x,s)ds.$$

If $(w * u_\infty)(x) > 0$, then

$$\lim_{t\to\infty} \int_0^t (w * u)(x,s)ds = +\infty$$

so that

$$\lim_{t\to+\infty} u(x,t) = +1.$$

Similarly, if $(w * u_\infty)(x) < 0$,

$$\lim_{t\to+\infty} u(x,t) = -1,$$

which proves (3.19). □

Let us turn now to analogous results relative to Equation (3.6). To avoid confusion, we let $\tilde{\gamma} u_0$ be the set $\{u(t)/t \in \mathbb{R}^+\}$, where $u(t)$ is the solution of (3.6), with initial condition u_0. The definition of $\omega^+ \tilde{\gamma} u_0$ is the analogue of that of $\omega^+ \gamma u_0$.

Theorem 3.4. *Assume* (3.8)-(3.10), *and let* u_0 *take its values in* $[-1,1]$; *then, either*

$$\omega^+ \tilde{\gamma} u_o = \{0\} \tag{3.20}$$

or

$$\begin{cases} 0 \notin \omega^+ \tilde{\gamma} u_o , \quad \Phi \textit{ is a positive constant on } \omega^+ \tilde{\gamma} u_o , \\ \textit{and for all } u \textit{ in } \omega^+ \tilde{\gamma} u_o, \ |u| = 1 \text{ a.e.}, \textit{ and } u(w * u) \geq 0 \text{ a.e.} \end{cases} \tag{3.21}$$

Proof. Assume that (3.20) does not hold; then, there exists a sequence t_n tending to infinity, such that

$$u_n \equiv u(.,t_n) \rightharpoonup u_\infty \quad \text{in } E ,$$

and u_∞ is not null. Thanks to (3.10),

$$w * u_\infty \neq 0 \quad \text{a.e. on } \Omega , \tag{3.22}$$

and

$$w * u_n \to w * u_\infty \quad \text{in } C^o(\Omega) \text{ (equipped with the uniform topology)} \tag{3.23}$$

According to (2.18), the derivative from the right $\frac{\partial^+ u}{\partial t}$ satisfies

$$\begin{cases} \frac{\partial^+ u}{\partial t}(x,t) = (w * u)(x,t) \quad \text{if } |u(x,t)| < 1 , \\ \frac{\partial^+ u}{\partial t}(x,t) = -u(x,t)((w * u)(x,t)u(x,t))^- , \text{ if } |u(x,t)| = 1, \end{cases} \tag{3.24}$$

with $r^- = -\min(0,r)$, for all real r. We have

$$\frac{d^+}{dt} \Phi(u(t)) = 2 \int_\Omega (w * u)(x,t) \frac{\partial^+ u}{\partial t}(x,t)dx,$$

and therefore, according to (3.24),

$$\frac{d^+}{dt} \Phi(u(t)) = 2\tilde{\Psi}(u(t))$$

with

$$\widetilde{\Psi}(u) = \int_{|u|<1} (w * u)^2 dx + \int_{|u| = 1} ((w * u)^-)^2 dx .$$

Clearly

$$\int_0^{+\infty} \widetilde{\Psi}(u(s))ds < +\infty. \tag{3.25}$$

Let us show that

$$\lim_{t\to\infty} \widetilde{\Psi}(u(t)) < +\infty.$$

From (3.25), for almost every x in Ω, by Fubini's theorem,

$$\int_0^{+\infty} |\frac{\partial^+ u}{\partial t}(x,t)|^2 dt < +\infty.$$

Assume that there is a sequence $t_m \to +\infty$ such that

$$|\frac{\partial^+ u}{\partial t}(x,t_m)| \geq \alpha > 0 .$$

Then, as

$$|u(x,t) - u(x,s)| \leq \int_0^t |u_t(x,\tau)|d\tau \leq \int_0^t |(w * u)(x,\tau)| d\tau$$

$$\leq |w|_{L^1}|t-s| = K|t-s|$$

and

$$|(w * u)(x,t) - (w * u)(x,s)| \leq K^2|t-s| ,$$

there exists an interval $[t_m-\beta,t_m]$ or $[t_m,t_m+\beta]$ such that

$$|\frac{\partial^+ u}{\partial t}(x,t_m)| \geq \frac{\alpha}{2} . \tag{3.26}$$

To see this, assume, for example, that

$$- 1 < u(x,t_m) \le 0$$

$$\frac{\partial^+ u}{\partial t}(x,t_m) \ge \alpha \quad .$$

Then, on some interval $[t_m-\gamma, t_m+\gamma']$, $|u(x,t\ )| < 1$; on such an interval,

$$\frac{\partial^+ u}{\partial t}(x,t) = (w * u)(x,t)$$

and thus

$$\alpha - K^2|t-t_n| \le \frac{\partial u}{\partial t}(x,t) \le K$$

so that, for $t \ge t_m$

$$u(x,t_m) + \alpha(t-t_m) - \frac{K^2}{2}(t-t_m)^2 \le u(x,t) \le u(x,t_m) + K(t-t_m).$$

We can see that we can take

$$\gamma' = \min(2\alpha K^{-2}, K^{-1}) \ ,$$

and on $[t_m, t_m+\gamma']$, we have

$$\frac{\partial u}{\partial t}(x,t) \ge \alpha - K^2(t-t_m)$$

which is not smaller than $\frac{\alpha}{2}$ if

$$t - t_m \le \frac{\alpha}{2K^2} \quad .$$

Therefore, if we let

$$\beta = \min(\alpha K^{-2}/2, K^{-1}) \ ,$$

(3.26) holds on $[t_m, t_m+\beta]$. In another case, for instance, if $1 > u(x,t_m) \ge 0$, $\frac{\partial^+ u}{\partial t} \ge \alpha$, then (3.26) holds on $[t_m-\beta, t_m]$; the two other cases are treated analogously.

Assume now that $u(x,t_m) = -1$; then, necessarily

$$\frac{\partial^+ u}{\partial t}(x,t_m) \geq \alpha ,$$

because $\frac{\partial^+ u}{\partial t}(x,t) \geq 0$ on $\{t | u(x,t) = -1\}$.

This case is treated as the case $-1 < u(x,t_m) \leq 0$, $\frac{\partial^+ u}{\partial t}(x,t_m) \geq \alpha$. The case $u(x,t_m) = +1$ is treated similarly. As β does not depend on m, we have a contradiction. Therefore, for almost every x in Ω,

$$\lim_{m\to\infty} \left| \frac{\partial^+ u}{\partial t}(x,t) \right| = 0 .$$

As $\frac{\partial^+ u}{\partial t}$ is estimated by K on Ω, we may apply Lebesgue's theorem, and

$$\lim_{t\to\infty} \tilde{\Psi}(u(t)) \equiv \lim_{t\to\infty} \int_\Omega \left|\frac{\partial^+ u}{\partial t}(x,t)\right|^2 dx = 0 . \tag{3.27}$$

Let now T be a positive, given number; if we define, for t_n such that $u(.,t_n) \rightharpoonup u_\infty$,

$$v_n(x,t) = u(x,t_n+t) ,$$

then, for all $t \in [0,T]$, $(v_n(.,t))_{t\geq 0}$ is relatively compact in E, the unit ball of $L^\infty(\Omega)$ equipped with the weak * topology, and $(v_n)_n$ is equicontinuous with values in E. Therefore, by Ascoli-Arzela's theorem, the sequence $(v_n)_n$ is relatively compact in $C^0([0,T];E)$. Extract a converging subsequence, whose limit is denoted by v_∞; then clearly

$$v_\infty(0) = u_\infty .$$

On the other hand, from (3.27), $\frac{\partial^+ v_n}{\partial t} \to 0$ in $L^1([0,T];E)$ which implies that,

$$\frac{\partial v_\infty}{\partial t} = 0 \text{ in the sense of distributions,}$$

and therefore,

$$v_\infty(t) = u_\infty \quad \forall\ t \in [0,T] .$$

Then, all the sequence v_n converges to v_∞, which is independent from t.

Let now A_ε^+ be the set

$$A_\varepsilon^+ = \{x \,|(w * u_\infty)(x) \geq 2\varepsilon > 0\} .$$

Then, given $T > 0$, there exists a $n(T)$ such that, for all $n \geq n(T)$, for all x in A_ε^+ and t in $[0,T]$,

$$(w * v_n)(x,t) \geq \varepsilon > 0 .$$

Assume that $v_n(x,0) < + 1$; then, on some interval $[0,t_n(x))$, $v_n(x,t) < + 1$, and thus

$$\frac{\partial v_n}{\partial t}(x,t) = (w * v_n)(x,t) \geq \varepsilon ,$$

so that

$$v_n(x,t) \geq - 1 + \varepsilon t.$$

Therefore, if $T \geq 0$, and $n \geq n(\varepsilon^{-1})$, for all x in A_ε^+, there exists a $t_n(x)$ such that

$$v_n(x,t_n(x)) = + 1.$$

One checks easily that

$$\frac{\partial^+ v_n}{\partial t}(x,t) \geq 0 \qquad \forall\, t \geq 0\,, \qquad \forall\, x \in A_\varepsilon^+\,,$$

and thus

$$v_n(x,t) = +1 \qquad \forall\, t \in [t_n(x),T]\,.$$

It follows that

$$v_n(x,T) = +1 \quad \text{a.e. on} \quad A_\varepsilon^+\,,$$

and therefore,

$$u_\infty(x) = \lim_{n\to\infty} v_n(x,T) = +1 \quad \text{on} \quad A_\varepsilon^+\,.$$

As ε is an arbitrary positive number,

$$|u_\infty(x)| = 1 \quad \text{and} \quad u_\infty(w * u_\infty) \geq 0 \quad \text{a.e. on} \quad \Omega,$$

and moreover, as

$$\int |v_n(x,T) - v_n(x,0)|\,dx \leq \sqrt{T} \quad \int_{t_n}^{t_n+T} \tilde{\Psi}(u(.,s))ds]^{1/2}\,,$$

$u_n(.,t)$ converges to u_∞ in $L^p(\Omega)$ strong, for all $p < +\infty$.

The rest of the proof follows easily. □

Remark 3.5. If, under assumptions (3.8)-(3.10), we compare the properties of equations (3.5) and (3.6), we can see that the set of stationary solutions of (3.5) which is the set of u's such that $|u(x)| = 1$ a.e. on Ω, is much larger than the set of stationary solutions of (3.6), as this last set can be easily proved to consist of $\{0\}$ and the u's such that

$$|u(x)| = 1 \quad \text{and} \quad u(w * u) \geq 0 \quad \text{a.e. on} \quad \Omega\,.$$

But it turns out that the set of stable stationary solutions of (3.5) is the same as the set of stationary solutions of (3.6). We shall denote

$$L = \{u \in L^\infty(\Omega)/|u| = 1 \text{ a.e. and } u(w * u) \geq 0 \text{ a.e.}\}$$

4. ON THE SET OF NON-ZERO LIMIT POINTS

What is needed is sufficient information on L, as defined in (3.28); we can firstly point out that as $w * u$ is analytic, the set of changes of sign of $w * u$ (which is the same as the set of changes of sign of u) is a locally finite union of analytic surfaces, as well as the set of zeroes of $w * u$.

In particular, in the case $\Omega =]0,1[\subset \mathbb{R}$, or $\Omega = \mathbb{T}^1$, $w * u$ may have only a finite number of changes of sign. Call these $a_1,\ldots,a_p$; then up to a multiplication by -1,

$$u(a,x) = \begin{cases} +1 & \text{on } (a_{2j},a_{2j+1}) \\ -1 & \text{on } (a_{2j+1},a_{2j+2}) \end{cases} \tag{4.1}$$

where we may of course set $a_o = 0$, $a_{p+1} = 1$. In the case $\Omega = \mathbb{T}^1$, p is necessarily even.

From the definition (3.24) of L, we deduce that

$$(w * u(a,.)(a_k) = 0 \qquad 1 \leq k \leq p. \tag{4.2}$$

The system (4.2) is a system of p equations with p unknowns $a_1,\ldots,a_p$. The equations are analytic in the p unknowns.
I conjecture that the solutions of (4.2) are isolated in the case $\Omega = (0,1)$, and that, in the case $\Omega = \mathbb{T}^1$, the set of solutions is a disjoint union of circles in $(\mathbb{T}^1)^p \ni a$.

This conjecture does not immediately imply that L is not an infinite union of circles or of points, in the respective cases $\Omega = \mathbb{T}^1$ and $\Omega = (0,1)$; but, we are interested only in a subset of L where $\Phi(u)$ is a given non-zero constant c, because, as $t \to \infty$, $\Phi(u(t))$ converges to $c \neq 0$ if for one subsequence $u_n \rightharpoonup u_\infty$, with $u_\infty \neq 0$. Then, the number of changes of sign of $u \in L$ such that $\Phi(u) = c$ is bounded. Suppose not; then, there exists a sequence $(u^p)_p$ such that $u^p \in L$, u^p has exactly p changes of sign and $\Phi(u^p) = c$. Observe first that if,

$$n_p = \sup_x \{\text{number of zeroes } a_j \text{ in } [x-1/\sqrt{p}, x+1/\sqrt{p}]\},$$

then

$$\lim_{p\to\infty} n_p = +\infty .$$

Let x_p be a sequence such that there are n_p zeroes of $w * u_p$ in $[x_p-1/\sqrt{p}, x_p+1/\sqrt{p}] \cap \overline{\Omega}$. By extracting a subsequence, one may assume that

$$x_p \to x_\infty \in \overline{\Omega} , \quad u_p \rightharpoonup u \quad \text{in } L^\infty(\Omega) \text{ weak } * .$$

Then, for any $k \leq n_p-1$, there exists, by the mean value theorem a point b_k^p in $[x_p-1/\sqrt{p}, x_p+1/\sqrt{p}]$ such that

$$\frac{d^k}{dx^k}(w * u_p)(b_k^p) = 0 .$$

Clearly, for each k,

$$b_k^p \to x_\infty .$$

As $\frac{d^k}{dx^k}(w * u_p)$ converges uniformly on $\overline{\Omega}$, for all k, we can see that

$$\frac{d^k}{dx^k}(w * u_\infty)(x_\infty) = 0 , \quad \forall k ,$$

and therefore

$$w * u_\infty \equiv 0 ,$$

by analyticity, which implies $u_\infty = 0$ a.e., and yields a contradiction.

If the above-mentioned conjecture is true, it will follow that

$$L \cap \{u \ / \int_\Omega |w * u| \ dx = c > 0\}$$

is a finite set if $\Omega = (0,1)$, and a finite union of disjoint circles if $\Omega = \mathbb{T}^1$. In particular, it will follow that in the case $\Omega = (0,1)$, the ω-limit set of a trajectory of (3.6) consists of one point only.
In general, I conjecture that the ω-limit set of trajectories of (3.6) is always reduced to one point, and I would expect the same result to hold for (3.5); these look like very challenging problems.

5. NUMERICAL EXPERIMENTS

An important step forward would be to be able to relate the spatial periodicity of elements of L, to features of w, and possibly of Ω. If we consider (3.5), and linearize around zero, we have

$$u_t = (w * u) \, 1_\Omega \tag{5.1}$$

which has of course a very simple asymptotic behavior for infinite time, because the operator $A : u \to (w * u) \, 1_\Omega$ is selfadjoint and completely continuous from $L^2(\Omega)$ to itself. Therefore, if u_o has components along eigenfunctions of A with positive eigenvalues, $|u|_{L^2(\Omega)}$ will tend to $+\infty$ as time goes to infinity, and the components corresponding to the eigenvalue of largest modulus will be vastly predominant over any other one. Swindale argues in [11] that, if $\Omega = \mathbb{R}^N$ or $\mathbb{T}^N$, then, with the nonlinearity $1-u^2$, the components corresponding to the largest value of $\hat{w}$ will be reinforced, together with all its harmonics. This argument, sensible as it is, has not yet given rise to a general theorem.

Therefore, together with E. Bienenstock, we were led to a systematic numerical study of the asymptotic behavior of (3.5)&(3.8), for different kinds of initial data, of w, and of Ω, with a specific interest in the effect of the boundary of Ω.

Most of these experiments were performed in one dimension, the remaining ones in two dimensions.

The discretization for (3.5) is

$$u_j^{n+1} = \Big(\sum_{k\in K} w_{j-k}\, u_k^n \Delta x\Big)(1-(u_j^n)^2)\Delta t + u_j^n \tag{5.2}$$

which is explicit, and the discretization for (3.6) is

$$u_j^{n+1} = T\Big(u_j^n + \Delta t \sum_{k\in K} w_{j-k} u_k^n \,\Delta x\Big) \tag{5.3}$$

where T is defined by

$$T(r) = \max(-1,\min(1,r)),\ r \in \mathbb{R}.$$

Of course, (5.3) is implicit in the constraint $-1 \le u \le 1$. Note that in these two discretizations the parameter $\Delta x \Delta t$ should not be too large to avoid an oscillatory asymptotic behavior. Moreover, introducing (3.6) was not motivated by a theoretician's love for maximal monotone operators,but by the expectation that the discretization (5.3) would be faster converging and thus more interesting than the discretization (5.2), in the neighborhood of states $|u_j| = 1$, $j = 1,\dots,J$: from the mere form of (3.5), one cannot expect more than an exponential convergence of $u(t)$ to a limit.

Influence of w

If $\hat{w}(\xi) \leq 0$ a.e. (example : $w(x) = -(1-|x|)^+$), then, for arbitrary initial data, the asymptotic state is zero.
If $\hat{w}$ reaches a strict global maximum at zero (example : $w(x) = 1$ on $[-a,a]$, 0 elsewhere), the asymptotic state, for arbitrary initial data, consists of blocks of +1 and -1, with no obvious spatial frequency.
Therefore, we shall assume that the range of $\hat{w}$ contains positive points, and that the maximum of $\hat{w}$ occurs only for non-vanishing ξ.

Random initial data

A "random" initial data is a sequence of J numbers given by the random number generator of the machine, where J is the number of discretization points. Under the above assumption, the experimental limiting state for (3.5) or (3.6), either in $\mathbb{T}^1$ or in $]0,1[$, is a square function $u_\infty = \pm 1$, and the periodicity of this function (or approximate periodicity in the case $]0,1[$) corresponds to the value ξ_m of the dual variable for which $\hat{w}$ attains its positive global maximum.

Periodic initial data, and close to periodic initial data

If we give an initial data to the system, of the form

$$u_o(x) = \alpha v_o(x) + \beta r(x)$$

where v_o is a periodic square function taking the values +1 and -1, α is strictly less than 1, β will be "small", and r will be random, and will take values between -1 and +1.

Then we observe interesting facts if v_o does not have the "ground" period i.e. the period corresponding to ξ_m .

Assume first that $\Omega = \mathbb{T}^1$. Then, if $\hat{w}(\xi')$ is positive for ξ' corresponding to the period of v_o , and if β is smaller than some $\beta_o(\alpha, v_o)$, then, consistently, the limiting state is a periodic square function which has the period of v_o ; this phenomenon occurs in both cases (3.5) and (3.6).

Assume now $\Omega =]0,1[$. Then, in the case (3.5), no interesting observation can be made, as the dynamics is asymptotically too slow; but, in case (3.6), if we start with $\beta = 0$, then, during the first iterations, u seem to grow, retaining the same sign as v_o , but in a second step, which can take 10 times the number of iterations of the first one, we can observe that the pattern of $+1$ and -1 changes slowly, starting at the ends of the segment, to reach the kind of pattern obtained from a random initial data, with approximate ground frequency.

To interpret this reorganization phenomenon, let us first point out that no true periodicity exists on a segment, and that the approximate frequencies of the eigenfunctions of $u \to (w * u)1_\Omega$ cannot be expected to be rational multiple of the lowest of them. Therefore, one could think that there are much less local minima of the Lyapunov function Φ given by (3.15) in the case of a segment than in the case of torus, or that the attraction basin of a square function which does not have the right period is much thinner in the case of a segment than in the case of a torus. Thus, the rounding error in the numerical computation would be enough to explain the above described phenomenon, after a large number of iterations.

6. CONCLUSION

This article does not give any definitive explanation for the appearance of striped asymptotic states in Swindale's model. It gives essentially hints, to a mathematical problem which is not very easy for several reasons :

* the operators involved are global, and not local.

* we considered a quadratic - but not necessarily convex - Lyapunov function on the unit ball of $L^{\infty}(\Omega)$; it can be shown easily that a non-zero minimum of such a Lyapunov function belongs to L , as defined by (3.24). Thus, a precise statement on the global minimum of Φ over the unit ball of $L^{\infty}(\Omega)$ is indeed a statement on the geometry of this ball, in Fourier variable, whenever $\Omega = \mathbb{T}^N$. This is not a well known area of mathematics.

I would like to emphasize the simplicity of the statement of the problem, together with its relevance to a number of different fields of science : modelling by a convolution a short distance positive interaction and a long distance negative interaction is, after all, a simple idea, and deserves more interest from mathematicians.

REFERENCES

[1] H. Brezis, Operateurs maximaux monotones et semi-groupes de contractions dans les espaces de Hilbert. North Holland (1973).

[2] D.J. Gates, O. Penrose, The Van der Waals limit for classical systems. I. A variational principle. Commun. Math. Phys. 15 (1969), 255-276.

[3] D.J. Gates, O. Penrose, The Van der Waals limit for classical systems. II. Existence and continuity of the canonical pressure. Commun. Math. Phys. 16 (1970), 231-237.

[4] D.J. Gates, O. Penrose, The Van der Waals Limit for classical systems. III. Deviation from the Van der Waals-Maxwell theory. Commun. Math. Phys. 17 (1970), 194-209.

[5] J. Neveu,Bases mathématiques du calcul des probabilités. Masson. Paris (1970).

[6] Y. Pomeau, Nonlinear pattern selection in a problem of elasticity. J. hysique - lettres 42 (1981), L1 - L4.

[7] Y. Pomeau, S. Zaleski, Wavelength selection in one-dimensional cellular structures. J. Physique 42 (1981), 515-528.

[8] Y. Pomeau, P. Manneville, Wavelength selection in axisymmetric cellular structures. J. Physique 42 (1981), 1067-1074.

[9] G. Ruget, Lecture at the Probability Theory Congress in Marseille, July 1982.

[10] M. Schatzman, A nonlinear evolution system with a convolution term arising in a biological model. Rapport interne n°74, Centre de Mathématiques Appliquées, Ecole Polytechnique (1981).

[11] N.V. Swindale, A model for the formation of ocular dominance stripes. Proc. R. Soc. Lond. B, 208 (1980), 243-264.

Michelle SCHATZMAN

Ecole Polytechnique
Centre de Mathématiques appliquées
Route de Saclay

91128 PALAISEAU CEDEX
FRANCE

R SENTIS

Equation de Riccati non classique, liée aux équations de transport

INTRODUCTION

Chandrasekhar [1] a souligné dès 1949 qu'à certains problèmes de transport de particules dans une bande (ou un demi-espace) homogène on pouvait associer des équations de Riccati. Il donne ainsi l'équation de Ricatti satisfaite par l'opérateur qui permet de calculer le flux sortant d'un demi-espace homogène en fonction du flux entrant (S. Chandrasekhar [1. chap. IV eq. 28]). Nous nous proposons de donner un cadre rigoureux et plus général pour étudier cette équation. Le cadre général que nous introduisons est d'un type nouveau: par rapport à Lions [3] ou Sorine [5] le terme quadratique de notre équation n'a pas le même signe; d'autre part l'opérateur G_1 de notre terme quadratique $P\, G_1\, P$ n'a pas la régularité supposée dans Tartar [6] , ni dans Kuiper-Shew [2](les hypothèses de cet article entraineraient que G_1 et G_2 soient des opérateurs bornés sur $L^2(K)$ ce qui impliquerait que la fonction $\gamma(k)$ soit strictement positive - ce qui n'est pas toujours le cas dans les problèmes de transport (voir exemple 2)).

. Le point capital est la résolution du système "aux deux bouts" suivant :

$$\begin{cases} \dfrac{dy}{dt} + L\, y = F\, y + G_1\, p & y(0) = f \qquad t \in [0,T] \\ \dfrac{dp}{dt} + L\, p = F\, p + G_2\, y & p(T) = 0 \end{cases} \qquad (*)$$

dans lequel L est un opérateur V-coercif associé à un triplet classique $V \subset H \subset V'$. Les outils de base sont l'introduction de cônes positifs dans V,

H et V' (tels que $H = H_+ - H_+$) et celle d'un sous espace E de H qui est du type espace L^∞ (pour une fonction remarquable γ on définit $E = \{f \in H / \exists \beta \geq 0 : \beta\gamma \leq f \leq \beta\gamma\}$). Et les opérateurs F et G_1, G_2 sont dans $\mathcal{L}(E,V')$. En faisant plusieurs hypothèses de positivité, en particulier $F\gamma + G_i\gamma \leq L\gamma$, on montre que le système (*) admet une solution unique grâce à un théorème de point fixe dans l'espace $L^\infty(0,T;E)$.

Nous donnons ensuite deux exemples auxquels nous pouvons appliquer nos résultats (dans le cadre du second exemple rentrent les problèmes de transport du type de ceux qui sont étudiés par Chandrasekhar [1]).

. Dans le §2, nous donnons l'interprétation du résultat précédent en terme de problème de contrôle. Et nous pouvons ainsi voir que la fonctionnelle suivante admet un point critique

$$J(u) = \frac{1}{2} \int_0^T (G_2 y_u(t), y_u(t)) - (G_1 u(t), u(t)) dt + (g, y_u(T))$$

où y_u est la solution d'une équation d'état classique associé au contrôle $u \in L^\infty(0,T;E)$.

Comme la fonctionnelle J, qui n'est ni convexe ni concave est cependant quadratique,on peut montrer directement qu'elle admet un point critique en utilisant les mêmes techniques que celles utilisées dans le §1.

. Dans le §3, nous étudions l'opérateur P_T défini de la façon suivante. Si (y,p) est solution de (*) on pose

$$p(0) = P_T f$$

On montre alors que si f est dans un sous espace E_0 de E, $P_T f$ dépend continuement de T . Et de plus P_T est dérivable par rapport à T en un sens faible et est solution de l'équation d'évolution de Riccati suivante :

$$\begin{cases} \dfrac{dP_T}{dT} = G_2 + (F-L)P_T + P_T^*(F-L) + P_T^* G_1 P_T \\ \\ P(0) = 0 \end{cases} \qquad (**)$$

qui est d'un type non classique, car G_1 n'a pas le "bon signe". (Les égalités dans (**) s'entendent au sens des opérateurs de $\mathcal{L}(E_o,E_o'))$. On donne enfin des corollaires dans le cadre de l'exemple 2 : la limite P_∞ de P_T (quand $T \to \infty$) vérifie une équation de Riccati stationnaire (qui est bien une généralisation de celle de Chandrasekhar [1. Chap. IV. équ (28)]). Et ce dernier résultat entraine une propriété de positivité de P_∞ qui est utilisée dans Sentis [4, Chap. 5] pour résoudre un problème d'interface pour des équations de transport. Je remercie M. Sorine pour les fructueuses conversations que j'ai eues avec lui sur ce problème.

1. CADRE GENERAL - SYSTEME AUX DEUX BOUTS

Soit V, H, V' un triplet d'espaces de Hilbert, quand H est identifié à son dual, V' est une réalisation du dual de V , avec :

$V \subset H \subset V'$ injections continues et denses.

On note $(.,.)_H$ le produit scalaire de H .

$(.,.)_V$ le produit de dualité (V,V').

$\|\cdot\|_V$, $\|\cdot\|$, $\|\cdot\|_*$ les normes de V, H et V'.

Soit H_+ une cône convexe fermé de H tel que

$$H = H_+ - H_+ \qquad (1)$$

$$H_+ = \{f \in H/(f,g)_H \geq 0, \ \forall\, g \in H_+\} \qquad (2)$$

(2) sera réalisée en particulier si $\forall f;g \in H_+$ on a $(f,g) \geq 0$ et $\forall f \in H$, $(f_+,f_-) = 0$ où f_+ et f_- sont des éléments de H_+ tels que $f_+ - f_- = f$] .

On note $V_+ = H_+ \cap V$, et V'_+ le cône dual de V_+ c'est-à-dire :

$$V'_+ = \{f \in V' \quad \text{t.q.} \quad (f,v)_V \geq 0 \quad \forall v \in V_+\}$$

On voit immédiatement que $V = V_+ - V_+$ et que : $H_+ = V'_+ \cap H$. (V'_+ est non vide).

Notons $f \leq g$ si et seulement si $g-f \in V'_+$, on a alors :

$$\begin{cases} \forall f, g \in H & f \leq g \quad \Leftrightarrow \quad g - f \in H_+ \\ \forall f, g \in V & f \leq g \quad \Leftrightarrow \quad g - f \in V_+ \end{cases}$$

D'autre part considérons $L \in \mathcal{L}(V,V')$ vérifiant

$$\exists \alpha > 0, (Lu,u)_V \geq \alpha \| u \|^2 \qquad \forall u \in V \tag{3}$$

$$\begin{cases} U_t(H_+) \subset H_+ \quad \forall t \text{, où } U_t \text{ est le semi-groupe de} \\ \qquad \text{type } C^0 \text{, engendré par } -L . \end{cases} \tag{4}$$

On se donne maintenant γ dans V_+ ($\gamma \neq 0$) et on définit l'espace E par :

$$E = \{f \in H \text{ t.q } \exists \beta \in \mathbb{R}_+ \ - \beta\gamma \leq f \leq \beta\gamma\} ; \| f \|_E = \inf\{\beta / - \beta\gamma \leq f \leq \beta\gamma\}$$

Muni de la norme $\| . \|_E$, E est un Banach. Notons $E_+ = E \cap \beta H_+$. Remarquons que l'injection de E dans H est continue et que :

Lemme 1 : *La norme* $\| . \|_E$ *est semi-continue inférieurement par rapport à* H.

En effet si $f_n \to f$ dans H et $\|f_n\|_E \leq \beta$, on a $- \beta\gamma \leq f_n \leq \beta\gamma$ et donc :

$$- \beta\gamma \leq f \leq \gamma\beta$$

Soit maintenant F, G_1 et G_2 trois opérateurs de $\mathcal{L}(E,V')$ pour lesquels on fait les hypothèses suivantes :

$$F\gamma + G_1\gamma \leq L\gamma \qquad F\gamma + G_2\gamma \leq L\gamma \tag{5}$$

$$F(E_+) \subset V'_+ \qquad G_i(E_+) \subset V'_+ \tag{6}$$

$$\forall f \in E_+ - \{0\} \qquad \exists C_f > 0 \quad \text{t.q.} \quad G_i f \geq C_f \, L\gamma \tag{7}$$

(On en déduit en particulier que $L\gamma \in V'_+$) .

Par la suite nous ferons des hypothèses supplémentaires, notées (17) (18) (32) et (33).

Donnons tout de suite les deux exemples de base :

Exemple 1.

Soit Ω un ouvert borné de frontière régulière dans $\mathbb{R}^d$. Avec les notations classiques prenons :

$$H = L^2(\Omega) \qquad V = H_0^1(\Omega) \qquad V' = H^{-1}(\Omega)$$

$$H_+ = \{f \in L^2(\Omega)/f(x) \geq 0 \quad \text{p.p.x}\} \; ; \; V_+ = \{f \in H_0^1(\Omega)/f(x) \geq 0 \quad \text{p.p.x}\}$$

$$(Lu,v) = \int_\Omega \sum_i \frac{\partial u}{\partial x_i} \frac{\partial v}{\partial x_i} \, dx \qquad (\text{Donc } L = -\Delta \text{ avec cond. de Dirichlet}).$$

On voit alors que (2) (3) et (4) sont satisfaits. Soit γ l'élément de V_+ tel que

$$L\gamma = 1 \qquad \gamma \in V$$

Alors $E = \{f \in H \text{ t.q. } \beta : -\beta\gamma \leq f \leq \beta\gamma\}$ est dense dans H. Définissons G_1 et G_2 par $(G_i f)(x) = \int \rho_i(x,x')f(x')dx'$

avec $\rho_s \geq \rho_i(x,x') \geq \rho_o > 0$; $\rho_i(x,x') = \rho_i(x',x) \forall x,x'$; $\rho_i \in C^2(\Omega,\Omega)$ [i = 1,2]

$$\int \rho_i(x,x')\, \gamma(x')dx' \leq 1 \qquad (i = 1,2)$$

et prenons $F = 0$. On voit alors que (5) et (6) sont vérifiés ainsi que (7) avec $C_f = \rho_o \int f(x')dx'$.

Exemple 2.

Soit K un compact de $\mathbb{R}^N$ muni d'une mesure de probabilité. Soit γ la fonction de K dans $\mathbb{R}$ définie par :

$$\gamma(k) = |k_1|^{1/2} \quad \forall\, k \in K \quad (k_1 \text{ étant la première composante de } k).$$

Soit $H = L^2(K)$ $\qquad H_+ = \{f \in L^2(K)\ ;\ f \geq 0\}$

$V = \gamma L^2(K) = \{\gamma f / f \in L^2(K)\}$, $\qquad V_+ = \gamma\, H_+$

$V' = \frac{1}{\gamma} L^2(K) = \{\frac{1}{\gamma} f / f \in L^2(K)\}$, $\qquad V'_+ = \frac{1}{\gamma} H_+$

$E = \gamma L^\infty(K) = \{\gamma f / f \in L^\infty(K)\}$

On muni V et E des normes $\| \cdot \|_V = \| \frac{\cdot}{\gamma} \|_{L^2}$, $\| \cdot \|_E = \| \frac{\cdot}{\gamma} \|_{L^\infty}$
On a ici $E \subset V$.

Soit $\sigma \in C(K)$, $\sigma_1 \in C(K \times K)$, $\sigma_2 \in C(K \times K)$ vérifiant :

$$\left\{\begin{array}{ll} \text{i)} & \exists \sigma_o \text{ t.q. } \; 0 < \sigma_o \leq \sigma \;,\; \sigma_o \leq \sigma_1(k,k') + \sigma_2(k,k'), \\ \text{ii)} & \int \sigma_1(k,k') + \sigma_2(k,k')dk' \leq \sigma(k), \\ \text{iii)} & \sigma_i(k,k') = \sigma_i(k',k) \;,\; \forall\, k,k' \; (i = 1,2) \;. \end{array}\right. \tag{8}$$

Alors on définit pour tout f de V' :

$$\begin{cases} Lf(k) = \sigma(k)\, f(k)/\gamma^2(k) \\ Ff(k) = \gamma(k)^{-1} \int \sigma_1(k,k')\, f(k')/\, \gamma(k')\, dk' \\ Gf(k) = \gamma(k)^{-1} \int \sigma_2(k,k')\, f(k')/\, \gamma(k')\, dk' \quad G_1 = G_2 = G\,. \end{cases} \tag{9}$$

On voit alors que (2),(3),(4),(5),(6),(7) sont satisfaits [avec $C_f = Cte. \int \frac{1}{\gamma(k)} f(k)\, dk$] .

Dans ce cadre rentre le cas provenant des équations de transport : On a alors : $K = \{k \in \mathcal{V} \text{ t.q. } k_1 \geq 0\}$ où $\mathcal{V}$ est une réunion compacte de sphères de $\mathbb{R}^N$ centrées en 0 .

Notons $\mathcal{V}_+ = K, \mathcal{V}_- = -K$. Alors si (y,p) est le couple solution du système (9') que nous résolvons ci-après et si on pose $u \in L^2(0,T;\mathcal{V})$ vérifiant :

$$u(t,.)|_{\mathcal{V}_+} = \gamma y\,; \qquad u(t,-.)|_{\mathcal{V}_-} = \gamma p\ .$$

alors u est la solution de l'équation de transport :

$$\begin{cases} v_1 \dfrac{\partial u}{\partial t} = Ku - \sigma_o u \\ \\ u(0,.)|_{\mathcal{V}_+} = f \qquad u(T,.)|_{\mathcal{V}_-} = 0 \end{cases} \tag{9'}$$

où $\sigma_o \in C(V)$ tel que : $\sigma_o|_{\mathcal{V}_+} = \sigma_o(-.)|_{\mathcal{V}_-} = \sigma$

$$Kf(v) = \int_{\mathcal{V}} \Sigma(v,w)\, f(w)\, dv\,;$$

$$\Sigma|_{\mathcal{V}_+ \times \mathcal{V}_+} = \sigma_1\,; \qquad \Sigma(-.,-.)|_{\mathcal{V}_- \times \mathcal{V}_-} = \sigma_1\,;$$

$$\Sigma(.,-.)|_{\mathcal{V}_+ \times \mathcal{V}_-} = \sigma_2\,; \qquad \Sigma(-.,.)|_{\mathcal{V}_- \times \mathcal{V}_+} = \sigma_2\,;$$

Nous allons maintenant énoncer le résultat fondamental qui est l'existence d'une solution pour un système "aux deux bouts". Soit $T > 0$ fixé. Il nous faut introduire les espaces (dépendant de T) dans lesquels nous allons travailler. Le premier espace est classique :

$$W_T = \{\psi \in L^2(0,T;V),\ \frac{d\psi}{dt} \in L^2(0,T;V')\} \quad .$$

Si on note $C(0,T;H)$ l'espace des fonctions continues de $[0,T]$ dans H, on sait que

$$W_T \subset C(0,T;H) \quad \text{avec injection continue .}$$

Définissons d'autre part :

$$S_T = L^\infty(0,T;E)\ ; \quad S_T^+ = \{\psi \in S_T / \ \psi(t) \in E_+ \quad \text{p.p.t}\}$$

<u>Proposition 1.</u>

i/ *Pour tout* f *de* E , *il existe une solution* (y,p) *unique dans* $[W_T \cap S_T]^2$ *pour le système aux deux bouts suivants :*

$$\begin{cases} \dfrac{dy}{dt} + Ly = Fy + G_1 p \\ -\dfrac{dp}{dt} + Lp = Fp + G_2 y \\ y(0) = f \quad p(T) = 0 \ . \end{cases} \tag{10}$$

ii/ *Si on note* P_T *l'application qui à* f *associe* $p(0)$:

$$p(0) = P_T f$$

alors : $P_T \in \mathcal{L}(E,E)$; $P_T(E_+) \subset E_+$; $\|P_T\|_{\mathcal{L}(E,E)} \leq 1$. □

Avant de faire la démonstration donnons des notations (T étant fixé) :
Pour tout $\psi \in L^2(0,T;V')$ notons $N\psi$ l'élément y de W_T solution de :

$$\begin{cases} \dot{y} + Ly = \psi \\ y(0) = 0 \end{cases}$$

et de même notons $\overline{N}\psi$ l'élément p de W_T solution de :

$$\begin{cases} -\dot{p} + Lp = \psi \\ p(T) = 0 \end{cases}$$

On a alors $(N\psi) \in C(0,T;H)$ et $(\overline{N}\psi) \in C(0,T,H)$ et :

Lemme 2 : *On a* :

$$U_t \ \gamma \leq U_\theta \gamma \leq \gamma \qquad \forall\ t,\theta \quad \theta \leq t \tag{11}$$

$$N(L\gamma)(t) = \gamma - U_t\,\gamma\ ; \ \overline{N}(L\gamma)(t) = \gamma - U_{T-t}\,\gamma \tag{12}$$

$$NF\ ,\ NG_i\ ,\ \overline{N}F\ ,\ NG_i \in \mathcal{L}(S_T,S_T) \tag{13}$$

NF , NG_i , $\overline{N}F$, $\overline{N}G_i$ *envoient* S_T^+ *dans* S_T^+ (14)

NF , $\overline{N}F$ *sont des contractions strictes sur* S_T ; *et*

$(I-NF)$, $(I-\overline{N}F)$ *sont inversibles dans* $\mathcal{L}(S_T,S_T)$. (15)

Démonstration.

Remarquons tout d'abord que si $\psi \in L^2(0,T,V')$ et $\psi(t) \in V'_+$ p.p.t alors $(N\psi)(t) \in H_+$.

Dans le cas où $\psi \in C(0,T;H)$ cela vient du fait que $N(t) = \int_0^t U_{t-s}\,\psi(s)ds$ et dans le cas général il suffit d'approcher ψ par $\psi_n \in C(0,T;H)$; $\psi_n(t) \in H_+$.

D'autre part soit z la solution de :

$$\begin{cases} \dot{z} + Lz = L\gamma \\ z(0) = 0 . \end{cases}$$

Alors on a $z(t) \in H_+$ et on remarque que $z(t) = \gamma - U_t\gamma$. D'oũ (12) et $U_t \gamma \leq \gamma$.

D'oũ (11) en appliquant la propriété de semi-groupe.

D'autre part d'après (5) et (6) on a :

$$N(F\gamma)(t) \leq N(L\gamma)(t) \qquad \forall t .$$

Donc

$$NF \in \mathcal{L}(S_T,S_T) \text{ et de même pour } NG_i , \overline{N}F \text{ et } \overline{N}G_i .$$

et (14) vient de la remarque faite au début de la démonstration. Enfin d'après (12) et les hypothèses (5) et (7) on a :

$$NF\gamma \leq NL\gamma - NG\gamma \leq (1-C_\gamma) NL\gamma \leq (1-C_\gamma)\gamma$$

Comme

$C_\gamma > 0$, on a (15). C.Q.F.D.

<u>Démonstration de la proposition 1.</u>

i/ D'après (11), $\forall f \in E$, la fonction $t \to U_t f$, notée $U.f$ est dans $S_T \cap W_T$.

Notons $\mathcal{A}_o$ l'opérateur de $(S_T \times S_T)$ dans $(S_T \times S_T)$ défini par :

$$\mathcal{A}_o \begin{pmatrix} z \\ q \end{pmatrix} = \begin{pmatrix} NF & NG_1 \\ NG_2 & NF \end{pmatrix} \begin{pmatrix} z \\ q \end{pmatrix} \quad \forall (z,q) \in S_T \times S_T .$$

Alors le système (10) est équivalent à trouver (y,p) dans $(S_T \times S_T)$ tel que :

$$\begin{pmatrix} y \\ p \end{pmatrix} = \mathcal{A}_o \begin{pmatrix} y \\ p \end{pmatrix} + \begin{pmatrix} U.f \\ 0 \end{pmatrix}$$

Notons $\|(z,q)\|_{S_T \times S_T} = \sup(\|z\|_{S_T}, \|q\|_{S_T})$; nous allons montrer qu'il existe $k_o > 0$ tel que

$$\|\mathcal{A}_o^2(z,q)\|_{S_T \times S_T} \leq (1-k_o)\ \|(z,w)\|_{S_T \times S_T} \quad \forall z,q \in S_T \ . \qquad (15)'$$

Posons $(\overline{z},\overline{q}) = \mathcal{A}_o^2(\gamma,\gamma)$. D'après (5), (11) et (12) on a

$$N(F\gamma+G_1\gamma) \leq \gamma - U_t\gamma \leq \gamma - U_T\gamma,\ \overline{N}(F\gamma+G_2\gamma) \leq \gamma - U_T\gamma$$

donc, grâce aux hypothèses (5) et (7) on a, en notant $k_o = C_{U_T\gamma}$:

$$0 \leq \overline{z} \leq N\ [F\gamma+G_1(\gamma-U_T\gamma)] \leq NL\gamma - k_oNL\gamma$$

$$0 \leq \overline{q} \leq N\ [F\gamma+G_2(\gamma-U_T\gamma)] \leq NL\gamma - k_oNL\gamma$$

Et grâce à (12) on voit que :

$$0 \leq \overline{z} \leq (1-k_o)\gamma \quad . \qquad 0 \leq \overline{q} \leq (1-k_o)\gamma \quad .$$

d'où (15)'. Donc la série $\sum_{n=0}^{\infty} \mathcal{A}_o^n$ est absolument convergente dans $(S_T \times S_T)$, [le terme général ayant une norme inférieure à $(1-k_o)^{n/2}$ x Cte] et on a :

$$(I-\mathcal{A}_o)^{-1} = \sum_{n=0}^{\infty} \mathcal{A}_o^n \ .$$

Et la solution de (10) est :

$$\begin{pmatrix} y \\ p \end{pmatrix} = (I-\mathcal{A}_o)^{-1} \begin{pmatrix} U.f \\ 0 \end{pmatrix}$$

ii/ Supposons maintenant que $0 \leq f \leq \gamma$. Comme $(NF+NG_1)\gamma \leq \gamma$ et $(\overline{N}G_2+\overline{N}F)\gamma \leq \overline{\gamma}$ on voit que $\mathcal{A}_o$ laisse invariant le convexe fermé :

$$\mathcal{C} = \{(\phi,\psi) \in S_T \times S_T / 0 \leq \phi(t) \leq \gamma \ ;\ 0 \leq \psi(t) \leq \gamma \quad \text{p.p.t}\}$$

Donc on a : $\mathcal{A}_o^n \begin{pmatrix} U.f \\ 0 \end{pmatrix} \in \mathcal{C}$ pour tout n et :

$$\begin{pmatrix} y \\ p \end{pmatrix} = \sum_n \mathcal{A}_o^n \begin{pmatrix} U.f \\ 0 \end{pmatrix} \in \mathcal{C}$$

D'autre part remarquons que :

$$\forall \xi \in S_T \cap C(0,T;H) \text{ on a : } \quad \xi(t) \in E \,,\ \|\xi(t)\|_E \leq \|\xi\|_{S_T} \quad \forall t \quad (16)$$

[En effet pour t quelconque il existe $t_p \to t$ tel que $\|\xi(t_p)\|_E \leq \|\xi\|_{S_T}$, comme $\xi(t_p) \to \xi(t)$ dans H, on obtient (16) d'après le lemme 1].

On a donc ici : $\quad 0 \leq y(t) \leq \gamma \qquad 0 \leq p(t) \leq \gamma \qquad \forall t$

Et en notant $p(0) = P_T f$ on a :

$$0 \leq P_T f \leq \gamma$$

Donc $P_T \in \mathcal{L}(E,E)$. C.Q.F.D.

<u>Remarque 1</u>. (<u>corollaire de (15)</u>). Pour tout $\phi \in L^2(0,T,V')$ et $f \in E$, il existe une solution unique dans $W_T \cap S_T$ de l'équation :

$$\begin{cases} \dot{y} + Ly = Fy + \phi \\ y(0) = f \end{cases}$$

qui est donnée par :

$$y = (I-NF)^{-1}(N\phi + U.f) \qquad U.f \text{ note la fonction : } t \to U_t$$

<u>Remarque 2</u>. Le système (10) est équivalent à trouver (y,p) vérifiant

$$\begin{pmatrix} y \\ p \end{pmatrix} = \mathcal{B}\begin{pmatrix} y \\ p \end{pmatrix} + \begin{pmatrix} (I-NF)^{-1}U.f \\ 0 \end{pmatrix} \text{ avec } \mathcal{B} = \begin{pmatrix} 0 & (I-NF)^{-1}NG_1 \\ (I-\overline{N}F)^{-1}NG_2 & 0 \end{pmatrix}$$

On aurait pu démontrer la proposition 1, en montrant que $\mathcal{B}^2$ est une contraction stricte c'est-à-dire que :

$$(I-NF)^{-1} NG_1(I-\overline{N}F)^{-1} NG_2 \gamma \leq (1-k_o)\gamma$$

(ce qui peut se faire à l'aide du Lemme 2 avec les mêmes méthodes que ci-dessus.

2. INTERPRETATION EN TERMES DE CONTROLE

D'après la remarque 1 ci-dessus, on sait que pour tout u dans S_T il existe une solution unique y_u dans $W_T \cap S_T$ de l'équation

$$\begin{cases} \dot{y} + Ly = Fy + G_1 u \\ y(0) = f \end{cases} \tag{17}$$

et l'application $u \to y_u$ est continue de S_T dans W_T et dans S_T.

Donnons maintenant deux hypothèses supplémentaires, qui seront faites dans toute la suite (et qui sont vérifiées dans les exemples 1 et 2).

$$(Lf,g)_V = (Lg,f)_V \quad ; \qquad \forall f,g \in V \tag{18}$$

$$(Ff,g)_V = (Fg,f)_V \quad ; \quad (G_i\, f,g)_V = (G_i g,f)_V \quad ; \qquad \forall\, f,g \in V \cap E \tag{19}$$

Nous donnons maintenant l'interprétation de la proposition 1 en terme de problème de contrôle.

Proposition 2. Soit f et g donnés dans E. Pour tout u de S_T posons :

$$J(u) = \frac{1}{2} \int_0^T (G_2 y_u, y_u)_V - (G_1 u,u)_V dt + (g,y_u(T))_H \tag{20}$$

où y_u est la solution de (17). Alors la fonctionnelle J admet un point critique u dans S_T. Ce point critique est donné par $u = p$ sachant que (y,p) est l'unique solution dans $(W_T \cap S_T)^2$ de

$$\begin{cases} \dot{y} + Ly = Fy + G_1 p \\ -\dot{p} + Lp = Fp + G_2 y \\ y(0) = f \; ; \quad p(T) = -g \end{cases} \tag{21}$$

D'autre part pour ce u critique on a :

$$J(u) = \frac{1}{2}(P_T f, f)_H + (g, y_u(T))_H \tag{22}$$

Démonstration. D'après la proposition 2 , J est continue sur S_T (car $G_1 y_u \in L^\infty(0,T;V'), y_u \in L^2(0,T;V)$). J est aussi différentiable et sa différentielle pour $u \in S_T$ est :

$$J'(u).v = \int_0^T (G_2 y_u, y'_u.v)_V - (G_1 u, v)_V \, dt + (g, (y'_u.v)(T))_H, \forall v \in S_T$$

où y'_u désigne la différentielle de y par rapport à u. Donc on a :

$$J'(u).v = \int_0^T (G_2 y_u, y_v - y_o)_V - (G_1 u, v)_V \, dt + (g, y_v(T) - y_o(T))_H, \forall v \in S_T \tag{23}$$

D'autre part d'après la proposition 1, on sait qu'il existe (y,p) dans $(W_T \cap S_T)^2$ solution de (21) . Choisissons alors $u = p$.
On a alors $y = y_u$. (Et u est égal à l'état adjoint associé au problème de contrôle (19),(21)). Et pour le u ainsi choisi on a :

$$\begin{aligned} \int_0^T (G_2 y_u, y_v - y_o) dt &= \int_0^T (-\dot{u} + Lu - Fu, y_v - y_o) dt \quad \forall v \in S_T \\ &= \int_0^T (u, \dot{y} + Ly_v - Fy_v - (\dot{y}_o + Ly_o - Fy_o))_V \, dt + (u(T), \; y_o(T) - y_v(T)) \\ &= \int_0^T (u, G_1 v - G_1 0)_V dt + (g, y_v(T) - y_o(T)) \end{aligned}$$

$$= \int_0^T (G_1u,v)_V dt + (g,y_v(T)-y_o(T)) \qquad \forall\, v \in S_T$$

En rapprochant cela de (23) on obtient bien : $J'(u) = 0$.

D'autre part on a :

$$J(u) = \frac{1}{2} \int_0^T (-\dot{u}+Lu-Fu,y_u)_V-(y+Ly_u-Fy_u,u)_V dt + (g,y_u(T))_H$$

$$= \frac{1}{2} \int_0^T - \frac{d}{dt}(u,y_u)_H dt + (g,y_u(T))_H = \frac{1}{2}(u(0),y_u(0))_H+(g,y_u(T))_H$$

D'où le résultat car :

$$(u(0),y_u(0))_H = (P_Tf,f)_H \tag{24}$$

C.Q.F.D.

Remarque 3. Montrons directement que le problème (17) (20) admet un point critique dans E , pour tout f fixé. (Nous remercions H. Berestycki de nous avoir donné l'idée de cette preuve). Nous n'utiliserons ici que les résultats du lemme 2. Posons donc :

$$B = (I-NF)^{-1} \qquad \overline{B} = (I-\overline{N}F)^{-1}$$

La solution de (17) pour u quelconque dans S_T est :

$$y_u = NBG_1u + BUf$$

Donc :

$$J(u) = \int_0^T \{(G_2BNG_1u,BNG_1u) + (G_2B\,U\,f,BNG_1u) + (G_2BNG_1u,B\,U\,f)$$

$$- (G_1u,u)\} dt + \int_0^T (G_2B\,U\,f,B\,U\,f)dt \quad .$$

Donc J est une forme quadratique en u dont la dérivée en u est :

$$\forall\, v \in S_T \qquad J'(u),v = \int_0^T (j_u(t),v(t))\, dt$$

avec $j_u = G_1[N^* B^* G_2 BNG_1 - I]\, u + G_1\, N^* B^* G_2 BUf$.

Comme G_1 est injectif, on a existence et unicité de u critique, si on montre que $M = N^* B^* G_2\, BNG_1$ est une contradiction stricte dans S_T.

On vérifie tout d'abord que : $N^* = \overline{N}$.

Puis on voit que :

$$\overline{N}\, B^* = \overline{N} \sum_{k=0}^{\infty} [(NF)^*]^k = \overline{N} \sum_{k=0}^{\infty} (F\overline{N})^k = \sum_{k=0}^{\infty} (\overline{N}F)^k\, \overline{N} = \overline{B}\, \overline{N}.$$

Posons donc $y = BNG_1 \gamma$, on voit que y est solution de :

$$\begin{cases} \dot{y} + Ly = Fy + G_1 \gamma \\ y(0) = 0 . \end{cases}$$

En posant $z = \gamma - y$ on a :

$$\begin{cases} \dot{z} + Lz = Fz + L\gamma - F\gamma - G_1\gamma \\ z(0) = \gamma . \end{cases}$$

D'après (5) on a $z \leq \tilde{z}$ où $\tilde{z}$ est la solution de :

$$\begin{cases} \dot{z} + L\tilde{z} = F\tilde{z} \\ \tilde{z}(0) = \gamma . \end{cases}$$

Donc $\tilde{z}(t) \geq 0$ et $\tilde{z}(t) \geq U_t\gamma$. D'où l'on tire grâce à (11) :

$$y(t) \leq \gamma - U_t\, \gamma \leq \gamma - U_T \gamma .$$

Donc on peut écrire en utilisant (7) avec $k_o = C_{U_T}$:

$$M\gamma = \overline{B}\, \overline{N}\, G_2\, B\, N\, G_1 \gamma \leq \overline{B}\, \overline{N}\, G_2 (\gamma - U_T \gamma) \leq \overline{B}\, N\, G_2 \gamma - k_o\, \overline{B}\, \overline{N}\, L\gamma .$$

De même que ci-dessus on vérifie que :

$$(\overline{B}\,\overline{N}\,G_2\gamma)(t) \leq \gamma - U_{T-t}\gamma \quad .$$

Et comme $\overline{B}\,g \geq g$; $\forall\, g \in S_T^+$, on a grâce à (12) :

$$(M\gamma)(t) \leq \gamma - U_{T-t}\gamma - k_o(\overline{N}\,L\gamma)(t) = (1-k_o)(\gamma - {}_{T-t}\gamma) \leq (1-k_o)\gamma .$$

Donc M est une contradiction stricte dans S_T .

C.Q.F.D.

La démonstration de la remarque 2 à la fin du §1 consiste également à montrer que M est une contraction stricte dans S_T .

3. EQUATION DE RICCATI

Afin de montrer que P_T satisfait une équation de Riccati, nous devons tout d'abord donner quelques propriétés de ces opérateurs P_T . Nous faisons toujours les hypothèses (2,3,4,5,6,7) et (17,18).

Proposition 3.

$$(P_T f, g) = (P_T g, f) \qquad f, g \in E$$

Démonstration. Soit f et g quelconques dans E. Soit (y,p) et (z,q) les uniques solutions dans $[W_T \cap S_T]^2$ des systèmes :

$$\begin{cases} \text{i/} \quad \dfrac{dy}{dt} = -\,Ly + Fy + G_1 p \\ \text{ii/} \quad \dfrac{dp}{dt} = -\,Lp + Fp + G_2 y \\ \text{iii/} \quad y(0) = f \; ; \quad p(T) = 0 \end{cases} \tag{25}$$

$$
\begin{cases}
i/ \quad \dfrac{dz}{dt} = -Lz + Fz + G_1 q \\
ii/ \quad -\dfrac{dg}{dt} = -Lg + Fq + G_2 z \\
iii/ \quad z(0) = g \; ; \qquad q(T) = 0
\end{cases}
\tag{26}
$$

On a alors :

$$p(0) = P_T f \; ; \qquad q(0) = P_T q \; .$$

En multipliant (25-ii) par z on obtient :

$$(p(0),z(0)) = \int_0^T - (Lp(s),z(s)) + (Fp(s),z(s)) + (G_2 y(s),z(s))ds \; .$$

De même en multipliant (26-i) par p, on obtient :

$$(p(0),z(0)) = -\int_0^T - (Lz(s),p(s)) + (Fz(s),p(s)) + (G_1 q(s),p(s))ds \; .$$

Donc en utilisant (17) et (18) on obtient :

$$(P_T f,g) = \frac{1}{2} \int_0^T (G_2 y(s),z(s)) - (G_1 q(s),p(s))ds \; .$$

De même en échangeant les rôles de (y,p) et (z,q) on obtient :

$$(P_T g,f) = \frac{1}{2} \int_0^T (G_2 z(s),y(s)) - (G_1 p(s),q(s))ds \; .$$

D'après (18) on obtient alors le résultat.

C.Q.F.D.

Proposition 4. Si $T \leq T_1$, on a alors :

$$P_T f \leq P_{T_1} f \qquad \forall\, f \in E_+ \; .$$

Démonstration. Soit (y,p) et (y_1,p_1) les solutions de (10) et de

$$\begin{cases} \dot{y}_1 + Ly_1 = Fy_1 + G_1p_1 \\ -\dot{p}_1 + Lp_1 = Fp_1 + G_2y_1 \\ y_1(0) = f \; ; \quad p_1(T_1) = 0 \; . \end{cases} \tag{27}$$

D'après la proposition 1-ii, on voit que $p_1(T) \in E_+$. Et donc, si on pose

$$y_2 = y-y_1 \; ; \quad p_2 = p-p_1$$

le couple (y_2,p_2) sera solution du système obtenu en remplaçant la dernière ligne de (10) par

$$y_2(0) = 0 \; ; \quad p_2(T) = p_1(T) \; .$$

Et en utilisant la proposition 1-ii) après avoir changé t en $T-t$, on voit que

$$p_2(0) = P_T f - P_{T_1} f \in E_+ \qquad \text{C.Q.F.D.}$$

Nous allons maintenant introduire de nouveaux espaces et faire quelques hypothèses techniques (que nous vérifierons dans les 2 exemples) afin de pouvoir démontrer le résultat de continuité de P_T par rapport à T puis d'écrire de façon convenable l'équation de Riccati satisfaite par P_T .

Notons B l'espace de Banach suivant

$$B = \{f \in V \quad \text{t.q.} \quad -\beta L\gamma \leq f \leq \beta L\gamma\}$$

$$\| f\|_B = \inf \; \{\beta > 0 \quad \text{t.q.} \quad -\beta L\gamma \leq f \leq \beta L\gamma\}$$

D'après (5) et (6) on voit immédiatement que :

$$F \in \mathfrak{L}(E,B) \qquad G_1,G_2 \in \mathfrak{L}(E,B) \; . \tag{28}$$

Notons maintenant E_o l'espace normé suivant :

$$E_o = \{f \in E \cap V \quad \text{t.q.} \quad L f \in B\}$$

$$\| f \|_{E_o} = \sup \{\|f\|_E , \|Lf\|_B\}$$

Cet espace non vide (car $\gamma \in E_o$) . Montrons le résultat suivant sur E_o .

<u>Lemme 4</u>. i/ *Il existe une constante* C_E *telle que*

$$\|f\|_V \leq C_E \|f\|_{E_o} \qquad \forall f \in E_o \tag{29}$$

et E_o *est un Banach.*

ii/ *La norme* $\|\cdot\|_{E_o}$ *est semi-continue inférieurement par rapport à* H .

<u>Démonstration</u>. i/ Comme L est V-coercif on a pour tout f de E_o :

$$\|f\|_V \leq \frac{1}{\alpha} \|Lf\|_* = \frac{1}{\alpha} \sup_{\|v\|_V=1} |(Lf,v)| \leq \frac{1}{\alpha} \|Lf\|_B \sup_{\|v\|=1} (L\gamma,v) = C_E \|Lf\|_B$$

avec $C_E = \frac{1}{\alpha} \|L\gamma\|_*$.

Donc $\|\cdot\|_{E_o} + \|\cdot\|$ est une norme équivalente à $\|\cdot\|_{E_o}$. Et E_o est bien un Banach.

ii/ Soit $\xi_n \in E_o$ tel que $\xi_n \to \xi$ dans H et $\|\xi_n\|_{E_o} \leq \beta$. En particulier $\|\xi_n\|_E \leq \beta$, d'après le lemme 1 on sait que $\xi \in E$ et que

$$\|\xi\|_E \leq \beta$$

D'autre part on a :

$$-\beta(L\gamma,v) \leq (L\xi_n,v) \leq \beta(L\gamma,v) \qquad \forall v \in V_+ \quad .$$

Et donc en faisant tendre n vers l'infini :

$$-\beta(L\gamma,v) \leq (L\xi,v) \leq \beta(L\gamma,v) .$$

D'oũ

$$\|L\xi\|_B \leq \beta \text{ . Donc } \xi \in E_o \text{ et } \|\xi\|_{E_o} \leq \beta$$

C.Q.F.D.

Remarquons que :

$$L \in \mathfrak{L}(E_o,B) \tag{30}$$

et E_o peut être interprété comme le domaine de L à valeurs dans B .

Faisons maintenant les dernières hypothèses :

$$NF, NG_1 , NG_2 \in \mathfrak{L}(S_T,L^\infty(0,T;E_o)) \tag{31}$$

$$\forall T > 0, \exists C_T \text{ , tel que } \forall f \in E_o \quad \| U_t f \|_{E_o} \leq C_T \|f\|_{E_o} \quad \forall t \in [0,T] \tag{32}$$

Vérifions que ces hypothèses sont bien satisfaites dans les deux exemples.

Exemple 1.

On a ici :

$B = L^\infty(\Omega)$ $\qquad E_o = E \cap W^{1,\infty}(\Omega)$ avec les notations des espaces de Sobolev.

En prenant $\psi \in S_T$ et $\phi = G_i\psi$, on voit que

$$\phi \in L^\infty(0,T;H)$$

donc

$$N\phi \in L^2(0,T;D(L)) \text{ et } \frac{d}{dt}(N\phi) \in L^2(0,T;H) .$$

On en déduit $N\phi \in C(0,T;V)$. Et comme :

$$\left|\frac{\partial^2\phi}{\partial x_j^2}(t,x)\right| = \left|\int \frac{\partial^2 p_i}{\partial x_j^2}(x,x')\psi(t,x')dx'\right| < C_1 \|\psi\|_T \qquad \forall(t,x)$$

on en déduit que $N\frac{(\partial^2\phi)}{\partial x_j^2}$ est borné dans $L^\infty([0,T]\times\Omega)$. D'où (32) car

$$\sum_j \frac{\partial^2}{\partial x_j^2}(N\phi) = N(\sum_j \frac{\partial^2}{\partial x_j^2}\phi) .$$

D'autre part si $f \in E_o$, on a $\frac{\partial^2 f}{\partial x_j^2} \in L^\infty(\Omega)$ et (33) vient du fait que

$$\left|\frac{\partial^2}{\partial x_j^2} U_t f\right|_{L^\infty} = \left|U_t \frac{\partial^2 f}{\partial x_j^2}\right|_{L^\infty} \leq Cte \qquad \forall t .$$

Exemple 2.

On a ici : $B = \frac{1}{\gamma} L^\infty(K)$ $\quad E_o = E = \gamma L^\infty(K)$ [les normes $\|\cdot\|_E$ et $\|\cdot\|_{E_o}$ sont équivalentes]. D'autre part (31) et (32) viennent immédiatement de (13) et (14).

Nous pouvons alors énoncer un premier résultat.

Proposition 5. *Pour tout* f *de* E_o, *l'application* $T \to P_T f$ *est continue dans* R^+ *dans* E.

Démonstration. Soit T fixé et h petit devant T. Considérons les solutions de (10) sur les intervalles $[0,T]$ et $[0,T+hT]$ avec $f \in E_o$.

Ramenons nous au même intervalle $[0,T]$. Cela revient à considérer (y,p) solution de (10) et (y_h,p_h) solution de

$$\begin{cases} \dot{y}_h + (1+h)Ly_h = (1+h)(Fy_h+G_1p_h) \\ -\dot{p}_h + (1+h)Lp_h = (1+h)(Fp_h+G_2y_h) \\ y_h(0) = f \qquad p_h(T) = 0 . \end{cases}$$

Les solutions y_h et p_h sont encore dans $W_T \cap S_T$. Il s'agit de montrer que :

$$p_h - p \to 0 \quad \text{dans } S_T \tag{33}$$

[En effet d'après la remarque (16) cela impliquera $p_h(0) - p(0) \to 0$ dans E]

Nous introduisons alors les solutions ξ_h et ζ_h des équations

$$\begin{cases} \dot{\xi}_h + L\xi_h = (1+h)(Fy_h+G_1p_h) & t \in [0,T] \\ \xi_h(0) = f \end{cases}$$

$$\begin{cases} -\dot{\zeta}_h + L\zeta_h = (1+h)(Fp_h+G_2y_h) & t \in [0,T] \\ \zeta_h(T) = 0 \end{cases}$$

Admettons un instant le lemme suivant :

<u>Lemme 5</u>. *Soit* $f \in E_o$ *et* $\phi,\psi \in S_T$. *Notons* w *et* w_h *les solutions dans* W_T *de* :

$$\dot{w} + Lw = F\phi + G_1\psi \qquad w(0) = f$$

$$\dot{w}_h + (1+h)Lw_h = F\phi + G_1\psi \qquad w_h(0) = f .$$

Alors il existe 2 constantes β_f *(indépendante de* ϕ,ψ*) et* β *(indépendante de* ϕ,ψ *et* f *) telles que :*

$$\| w_h-w \|_{S_T} \le \frac{h}{1+h}(\beta_f+\beta\|\phi\|_{S_T}+\beta\|\psi\|_{S_T})$$

Appliquons ce lemme avec $(w,w_h) = (\xi_h,y_h)$ et $(\phi,\psi) = (y_h,p_h)$.
D'après (13) comme y_h et p_h sont majorées dans S_T par $\| f \|_E$ on a, quand $h \to 0$:

$$y_h - \xi_h \to 0 \ ; \quad NF(y_h-\xi_h) \to 0 \ ; \quad NG_i(y_h-\xi_h) \to 0 \ \text{ dans } S_T \tag{34}$$

De même on montre que, lorsque $h \to 0$, on a :

$$p_h - \zeta_h \to 0 \ ; \quad \overline{N}F(p_h-\zeta_h) \to 0 \ ; \quad \overline{N}G_i(p_h-\zeta_h) \to 0 \ \text{ dans } S_T \tag{34'}$$

Posons maintenant

$$z_h = \xi_h - y \qquad\qquad q_h = \zeta_h - p \ .$$

On a alors

$$\left\{\begin{array}{l} z_h = N(Fz_h-G_1q_h) + x_{1h} \text{ avec } x_{1h} = N\,F(y_h-\xi_h) + G_1(p_h-\zeta_h) \\ \qquad\qquad\qquad\qquad\qquad\qquad + hN(Fy_h+G_1p_h) \\ q_h = \overline{N}(Fq_h-G_2z_h) + x_{2h} \text{ avec } x_{2h} = \overline{N}\,F(p_h-\zeta_h) + g_2(y_h-\xi_h) \\ \qquad\qquad\qquad\qquad\qquad\qquad + h\overline{N}(Fp_h+G_2y_h) \end{array}\right. \tag{35}$$

Or en reprenant l'opérateur $\mathcal{A}_o$ défini par

$$\mathcal{A}_o(z,q) = (Z,Q) \ ; \quad Z = N(Fz+G_1q); \quad Q = \overline{N}(Fq+G_2z)$$

le système (36) peut se réécrire :

$$\begin{pmatrix} z_h \\ q_h \end{pmatrix} = \mathcal{A}_o \begin{pmatrix} z_h \\ q_h \end{pmatrix} + \begin{pmatrix} x_{1h} \\ x_{2h} \end{pmatrix} \text{ et } \begin{pmatrix} z_h \\ q_h \end{pmatrix} = \mathcal{A}_o^2 \begin{pmatrix} z_h \\ q_h \end{pmatrix} + (\mathcal{A}_o + I) \begin{pmatrix} x_{1h} \\ x_{2h} \end{pmatrix} \tag{36}$$

Or on sait que $\mathcal{A}_o$ est dans $\mathfrak{L}((S_T)^2 , (S_T)^2)$ et d'après (15)' que $(\mathcal{A}_o)^2$

est une contraction stricte dans $(S_T)^2$. Donc $I - (\mathcal{A}_o)^2$ est un opérateur inversible dans $(S_T)^2$. D'autre part d'après (34), (34)' et, comme $\| F_{y_h} + G_1 p_h \|_{S_T}$ et $\| F p_h + G_2 y_h \|_{S_T}$ sont majorés par $\| f \|_E$ on voit que :

$$x_{1h} \to 0 ; \qquad x_{2h} \to 0 \qquad \text{dans } S_T \quad \text{quand } h \to 0 .$$

Donc l'on tire de (37) que :

$$z_h \to 0 ; \qquad q_h \to 0 \qquad \text{dans } S_T$$

Cela joint à (34)' implique (33).

C.Q.F.D.

Démonstration du lemme 5. Notons $z_h = w_h - w$ et $Z_h(t) = z_h(\frac{t}{1+h})$ donc $Z_h \in L^\infty(0,T+Th;E)$.

On a alors :

$$\dot{z}_h + (1+h)\, L z_h = -h\, L\, w \qquad z_h(0) = 0$$

$$\dot{Z}_h + L\, Z_h(t) = \frac{-h}{1+h} L\, w(\frac{t}{1+h}) \qquad Z_h(0) = 0$$

Donc $Z_h = \frac{-h}{1+h} N[Lw(\frac{\cdot}{1+h})]$. Comme $w(t) = N(F\phi + G_1\psi) + U_t f$ d'après les hypothèses (32) et (33) on en déduit que $Lw(\frac{\cdot}{1+h}) \in L^\infty(0,T;B)$ et : il existe β_f indépendant de (ϕ,ψ) tel que :

$$\| Lw \|_B \leq \beta_f + \beta(\|\psi\|_{S_T} + \|\psi\|_{S_T}) \quad \text{avec } \beta = \sup(\| LNF \|, \| LNG_1 \|).$$

Le lemme est alors une conséquence de (12).

C.Q.F.D.

Proposition 8. i/ *Pour tout* $f \in E_o$ *la solution* (y,p) *de* (10) *est dans* $L^\infty(0,T;E_o)^2$.

ii/ *On a :* $P_T \in \mathcal{L}(E_o,E_o) \quad \forall\, T > 0$

Démonstration. i/ On voit d'après (32) et (33) que :

$$y = N\,(Fy+G_1p) + U.f \in L^{\infty}(0,T;E_o)$$

$$p = \overline{N}\,(Fp+G_2y) \in L^{\infty}(0,T;E_o)$$

ii/ Plus précisément il existe une constante β_T telle que

$$\|p\|_{L^{\infty}(0,T;E_o)} \leq \beta_T \,.$$

Donc il existe une suite t_n tendant vers 0 telle que :

$$p(t_n) \in E_o \,,\ \|p(t_n)\|_{E_o} \leq \beta_T \,.$$

Le résultat est alors une conséquence du point ii/ du lemme 4.

C.Q.F.D.

Notons E'_o le dual de E_o. Nous pouvons maintenant énoncer le résultat principal affirmant que P_T satisfait une équation de Riccati au sens des opérateurs de $\mathcal{L}(E_o,E'_o)$. L'espace E_o joue le rôle de l'espace $D(L)$ pour les équations de Riccati intervenant pour les problèmes de contrôles dans un cadre variationnel(l'équation de Riccati étant satisfaite dans $\mathcal{L}(D(L),D(L)')$; cf. Lions [1] ou Sorine [1]).

Théorème. *Pour tout* $T > 0$, P_T *admet une dérivée au sens faible* $\frac{dP_T}{dT}$ *dans* $\mathcal{L}(E_o,E'_o)$ *[c'est-à-dire* $\frac{1}{h}(P_{T+h}f-P_Tf) \underset{h\to 0}{\rightarrow} \frac{dP_T}{dt} f$ *dans* E'_o *faible * pour tout* f *de* E_o*] et on a :*

$$\frac{dP_T}{dT} = G_2 + (F-L)P_T+P^*_T(F-L) + P^*_TG_1P_T \tag{37}$$

égalité dans $\mathcal{L}(E_o,E'_o)$ *où* P^*_T *est l'opérateur de* $\mathcal{L}(E'_o,E'_o)$ *dual de* P_T.

De plus, $\forall f,g \in E_o$ *on a :*

$$(P_T f,g) = \int_0^T ([G_2+(F-L)P_t]\, f,g) + ([(F-L)+G_1 P_t]f,P_t g)dt \qquad (38)$$

Avant de faire la démonstration donnons un lemme.

Lemme 6. i/ $\forall\ \xi \in E \cap V\ ,\quad \forall\ \eta \in B\ ,\ |(\xi,\eta)| \le \|\xi\|_E\ \|\eta\|_B$ Cte

ii/ *Soit* ξ_n *une suite bornée dans* E_o . *Si on a* $\xi_n \to \xi$ *dans* H *alors on a :*

$$(\xi_n\ ,\eta)\ \to (\xi,\eta) \qquad \forall\ \eta \in V'\ .$$

Démonstration du lemme. Le point i/ est immédiatement vérifié dans le cas où $\eta \in B \cap V'_+$ et dans le cas général il suffit d'écrire $\eta = \eta_+ - \eta_-$ avec $\eta_+,\eta_- \in B \cap V'_+$.

ii/ D'après le lemme 4-ii/ on sait que $\xi \in E_o$, donc $\xi \in V$.

D'après le lemme 4-i/, ξ_n est borné dans V , donc converge faiblement dans V; nécessairement vers ξ .

C.Q.F.D.

Démonstration du théorème. Considérons (y,p) l'unique solution de (10) avec $f \in E_o$. Soit $h > 0$ (h petit devant T). D'après (16) on a :

$$y(h) \in E$$

D'après l'unicité de la solution de (10) sur [h,T] on a :

$$p(h) = P_{T-h}\ y(h) \qquad (39)$$

D'autre part, pour tout g et ϕ , éléments de E_o , on a (en utilisant le fait que (y,p) est solution au sens faible et le fait que $E_o \subset V$) :

$$(y(h),\phi)_H - (f,\phi)_H = \int_0^h (Fy(t) + G_1p(t),\phi) - (Ly(t),\phi)\, dt \qquad (40)$$

$$(p(0),g)_H - (p(h),g)_H = \int_0^h (Fp(t) + G_2y(t),g) - (Lp(t),g)dt \qquad (41)$$

En utilisant la proposition précédente 8.ii/ on voit que l'on peut prendre

$$\phi = P_{T-h}g$$

et d'après (40) et la proposition 5 on a :

$$(y(h),\phi) = (P_{T-h}y(h),g) = (p(h),g)$$

Comme $(p(0),g)_H - (f,\phi)_H = (P_Tf-P_{T-h}f,g)_H$, en additionnant (40) et (41) on obtient :

$$\begin{aligned}(P_Tf-P_{T-h}f,g) &= \int_0^h (Fp(t)+G_2y(t)-Lp(t),g)+(G_1p(t)-Ly(t),P_{T-h}g)dt \\ &= \int_0^h (Fg-Lg+G_1P_Tg,p(t))+(G_2g+FP_Tg-LP_Tg,y(t))dt \qquad (42) \\ &\quad + \int_0^h (G_1p(t)+Fy(t)-Ly(t),P_Tg-P_{T-h}g)dt \,.\end{aligned}$$

D'après le lemme 6.i/ et la proposition 8.i/ quand $h \to 0$, on a :

$$|(G_1p(t)+Fy(t)-Ly(t),P_Tg-P_{T-h}g)| \leq C\,te \|P_Tg-P_{T-h}g\|_E \to 0 \quad \text{p.p.t}$$

donc la 2ème intégrale du second membre de (42) tend vers 0 quand $h \to 0$. Et grâce au lemme ii/ , comme $p(t)$ et $y(t)$ sont bornés dans E_0 et comme $p(t) \underset{H}{\to} P_Tf$ et $y(t) \underset{H}{\to} f$, on voit que lorsque $h \to 0$ on a :

$$\begin{aligned}\frac{1}{h}(P_Tf-P_{T-h}f,g) &\to (Fg-Lg+G_1P_Tg,P_Tf) + (G_2g+FP_Tg-LP_Tg,f) \\ &= ((G_2+(F-L)P_T)f,g) + (((F-L)+G_1P_T)\, f,P_Tg) \,.\end{aligned}$$

De la même façon que ci-dessus on obtient quand $h \to 0$:

$$\frac{1}{h}(P_{T+h}f-P_Tf,g) \to ((G_2+(F-L)P_T)f,g)_V + (((F-L)+GP_T)f,P_Tg)_V \;.$$

Donc on voit que pour tout f et g dans E_o on a :

$$\frac{d}{dT}(P_Tf,g) = ([G_2+(F-L)P_T+P_T^*(F-L)+P_T^*G_1P_T]f,g) \;. \qquad (43)$$

d'où (37), et on obtient (38) en intégrant (43) en T, sur $[0,T_o]$.

Grâce à $(22)^{bis}$, on a en effet $(P_Tf,g) \to 0$ quand $T \to 0$).

C.Q.F.D.

Dans le cadre de l'exemple 2, en utilisant le fait que $f \in E_+$, la famille P_tf est croissante et majorée donc convergente dans V , (quand $t \to \infty$) on montre que (pour les démonstrations voir Sentis [1, Ch. VI]) :

Corollaire 1. *Il existe un opérateur* P_∞ *de* $\mathcal{L}(E,\mathcal{V})$ *tel que :*

$$P_Tf \to P f \quad \textit{dans } \mathcal{V} \qquad \forall f \in E$$

et P_∞ *vérifie l'équation (où* P_∞^* *est le dual de* P_∞ *)*

$$G_2+(F-L)P_\infty+P_\infty^*(F-L)+P_\infty^* G_1P_\infty= 0 \quad \textit{égalité dans} \quad \mathcal{L}(E,E').$$

Corollaire 2. *Pour tout* $f \in E_+$, *on a :*

$$P^*(\frac{\sigma}{\gamma^2} f) + \frac{\sigma}{\gamma^2}(P_\infty f) \geq \frac{Cte}{\gamma}\int \frac{1}{\gamma(k)} f(k)\, dk \;.$$

REFERENCES

[1] S. Chadrasekhar, "Radiative transfer". Dover, New-York 1950.

[2] H. Kuiper, S. Shew, Strong solution for infinite dimensional Riccati equations arising in transport theory . Siam. J. Math. Anal. 11 (1980), 211-222.

[3] J.L. Lions, Contrôle optimal des systèmes... . Dunod, Paris, 1968.

[4] R. Sentis, Analyse asymptotique d'équations de transport (Partie A). Thèse d'Etat, Université Paris IX, 1981.

[5] M. Sorine, Un résultat d'existence et d'unité pour l'équation de Riccati stationnaire . Rapport Inria. n°55 (Février 1981).

[6] L. Tartar, Sur l'étude directe d'équations non linéaires . J. Functional Analysis 6 (1974), 1-47.

Remi SENTIS

CEA
Centre d'Etudes de Limeil
BP n°27
94190 VILLENEUVE ST-GEORGES

FRANCE

H J SUSSMAN

On the spatial differential equations of nonlinear filtering

In this lecture I will discuss the problem of constructing "robust" solutions for the equations of nonlinear filtering. These equations involve an arbitrary continuous function y, and one wants to solve them for all such y's, or at least for all y's in a set of Wiener measure one. Since the equations contain the time-derivative $\dot{y}$, some reinterpretation is needed before we can actually talk about "solutions". One such interpretation is to regard the equations as stochastic, and to look for solutions which are stochastic processes. The work of E. Pardoux ([1]) is a beautiful example of how to pursue this direction.

Here we shall outline the "pathwise" approach, which consists of working with an individual path $y(.)$. The work discussed here was originally motivated by a desire to prove rigorously a result which was suggested by algebraic calculations of Hazenwinkel and Marcus, namely, that for the "cubic sensor problem" ($dx = dw$, $dy = x^3 dt + dv$) there exist no conditional statistics of the state x_t given the observations y_s, $0 \leq s \leq t$, which can be computed by a finite-dimensional filter. This was actually done (cf. [2]), and the more general development of the theory is presented in [3], to which the reader is referred for the details of the results and proofs sketched here.

We shall proceed in the following order :

1) precise statement of the problem,

2) reduction of the problem to that of proving certain L^1 estimates for functions of Wiener paths,

3) methods for proving the L^1 estimates, and

4) examples.

1. STATEMENT OF THE PROBLEM

Let $X_o, \ldots, X_m$ be C^∞ vector fields on $\mathbb{R}^n$, and let L be the second-order operator given by

$$L = X_o + \frac{1}{2} \sum_{i=1}^{m} X_i^2 . \tag{1}$$

(As usual, a vector field X on $\mathbb{R}^n$ is regarded indistinctly as a vector-valued function on $\mathbb{R}^n$ or as a first-order differential operator).

Let $h : \mathbb{R}^n \to \mathbb{R}$ be a C^∞ function. If $T > 0$, let C_T^o denote the space of all continuous functions $y(.) : [0,T] \to \mathbb{R}$ such that $y(0) = 0$. If $0 < \alpha \leq 1$, let C_T^α denote the set of those $y(.) \in C_T^o$ that satisfy a Hölder condition with exponent α.

The expression

$$\frac{\partial u}{\partial t} = L^* u + (\dot{y}h - \frac{h^2}{2})u \tag{2.a}$$

is called the Duncan-Mortensen-Zakai (DMZ) equation that corresponds to $X_o, \ldots, X_m$ and h. The variable u stands for a function or distribution on $[0,T] \times \mathbb{R}^n$, and $\dot{y}$ means dy/dt. The problem to be discussed here is that of the existence, uniqueness, and continuous dependence on y and on the initial condition ϕ, of the solutions of the Cauchy problem that consists of (2.a) together with the initial condition

$$u(0,x) = \phi(x) \tag{2.b}$$

One wants to prove this for large classes of ϕ's and y's.
More precisely, one wants to find "good" spaces U_o, $U, \mathcal{Q}_T$ such that U_o, U are spaces of distributions on $\mathbb{R}^n$ and on $[0,T] \times \mathbb{R}^n$, respectively, that $\mathcal{Q}_T$ is a subspace of C_T^o, and that

(I) *For every* $\phi \in U_o$, $y \in \mathcal{Q}_T$, *there exists one and only solution* $u \in U$ *of* (2.a,b), *and this solution depends continuously on* ϕ *and* y.

Our problem arises from the theory of nonlinear filtering. As explained in the Appendix, if ϕ is a probability measure, and y is a path of the observation process, then the solution u of (2.a,b) is a function of $t \in [0,T]$ with values finite Borel measures on $\mathbb{R}^n$, such that, for each t, the measure $(\int_{\mathbb{R}^n} u(t,x)dx)^{-1}u(t,.)$ is a probability measure on $\mathbb{R}^n$. Moreover, the observation process is "almost" a Wiener process (cf. the Appendix) and, in particular, if P_T denotes standard Wiener measure on C_T^o, then a set S of y paths has probability one iff $P_T(S) = 1$. So, if we want to solve (2.a,b) for almost all paths y, we need $\mathcal{Q}_T$ to be so large that $P_T(\mathcal{Q}_T) = 1$. If we let Bor $(\mathbb{R}^n)$ denote the set of all finite Borel measures on $\mathbb{R}^n$, it follows that the case of interest for nonlinear filtering is when

(II) $U_o \subseteq \mathrm{Bor}(\mathbb{R}^n)$, U *is a space of functions from* $[0,T]$ *to* $\mathrm{Bor}(\mathbb{R}^n)$, *and* $\mathcal{Q}_T$ *is a subset of* C_T^o *which satisfies.*

$$P_T(\mathcal{Q}_T) = 1. \tag{3}$$

Notice that $P_T(C_T^\alpha) = 0$ if $\frac{1}{2} \leq \alpha$, but $P_T(C_T^\alpha) = 1$ if $0 \leq \alpha < \frac{1}{2}$. Therefore, a good choice of $\mathcal{Q}_T$ would be C_T^α, if $\alpha < \frac{1}{2}$. (But $\alpha \geq \frac{1}{2}$ would not be satisfactory). Notice also that, if u is a $\mathrm{Bor}(\mathbb{R}^n)$-valued function of $t \in [0,T]$, and if $y \in C_T^o$, then it is not a priori clear how to define the product of the distributions u and $\dot{y}$. Therefore, it is not completely clear what exactly is meant by u being a "solution" of (2.a). (It would be clear if we limited ourselves to $y \in C_T^1$ but this would not suffice to get a $\mathcal{Q}_T$ for which (3) holds. It would also be clear if u were C^1 as a function of t, but it is easy to see that, in general, one cannot expect this to be the case). So, before we can even state (I), we must solve the preliminary problem.

(I-) *define what it means for a* $u \in U$ *to be a solution of* (2.a) *for a given* $y \in \mathcal{Q}_T$.

So, summarizing, we can now state our problem as follows :

Find spaces U_o, $U, \mathcal{Q}_T$ *such that* (II) *holds, and that one can solve* (I-) *in such a way that* (I) *holds.*

Naturally, the larger U_o and $\mathcal{Q}_T$ are, and the smaller U is for a given $U_o, \mathcal{Q}_T$, the better our theorem will be.

We shall make the following hypothesis on $X_o, \ldots, X_m$:

(H1) *All the derivatives of the coefficients of the* X_i *, of all orders* ≥ 1 *, are bounded.*

(Notice that the X_i are allowed to have linear growth). However, *we do not want to make restrictive hypotheses on the growth of* h.

Hypothesis (H1) guarantees the existence and uniqueness of the solutions of (2.a,b) if $\phi \in \mathcal{D}(\mathbb{R}^n)$ and $y \in C_T^1$. (Because, if y is Lipschitz with constant K, then the function $(t,x) \mapsto \dot{y}(t)h(x) - \frac{h(x)^2}{2}$ is bounded above by $\frac{1}{2}K^2$). So there is a well defined "solution map" $(\phi,y) \to u_{\phi,y}$ from $\mathcal{D}(\mathbb{R}^n) \times C_T^1$. Suppose that U_o, U, $\mathcal{Q}_T$ are spaces for which (II) holds, and which are such that $\mathcal{D}(\mathbb{R}^n) \times C_T^1$ is dense in $U_o \times \mathcal{Q}_T$.
Then, if we manage to solve (I-) and then to prove (I), we will have cons-

tructed a continuous extension $\tilde{u}$ of u to $U_o \times \mathcal{Q}_T$. Since such a continuous extension, if it exists, is unique, we could actually use it to circumvent (I-). We could *define* "solution" of (2.a,b) by saying that, for a given $\bar{\phi} \in U_o$, $\bar{y} \in \mathcal{Q}_T$, the distribution $\bar{u}$ is a *solution* of (2.a,b) if $\bar{u} = \tilde{u}_{\bar{\phi},\bar{y}}$, where $\tilde{u}$ is a continuous extension, to some neighborhood of $(\bar{\phi},\bar{y})$ in $U_o \times \mathcal{Q}_T$, of the solution map $(\phi,y) \to u_{\phi,y}$. We shall refer to this as the "continuous extension approach" (CEA). Notice that, if we follow the CEA, then the uniqueness of solutions is trivial.

There is another approach, called the REA (robust equation approach), which does not involve a definition by continuous extension. We make the transformation

$$u(t,x) = e^{y(t)h(x)}\, v(t,x) \,, \tag{4}$$

which gives the equation

$$\frac{\partial v}{\partial t} = \tilde{L}_t^* v + g_y v \,, \tag{5}$$

where

$$\tilde{L}_t = L + y(t) \sum_{i=1}^{m} (X_i h) X_i \tag{6}$$

and

$$g_y(t,x) = y(t)(Lh)(x) + \frac{y(t)^2}{2} \sum_{i=1}^{m} (X_i h)^2(x) - \frac{h(x)^2}{2} \,. \tag{7}$$

Since equation (6) does not contain $\dot{y}$, one can *define* a concept of "solution" of (2.a) by declaring a $\mathrm{Bor}(\mathbb{R}^n)$-valued function $t \to u(t,.)$ to be a solution of (2.a) if the v defined by (4) is a solution of (5) and of $v(0,x) = \phi(x)$. (Notice that the various multiplications by y that are required to make sense of (5) are well defined, since one is always multiplying functions from $[0,T]$ to $\mathcal{D}'(\mathbb{R}^n)$ by functions $y \in C_T^o$.

Equation (5) is the *robust* DMZ *equation*. If we follow the REA, we must prove existence, uniqueness and continuous dependence on (ϕ,y) of the solutions of (5), in appropriate spaces. Once this is done, the transformation (4) can be used to construct the continuous extension $\tilde{u}$, i.e. to carry out successfully the CEA. On the other hand, the CEA by itself does not suffice to obtain the uniqueness of the solutions of (5) or, equivalently, of the solutions of (2.a) in the REA sense. Thus the REA, when it works, gives a better result than the CEA.

In this note we shall concentrate on the CEA, and we will use the Feynman-Kac formula to get the continuous extension. It turns out that, when this method works, then one can also prove uniqueness for the robust equation (and thus actually carry out the REA) under some minor extra technical conditions (amounting, essentially, to the requirement that one can prove existence for the adjoint equation as well, and thus uniqueness for the original equation).

2. REDUCTION TO THE PROOF OF L^1 BOUNDS

In order to apply the Feynman-Kac formula, we let $\mathfrak{W}$ denote the space of continuous functions $w = [0,\infty[\to \mathbb{R}$ such that $w(0) = 0$, and we let $\mathfrak{W}^m$ denote the product of m copies of $\mathfrak{W}$. If $w = (w^1,\dots,w^m) \in \mathfrak{W}^m$, and $t \in [0,\infty[$, we write $W_t^i(w) = w^i(t)$. If P^m denotes Wiener measure on $\mathfrak{W}^m$, (and $\mathcal{A}_t$ is the σ-algebra generated by the W_s^i, $0 \le s \le t$, $i = 1,\dots,m$), then $W = (W^1,\dots,W^m)$ is a standard m-dimensional Wiener process adapted to the σ-algebras $\mathcal{A}_t$, on the probability space $(\mathfrak{W}^m,\mathcal{A}_\infty,P^m)$.
For each $z \in \mathbb{R}^n$, let $\xi^z = \{\xi_t^z \mid t \ge 0\}$ be the stochastic process which

is the solution of the stochastic differential equation

$$d\xi_t = X_o(\xi_t)dt + \sum_{i=1}^{m} X_i(\xi_t)dW_t^i \tag{8.a}$$

with the initial condition

$$\xi_o = z \ . \tag{8.b}$$

(Equation (8.a) is interpreted in the Fisk-Stratonovich sense. We will write $d, \int$ when we talk about Stratonovich differentials or integrals, and $\tilde{d}, \tilde{\int}$ when we mean the differentials or integrals in the Ito sense. For those who prefer the Ito formalism, equation (8.a) is equivalent to the Ito equation

$$\tilde{d}\xi_t = \tilde{X}_o(\xi_t)dt + \sum_{i=1}^{m} X_i(\xi_t)dW_t^i \tag{9}$$

where

$$\tilde{X}_o(x) = X_o(x) + \frac{1}{2} \sum_{i=1}^{m} [(\frac{\partial X_i}{\partial x})X_i] (x) \ . \tag{10}$$

Here the X_i are written as column vectors and $\frac{\partial X_i}{\partial x}$ is the square matrix with columns $\frac{\partial X_i}{\partial x_j}$, $j = 1,\dots,n)$.

If $\mu \in \mathrm{Bor}(\mathbb{R}^n)$, $y \in C_T^1$, $0 \le t \le T$, and ψ is a bounded Borel function on $\mathbb{R}^n$, define

$$B_t^y(\mu,\psi) = \int_{\mathbb{R}^n} [E(\psi(\xi_t^z)U_t^{z,y})] \, d\mu(z) \ , \tag{11}$$

where

$$U_t^{z,y} = e^{\int_0^t \dot{y}(s)h(\xi_s^z)ds - \frac{1}{2}\int_0^t h(\xi_s^z)^2 ds} , \tag{12}$$

and E denotes expectation with respect to Wiener measure.

The function $U_t^{z,y}$ from $\mathcal{W}^m$ to $\mathbb{R}$ is bounded above.
Indeed, we have

$$U_t^{z,y}(w) \leq e^{\frac{1}{2}\int_0^t \dot{y}(s)^2 ds} \qquad \text{for all } w \in \mathcal{W}^m . \tag{13}$$

So we have $|B_t^y(\mu,\psi)| \leq c\|\mu\|_{TV} \ \|\psi\|_{L^\infty}$ where c is a y-dependent constant and $\|\mu\|_{TV}$ is the total variation of μ. Therefore there are measures $\beta_t^y(u) \in \text{Bor}(\mathbb{R}^n)$ such that $<\beta_t^y(\mu),\psi> = B_t^y(\mu,\psi)$ for ψ Borel-measurable and bounded on $\mathbb{R}^n$. The Feynman-Kac formula is, essentially, the assertion that the map $t \to \beta_t^y(\mu)$ is the solution of (2.a) with initial condition μ. (The verification that $t \to \beta_t^y(\mu)$ solves (2.a) is by a standard stochastic differentiation. The sense in which the initial condition holds is also clear, since $\beta_t^y(\mu)$ depends continuously on t, in the sense that $t \to <\beta_t^y(\mu),\psi>$ is continuous for each bounded continuous $\psi : \mathbb{R}^n \to \mathbb{R}$).

The preceding construction, valid if $y \in C_T^1$, also leads to a method for constructing solutions for more general y's. The key observation, due to J.M.C. Clark, is that $U_t^{z,y}$ actually makes perfect sense for arbitrary $y \in C_T^o$, if $0 \leq t \leq T$. Indeed, one can rewrite the first integral of the exponent of (12) by an integration by parts, thereby obtaining :

$$\begin{cases} \int_0^t \dot{y}(s)h(\xi_s^z)ds = y(t)h(\xi_t^z) - \tilde{\int_0^t} y(s)\tilde{d}[h(\xi_s^z)] \\ \qquad\qquad\qquad\quad = y(t)h(z) + A(y,t,z) \end{cases} \tag{14}$$

where

$$A(y,t,z) = \tilde{\int_0^t} [y(t)-y(s)] \ \tilde{d}[h(\xi_s^z)] . \tag{15}$$

So

$$U_t^{z,y} = e^{y(t)h(z)+A(y,t,z) - 1/2\int_0^t h(\xi_s^z)^2 ds} \tag{16}$$

This formula, which was derived for $y \in C_T^1$, enables us to assign a meaning to $U_t^{z,y}$ for arbitrary $y \in C_T^o$, $t \in [0,T]$, $z \in \mathbb{R}^n$. However, we no longer know that $U_t^{z,y}$ is bounded above, and so we cannot define $B_t^y(\mu,\psi)$ and $\beta_t^y(\mu)$ as was done before when y was in C_T^1. On the other hand, requiring $U_t^{z,y}$ to be bounded is certainly too much, since all that is needed to define the $\beta_t^y(\mu)$ is an L^1 estimate. It turns out that this is actually all that is needed : once we get an L^1 estimate for $U_t^{z,y}$, then everything else follows easily. Before we embark upon the really crucial task of finding the L^1 estimates, we prove a simple Lemma which formalizes our assertion that the estimates are all we need. We use the following notation : a) $B(\rho)$ denotes the ball of radius ρ in $\mathbb{R}^n$, centered at 0, b) $U^{\#}(\rho)$ is the space of all Borel measures in $\mathbb{R}^n$ whose support is contained in $B(\rho)$, equipped with the total variation norm, c) $U_o^{\#}$ is the space of all Borel measures on $\mathbb{R}^n$ which have compact support, topologized as the inductive limit of the $U_o^{\#}(\rho)$, d) $U_T^{\#}$ is the space of all continuous functions $\sigma : [0,T] \to \mathrm{Bor}(\mathbb{R}^n)$ ("continuity" means that $t \to \langle \sigma(t),\psi \rangle$ is continuous for every bounded continuous $\psi : \mathbb{R}^n \to \mathbb{R}$); the space $U_T^{\#}$ is equipped with the topology defined by the seminorms $|\sigma|_\psi = \sup_{0 \le t \le T} |\langle \sigma(t),\psi \rangle|$, for $\psi : \mathbb{R}^n \to \mathbb{R}$ continuous and bounded, e) an *admissible space* on $[0,T]$ is a space $\mathcal{Q}_T$ of functions on $[0,T]$, such that $C_T^1 \subseteq \mathcal{Q}_T \subseteq C_T^o$, which is equipped with a norm $\|\ldots\|_T$ in such a way that the inclusions $C_T^1 \subset \mathcal{Q}_T$ and $\mathcal{Q}_T \subset C_T^o$ are continuous, and C_T^1 is dense in $\mathcal{Q}_T$, f) if $\mathcal{Q}_T$ is an admissible space on $[0,T]$, and $\eta > 0$, $\mathcal{Q}_T(\eta)$ is the ball of radius η in $\mathcal{Q}_T$, with center 0.

We say that *local* L^1 bounds *hold for* y *in* $\mathcal{Q}_T$, if

(i) the function $U_t^{z,y}$ belongs to $L^1(\mathcal{W}^m, \mathcal{A}_\infty, P^m)$ for each $y \in \mathcal{Q}_T$,

and

(ii) for each ρ,η, the quantity

$$K(\rho,\eta) = \sup \{ \| U_t^{z,y} \|_{L^1} : z \in B(\rho),\ y \in \mathcal{Q}_T(\eta),\ t \in [0,T] \}$$

is finite.

<u>Lemma</u>. *Let* $\mathcal{Q}_T$ *be an admissible space on* $[0,T]$ *, and suppose that local* L^1 *bounds hold for* $y \in \mathcal{Q}_T$ *. Then there exists a continuous map* λ *from* $U_o^{\#} \times \mathcal{Q}_T$ *to* $U_T^{\#}$ *which, to each* $\mu \in U_o^{\#}$, $y \in \mathcal{Q}_T$ *, assigns a continuous map* $t \to \beta_t^y(\mu)$ *from* $[0,T]$ *to* $\mathrm{Bor}(\mathbb{R}^n)$ *, in such a way that, if* $y \in C_T^1$ *, then* $t \to \beta_t^y(\mu)$ *is the solution of the* DMZ *equation for* y *with initial condition* μ .

<u>Remark</u>. The hypotheses of the Lemma are "minimal", in the sense that it is hard to figure out any weaker hypotheses under which one might get any conclusion at all. Indeed, the L^1 norm of $U_t^{z,y}$ is clearly the total measure of $\mathbb{R}^n$ under the measure $\beta_t^y(\delta_z)$ (where δ_z is a Dirac delta at z). The condition that $\| U_t^{z,y} \|_{L^1}$ is finite, for a given y , is therefore necessary if we want the DMZ equation for that y to have a fundamental solution which, for each t , is a finite measure. The Lemma only assumes, in addition to this necessary condition, that the L^1 norms of the $U_t^{z,y}$ remain bounded when z,y are bounded, and it is hard to imagine how this could fail to hold if condition (i) holds. Not surprisingly, the conclusion of the Lemma is also "minimal", in that it only gives solutions for an initial condition with compact support, and the solution is only asserted to belong to a very large class of distributions. Under more restrictive assumptions one can prove better results (cf. Sussmann[3] for the details). For instance, suppose that, instead of assuming local L^1 bounds, we assume local L^p bounds for

all $p \in [1,\infty[$. Suppose also that h and all its derivatives grow not faster than polynomially. Then in the expression (11), if one computes $B_t^y(\mu, \frac{\partial\psi}{x_i})$ instead of $B_t^y(\mu,\psi)$, one can write $\frac{\partial\psi}{x_i}(\xi_t^z)$ as a differential operator in the z variables, acting on $\psi(\xi_t^z)$. If we let $M^z = \{M_t^z\}$ be the square-matrix-valued process which solves

$$dM_t^z = \frac{\partial X_o}{\partial x}(\xi_t^z)M_t^z\,dt + \sum_{i=1}^{m} \frac{\partial X_i}{\partial x}(\xi_t^z)M_t^z\,dW_t^i \tag{17.a}$$

$$M_o^z = \text{identity} \tag{17.b}$$

then it is easy to see that, if

$$\zeta_t^{j,z} = \frac{\partial}{\partial z_j}(\xi_t^z), \text{ then } \zeta_t^{j,z} = M_t^z e_j \text{ , where } e_j = (0,\dots,0,1,0,\dots,0)$$

the 1 being in the j-th place. We have

$$\text{grad}_z\,[\psi(\xi_t^z)] = [(\text{grad}_x\psi)(\xi_t^z)]\,M_t^z\ . \tag{18}$$

The matrix M_t^z is almost surely nonsingular, since $(M_t^z)^{-1}$ satisfies an equation similar to (17.a), which it is easy to write down. So

$$(\text{grad}_x\psi)\,(\xi_t^z) = \text{grad}_z\,[\psi(\xi_t^z)\,](M_t^z)^{-1} \tag{19}$$

and

$$\frac{\partial\psi}{\partial x_i}(\xi_t^z) = \{\text{grad}_z[\psi(\xi_t^z)]\}\,(M_t^z)^{-1}e_i\ . \tag{20}$$

Therefore, if $\psi \in \mathfrak{D}(\mathbb{R}^n)$, we have

$$\begin{aligned} \langle\beta_t^y(\mu)\,,\frac{\partial\psi}{\partial x_i}\rangle &= \int_{\mathbb{R}^n} E(\frac{\partial\psi}{\partial x_i}(\xi_t^z)U_t^{z,y})\,d\mu(z) \\ &= \int_{\mathbb{R}^n} E[(\{\text{grad}_z[\psi(\xi_t^z)]\}\,(M_t^z)^{-1}e_i)U_t^{z,y}]\,d\mu(z)\ . \end{aligned} \tag{21}$$

The z-differentiation can then be shifted to the other factors, provided that μ is a C^1 function, i.e. that $d\mu(z) = \phi(z)dz$, $\phi \in C^1(\mathbb{R}^n)$. This yields an expression for $B_t^y(\mu, \frac{\partial\psi}{\partial x_i})$ of the form

$$\int_{\mathbb{R}^n} E[\psi(\xi_t^z)(a_{i,t}^z\phi(z)+ \psi\ b_{ij,t}^z\ \frac{\partial\phi}{\partial z_j}(z))U_t^{z,y}]\,dz\ . \tag{22}$$

The $a_{i,t}^z$, $b_{ij,t}^z$ can be shown to be in $L^p(\mathcal{W}^m, \mathcal{A}_\infty, P^m)$ for $p \in [1,\infty[$ and, since the same is true for $U_t^{z,y}$, one gets an estimate

$$|B_t^y(\mu, \frac{\partial\psi}{\partial x_i})| \leq \text{constant} \times |\psi|_{L^\infty}. \tag{23}$$

This can be iterated, and one ends up proving that all the distributional derivatives of the $\beta_t^y(\mu)$, of all orders, are finite Borel measures, so that the $\beta_t^y(\mu)$ are in fact smooth functions, if μ is.

As a second example, if the operator $\frac{\partial}{\partial t} - L^*$ satisfies Hormander-like conditions (precisely, if the linear span of $X_1(x),\ldots,X_m(x)$, and of all brackets of 2 or more X_i's , $0 \leq i \leq m$, is all of $\mathbb{R}^n$ at each x) then one can get smoothness of $\beta_t^y(\mu)$ for $t > 0$, even if μ itself is not smooth, by means of the Malliavin calculus.

As a third example, if one has bounds which are more global (e.g. if $K(\rho,\eta)$ remains bounded as $\rho \to \infty$), then one can construct $\beta_t^y(\mu)$ even when μ does not have compact support.

<u>Proof of the Lemma</u>. Let $\mu \in U_o^\#$, $y \in \mathcal{Q}_T$. We define $B_t^y(\mu,\psi)$ for a bounded Borel function $\psi : \mathbb{R}^n \to \mathbb{R}$ by equation (11). If $\mu \in U_o^\#(\rho)$, and $y \in \mathcal{Q}_T(\eta)$, then we have the bound

$$|B_t^y(\mu,\psi)| \leq K(\rho,\eta)\, \|\mu\|_{TV}\, \|\psi\|_{L^\infty}. \tag{24}$$

So we can define measures $\beta_t^y(\mu) \in Bor(\mathbb{R}^n)$ by

$$< \beta_t^y(\mu),\psi > = B_t^y(\mu,\psi) \quad , \tag{25}$$

and we have the bound

$$\| \beta_t^y(\mu) \|_{TV} \leq K(\rho,\eta) \|\mu\|_{TV} \quad . \tag{26}$$

Now let y, y' belong to $\mathcal{Q}_T(\eta)$. We can write

$$U_t^{z,y'} - U_t^{z,y} = \int_0^1 \frac{\partial}{\partial\theta} [U_t^{z,y_\theta}] d\theta \tag{27}$$

where

$$y_\theta = \theta y' + (1-\theta) y \ . \tag{28}$$

Then

$$\frac{\partial}{\partial\theta} [U_t^{z,y_\theta}] = DU_t^{z,y_\theta} \tag{29}$$

where $D = D_1 + D_2$, and

$$D_1 = [y'(t)-y(t)] \ h(z) \tag{30.a}$$

$$D_2 = \tilde{\int}_0^t (y'(t)-y(t)- [y'(s)-y(s)]) \ \tilde{d}h(\xi_s^z) \ . \tag{30.b}$$

Let v be the function $\|y'-y\|_T^{-1}(y'-y)$, so that $v \in \mathcal{Q}_T(1)$.
We have :

$$D_2 = \|y'-y\|_T \ \tilde{\int}_0^t [v(t)-v(s)] \ \tilde{d}h(\xi_s^z) \tag{31}$$

and

$$| \tilde{\int}_0^t [v(t)-v(s)] \ \tilde{d}h(\xi_s^z)| \leq Cos\ h \quad \tilde{\int}_0^t [v(t)-v(s)] \ \tilde{d}h(\xi_s^z) \tag{32}$$

so that

$$|D_2 U_t^{z,y_\theta}| \le \frac{1}{2} \|y'-y\|_T \ (U_t^{z,y_\theta+v} + U_t^{z,y_\theta-v}) \tag{33}$$

and then

$$\| D_2 U_t^{z,y_\theta} \|_{L^1} \le \|y'-y\|_T \ K(\rho,1+\eta) \tag{34}$$

since the functions $y_\theta \pm v$ are in $\mathcal{Q}_T(1+\eta)$. On the other hand, if γ is a constant such that $\|f\|_{L^\infty} \le \gamma \|f\|_T$ for $f \in \mathcal{Q}_T$, then we have $|D_1| \le \gamma\, \nu(\rho)\ \|y'-y\|$, where $\nu(\rho) = \sup\{|h(z)| : z \in B(\rho)\}$. So

$$\| D_1 U_t^{z,y_\theta} \|_{L^1} \le \gamma\nu(\rho)K(\rho,\eta)\ \| y'-y\|_T \ .$$

Therefore

$$\| U_t^{z,y'} - U_t^{z,y} \|_{L^1} \le K^{\#}(\rho,\eta)\ \|y'-y\|_T \ , \tag{35}$$

where $K^{\#}(\rho,\eta) = \gamma\nu(\rho)K(\rho,\eta) + K(\rho,1+\eta)$.

So

$$\| \beta_t^{y'}(\mu) - \beta_t^{y}(\mu) \|_{TV} \le K^{\#}(\rho,\eta)\ \|y'-y\|_T \ . \tag{36}$$

Using (36) we can prove, first of all, that the map $t \to \beta_t^y(\mu)$ is continuous for each $y \in \mathcal{Q}_T$, $\mu \in U_o^{\#}$. We must show that $t \to \langle \beta_t^y(\mu),\psi\rangle$ is continuous if $\psi : \mathbb{R}^n \to \mathbb{R}$ is continuous and bounded. Since C_T^1 is dense in $\mathcal{Q}_T$, we can pick a sequence $\{y_k\}$ of elements of C_T^1 that converges to y in $\mathcal{Q}_T$. The bound (36) then implies that the functions $t \to \langle \beta_t^{y_k}(\mu),\psi\rangle$ converge uniformly to $t \to \langle \beta_t^y(\mu),\psi\rangle$. So it suffices to prove that $t \to \langle \beta_t^y(\mu),\psi\rangle$ is continuous if $y \in C_T^1$. But, in this case, the continuity is trivial. Indeed, the functions $|\psi(\xi_t^z)U_t^{z,y}|$ (from $\mathbb{R}^n \times \mathcal{W}^m$ to $\mathbb{R}$) are bounded by $e^{\frac{1}{2}\int_0^T \dot{y}(s)^2 ds} \| \psi \|_{L^\infty}$, which does not depend on z, w, or t.

For fixed z, $\psi(\xi_t^z)U_t^{z,y}$ is almost surely continuous in t, because ξ_t^z has continuous sample paths, and ψ is continuous. So, for each z, $t \to E(\psi(\xi_t^z)U_t^{z,y})$ is continuous in t. Since $|E(\psi(\xi_t^z)U_t^{z,y})|$ is also bounded by a constant, independent of z,t, we conclude that $<\beta_t^y(\mu),\psi>$ is continuous in t, as stated.

For $\mu \in U_o^\#$, $y \in \mathcal{Q}_T$, let $\lambda(\mu,y)$ be the map $t \to \beta_t^y(\mu)$. Then we have proved that $\lambda(\mu,y)$ is in $U_T^\#$. Moreover, if ψ is continuous and bounded on $\mathbb{R}^n$, the bounds (26) and (36) give, if $\mu_i \in U_o^\#(\rho)$, $y_i \in \mathcal{Q}_T(\eta)$, $i = 1,2$:

$$|\lambda(\mu_1,y_1)-\lambda(\mu_2 y_2)|_\psi \leq K^\#(\rho,\eta)\,\|y_1-y_2\|_T + K(\rho,\eta)\|\mu_1-\mu_2\|_{TV} \qquad (37)$$

This shows that λ is indeed continuous from $U_o^\# \times \mathcal{Q}_T$ to $U_T^\#$, and *completes the proof of the Lemma.*

3. VERIFYING THE L^1 BOUNDS

We now turn to the most interesting question, which is that of finding conditions under which local L^1 bounds can be shown to hold for some reasonable spaces $\mathcal{Q}_T$. We shall present several methods for doing so.

First method. If $v : [0,T] \to \mathbb{R}$ is a continuous function, let

$$V_t^{z,v} = e^{\tilde{\int}_0^t v(s)\tilde{d}h(\xi_s^z) - \frac{1}{2}\tilde{\int}_0^t h(\xi_s^z)^2 ds} \qquad (38)$$

The stochastic differential of $V_t^{z,v}$ with respect to t (in Ito's sense) is

$$\tilde{d}V_t^{z,v} = V_t^{z,v}\,[\,v(t)\tilde{d}h(\xi_t^z) + \frac{v(t)^2}{2}\,[\tilde{d}h(\xi_t^z)]^2 - \frac{1}{2}h(\xi_t^z)^2 dt\,] \tag{39}$$

The coefficient of dt in this formula is $g_v(t\,,\xi_t^z)$, where g is precisely the function defined in (7). Suppose that :

(C1) for each $\varepsilon > 0$ there is an $\alpha(\varepsilon)$ such that

$$\sum_{i=1}^{m}(X_i h)^2 + |Lh| \leq \varepsilon h^2 + \alpha(\varepsilon) \tag{40}$$

Then $g_v(t,\xi_t^z)$ is bounded above, for each v, by a constant K which can be chosen independently of z, v and t, as long as $t \in [0,T]$, and that $\| v \|_{L^\infty}$ stays bounded. One can write $V_t^{z,v}$ as a stochastic integral

$$V_t^{z,v} = 1 + \int_0^t g_v(s,\xi_s^z)V_s^{z,v}ds + \sum_{i=1}^{m}\int_0^t v(s)(X_i h)(\xi_s^z)V_s^{z,v}\,dW_s^i \tag{41}$$

Proceeding as if the Ito integrals in (41) were martingales, we get

$$E(V_t^{z,v}) \leq 1 + K\int_0^t E(V_s^{z,v})ds\ , \tag{42}$$

which gives the bound

$$E(V_t^{z,v}) \leq e^{tK} \tag{43}$$

Naturally, we do not know that the Ito integrals in (41) are martingales. However, since $V_t^{z,v}$ is positive, a simple argument (using stopping times and Fatou's Lemma) yields the bound (43) rigorously. Since

$$U_t^{z,y} = e^{y(t)h(z)}\,V_t^{z,v_t(y)}\qquad , \tag{44}$$

where $v_t(y)$ is the function $s \to y(t) - y(s)$, the local L^1 bounds follow from (43).

So we have established :

Proposition 1. *The local* L^1 *bounds hold for* $y \in C_T^0$ *if condition* (C1) *holds.*

(Not surprisingly, condition (C1) is precisely what is needed for the zeroth-order coefficient of the robust equation to be bounded above. Thus, in this case, one could equally well have studied the problem by PDE methods).

Second method. One tries to estimate the expectation of

$$Z_t^{z,v} = e^{\zeta_t^{z,v}} \tag{45}$$

where

$$\zeta_t^{z,v} = \tilde{\int_0^t} v(s)\tilde{d}h(\xi_s^z) \tag{46}$$

If one can prove that $E(Z_t^{z,v})$ is finite, in a suitably uniform way, then we get the local L^1 bounds, since $V_t^{z,v} \leq Z_t^{z,v}$. To estimate $E(Z_t^{z,v})$ we use the elementary observation that, if f is in L^p for all finite p, and satisfies

$$\| f \|_{L^p} \leq Cp^{\alpha} \tag{47}$$

for some $C > 0$, $\alpha < 1$, and for all $p \in [1,\infty[$, then e^f is in L^1, and the L^1 norm of e^f is bounded by a constant which only depends on C and α.

So we must prove that the L^p norms of $\zeta_t^{z,v}$ do not grow too fast as $p \to \infty$. Since

$$\zeta_t^{z,v} = \int_0^t v(s)(Lh)(\xi_s^z)ds + \sum_{i=1}^m \int_0^t v(s)(X_i h)(\xi_s^z)dW_s^i \;, \tag{48}$$

the Burkholder-Hunt-Gundy inequalities give the estimate

$$\| \zeta_t^{z,v} \|_{L^p} \le \| v \|_{L^\infty} [\int_0^t \| Lh(\xi_s^z) \|_{L^p} ds + \tag{49}$$

$$+ p^{1/2} \sum_{i=1}^m (\int_0^t \| X_i h(\xi_s^z) \|_{L^p}^2 \, ds)^{1/2}$$

Now suppose that the following condition holds :

(C2,α) for each ρ there is a C such that

$$\| Lh(\xi_t^z) \|_{L^p} \le Cp^\alpha \tag{50}$$

and

$$\| X_i \, h \, (\xi_t^z) \|_{L^p} \le Cp^{\alpha - 1/2} \quad \text{for} \quad i = 1,\dots,m \tag{51}$$

for all $z \in B(\rho)$, $t \in [0,T]$.

One then gets an estimate of the form

$$\| \zeta_t^{z,v} \| \le C' p^\alpha \, . \tag{52}$$

If $\alpha < 1$, then we obtain the L^1 bound for $Z_t^{z,v}$. This proves :

Proposition 2. *Suppose that* (C2,α) *holds for some* $\alpha < 1$. *Then the local* L^1 *bounds hold for* $y \in C_T^0$.

The preceding method does not exploit at all the presence of the term $-\frac{1}{2}\int_0^t h(X_s)^2 ds$ in the definition of $U_t^{z,y}$. We now present an improved version which does make use of this term.

Third method. Let

$$\bar{\zeta}_t^{z,y} = \int_0^t \dot{y}(s) h(\xi_s^z) ds \tag{53}$$

so that

$$\bar{\zeta}_t^{z,y} = y(t)h(z) + \zeta_t^{z,v_t}(y) \qquad (54)$$

where $\quad v_t(v)(s) = y(t) - y(s)$.

Let $\bar{P}_z$ be the measure

$$d\bar{P}_z = e^{-\frac{1}{2}\int_0^t h(\xi_s^z)^2 ds} dP^m \qquad (55)$$

We estimate the L^p norms of $\bar{\zeta}_t^{z,y}$ when $y \in C_T^1$, relative to the measure $\bar{P}_z$. Since

$$|\bar{\zeta}_t^{z,y}| \leq \left(\int_0^t \dot{y}(s)^2 ds\right)^{1/2} \left(\int_0^t h(\xi_s^z)^2 ds\right)^{1/2} \qquad (56)$$

we have

$$\int_{\mathcal{W}^m} |\bar{\zeta}_t^{z,y}|^p \, d\bar{P}_z \leq \|\dot{y}\|_{L^\infty}^p \int_{\mathcal{W}^m} \Phi^p \, e^{-\frac{1}{2}\Phi^2} \, dP^m \qquad (57)$$

where Φ is the function $\left(\int_0^t h(X_s)^2 ds\right)^{1/2}$.

The upper bound $\Phi^p \, e^{(-1/2)\Phi^2} \leq (p/e)^{p/2}$ gives

$$\int_{\mathcal{W}^m} |\bar{\zeta}_t^{z,y}|^p \, d\bar{P}_z \leq \|\dot{y}\|_{L^\infty}^p (p/e)^{p/2} \qquad (58)$$

so that

$$\|\bar{\zeta}_t^{z,y}\|_{L^p(\mathcal{W}^m, \mathcal{A}_\infty, \bar{P}_z)} \leq \|y\|_{C_T^1} \left(\frac{p}{e}\right)^{1/2}. \qquad (59)$$

Now suppose that $(C2,\alpha)$ holds for some α (not necessarily < 1). Then we get estimates

$$\|\bar{\zeta}_t^{z,y}\|_{L^p(\mathcal{W}^m, \mathcal{A}_\infty, \bar{P}_z)} \leq \text{constant} \times p^\alpha \times \|y\|_{C_T^o} \qquad (60)$$

By interpolation, we get estimates

$$\| \bar{\zeta}_t^{z,y} \|_{L^p(\mathcal{W}^m, \mathcal{A}_\infty, \bar{P}_z)} \leq \text{constant} \times \|y\|_{C_T^\beta} \times p^{\frac{\beta}{2} + (1-\beta)\alpha} \tag{61}$$

if y is in the Hölder space C_T^β.

If

$$\frac{\beta}{2} + (1-\beta)\alpha < 1 \tag{62}$$

then the estimate (61) implies that the exponential of $\bar{\zeta}_t^{z,y}$ is integrable with respect to the measure $\bar{P}_z$, i.e. that $E(U_t^{z,y})$ is finite. The required uniformity also follows, and we get

<u>Proposition 3</u>. *Suppose that* (C2,α) *holds for some* $\alpha \geq 1$. *Then the local* L^1 *bounds hold for* $y \in C_T^\beta$, *if*

$$\beta > 1 - \frac{1}{2\alpha - 1}. \tag{63}$$

Notice that Proposition 3 gives a "good" result (i.e. a $\beta < \frac{1}{2}$) as long as $\alpha < \frac{3}{2}$, which is an improvement over the requirement $\alpha < 1$ of Proposition 2.

Finally, we combine all the cases considered before into a

Fourth method. Suppose we can break up the functions Lh, X_ih as follows

$$Lh = (Lh)_1 + (Lh)_2 \tag{64}$$

$$X_ih = (X_ih)_1 + (X_ih)_2 \tag{65}$$

and that the $(X_ih)_1$, $(Lh)_1$ satisfy the bounds for the first method, and the $(X_ih)_2$, $(Lh)_2$ those of the second or third method. That is, suppose that

(C4) *For each* $\varepsilon > 0$ *there is a* $K(\varepsilon)$ *such that*

$$\sum (X_i h)_1^2 + |(Lh)_1| \le \varepsilon h^2 + K(\varepsilon) \tag{66}$$

and that

(C4,α) *there exists, for each* ρ *, a constant* C *such that*

$$\| (X_i h)_2(\xi_t^z) \|_{L^p} \le C p^{\alpha - \frac{1}{2}} \quad \text{for } i = 1,\dots,m \tag{67}$$

and

$$\| (Lh)_2(\xi_t^z) \|_{L^p} \le C p^{\alpha} \ ,$$

for all $t \in [0,T]$, $z \in B(\rho)$, $p \in [1,\infty[$.

Then we can break up $U_t^{z,y}$ into a product $(U_t^{z,y})_1 \, (U_t^{z,y})_2$, where

$$(U_t^{z,y})_j = e^{\int_0^t v(s)\,[(Lh)_j(\xi_s^z)ds + \sum_{i=1}^m (X_i h)_j(\xi_s^z)dW_s^i] - \frac{1}{4}\int_0^t h(\xi_s^z)^2 ds} \tag{68}$$

The factor $(U_t^{z,y})_1$ can be handled by the first method, while $(U_t^{z,y})_2$ is handled by the second method if $\alpha < 1$, or by the third method if $\alpha \ge 1$. For both factors, the corresponding method applies equally well to produce L^p bounds for any p, and the fact that the factor $\frac{1}{2}$ has been replaced by $\frac{1}{4}$ makes no difference. In particular, we can get L^2 estimates for the $(U_t^{z,y})_j$, j = 1,2, and then obtain the desired L^1 estimate for $U_t^{z,y}$. So, we have proved :

<u>Theorem</u>. *Suppose that the functions* Lh, $X_i h$ *have decompositions* (64) (65), *such that* (C4) *holds, and that* (C4,α) *holds for an* $\alpha \ge 0$. *Then the local* L^1 *bounds hold for* $y \in C_T^\beta$, *provided that either* $\alpha < 1$ *and* $0 \le \beta \le 1$, *or* $\alpha \ge 1$ *and* $1 \ge \beta > 1 - \frac{1}{2\alpha - 1}$.

4. EXAMPLES

In order to verify conditions of the type $(C2,\alpha)$ or $(C4,\alpha)$, we shall use repeatedly the fact that, if G is a Gaussian random variable, then the L^p norms of G behave like $p^{1/2}$ as $p \to \infty$.

Example 1. $n = 1$, $m = 1$, $X_o = 0$, $X_1 = \frac{d}{dx}$, h a polynomial. Then $X_1 h = h'$, $Lh = \frac{1}{2} h''$, and the DMZ equation is

$$\frac{\partial u}{\partial t} = \frac{1}{2} \frac{\partial^2 u}{\partial x^2} + (\dot{y}h - \frac{h^2}{2})u . \tag{69}$$

Since h' and h'' are dominated by h, condition (C1) holds. So the first method works and the solution map $(\phi, y) \to u_{\phi,y}$ has a continuous extension to $y \in C_T^o$.

Example 2. (Linear filtering). Here n and m are arbitrary, X_o is a linear vector field, $X_1, \ldots, X_m$ are constant vector fields, and h is a linear function. Then the $X_i h$, $i = 1, \ldots, m$ are constant functions, and Lh is linear. The processes ξ^z are Gaussian, and so $(C2,\alpha)$ is satisfied with $\alpha = \frac{1}{2}$.

Example 3. $n = 2$, $m = 2$, $X_o = 0$, $X_i = \frac{\partial}{\partial x_i}$ for $i = 1,2$, $h(x_1, x_2) = ax_1^k + bx_2$. Here we encounter for the first time the main source of difficulties for this type of problem, namely, the fact that the derivatives of h need not be nominated by h. We have $X_1 h = ak\, x_1^{k-1}$, $X_2 h = b$, $Lh = \frac{ak(k-1)}{2} x_1^{k-2}$. The process $\xi^z = (\xi^{z,1}, \xi^{z,2})$ is Gaussian, but this only gives us the bound $\| X_1 h(\xi_t^z) \|_{L^p} \leq \text{constant } p^{k/2-1/2}$, which does not suffice to satisfy $(C2,\alpha)$ with an $\alpha < 1$, if $k \geq 2$. However, the fourth method works. Assume $a \neq 0$, $b \neq 0$. (The case when $ab = 0$ is trivial).

Let S be the set of points $(x_1,x_2) \in \mathbb{R}^2$ such that $|ax_1^k| > 2|bx_2|$. Break up X_1h and Lh into $X_1h = (X_1h)_1 + (X_1h)_2$, $Lh = (Lh)_1 + (Lh)_2$, where $(X_1h)_1 = \chi X_1h$, $(Lh)_1 = \chi Lh$, and χ is the indicator function of S. Then we have $|ax_1^k + bx_2| \geq \frac{|ax_1^k|}{2}$ for $(x_1,x_2) \in S$. From this we get the bound $|x_1^{k-1}| \leq c|h(x_1,x_2)|^{\gamma}$, for some c, γ with $\gamma<1$, if $(x_1,x_2) \in S$. This implies that, if $\varepsilon > 0$, the functions $[(X_1h)_1]^2 - \varepsilon h^2$ and $|(Lh)_1| - \varepsilon h^2$ are bounded above. On the other hand, if $(x_1,x_2) \notin S$, we get

$$|x_1^{k-1}| \leq c|x_2|^{1-1/k} \tag{70}$$

for some c . So we have

$$\| (X_1h)_2(\xi_t^z)\|_{L^p} \leq c \| |\xi_t^{z,2}|^{1-1/k}\|_{L^p} \tag{71}$$

and

$$\| (Lh)_2(\xi_t^z) \|_{L^p} \leq c \| |\xi_t^{z,2}|^{1-2/k} \|_{L^p} \tag{72}$$

Since $\xi_t^{z,2}$ is Gaussian, we get $\|(X_1h)_2(\xi_t^z)\|_{L^p} \leq \text{constant} \times p^{\alpha-1/2}$ with $\alpha = 1-1/2k$. (For $(Lh)_2(\xi_t^z)$ we get an even stronger bound than the one we need). Since $\alpha < \frac{1}{2}$, the fourth method works and we get solutions for all $y \in C_T^o$.

Example 4. n, m, X_o, X_1, X_2 as in Example 3, but $h(x_1,x_2) = ax_1^k + bx_2^2$, with $ab \neq 0$. One proceeds exactly as in Example 3, except that now, in the bounds (70), (71), (72), x_2 and $\xi_t^{z,2}$ are replaced by their squares. We then get condition $(C4,\alpha)$ with $\alpha = \frac{3}{2} - \frac{1}{k}$, so that we obtain solutions of the DMZ equation for $y \in C^{\beta}$, $\beta > \frac{1}{2} - \frac{1}{2k-2}$.

Example 5. Same as Example 4, but with $h(x_1,x_2) = x_1^3 + x_2^3$. In this case, the third method works with α exactly equal to 3/2, and then one gets solutions for $y \in C_T^\beta$, $\beta > \frac{1}{2}$. As explained before, this is not good enough.

Example 6. $n = 2$, $m = 2$, $X_o = 0$, $X_1 = \frac{\partial}{\partial x_1}$, $X_2 = x_1 \frac{\partial}{\partial x_2}$, $h = x_2$. The DMZ equation is

$$\frac{\partial u}{\partial t} = \frac{1}{2} \left(\frac{\partial^2 u}{\partial x_1^2} + x_1^2 \frac{\partial^2 u}{\partial x_2^2} \right) + \left(\dot{y} x_2 - \frac{x_2^2}{2}\right) u \, . \tag{73}$$

We have $X_1 h = 0$, $X_2 h = x_1$, $Lh = 0$. Since $\xi^{z,1}$ is a Gaussian process, $(C2,\alpha)$ holds with $\alpha = 1$. So (74) has solutions for $y \in C_T^\beta$, $\beta > 0$ arbitrary.

Clearly, the critical question at the moment is to find new techniques which will make it possible to improve the Hölder exponents in situations such as that of Example 4. It would also be desirable to find some examples of problems where, for some $\beta \geq 0$, it is not true that the L^1 bounds hold for $y \in C_T^\beta$. We suspect that such examples exist, but we do not know any.

APPENDIX

Let $x = \{x_t, t \geq 0\}$ be a stochastic process with values in $\mathbb{R}^n$, which satisfies a Stratonovich equation

$$dx = X_o(x)dt + \sum_{i=1}^{m} X_i(x)\, dW^i \, ,$$

where $W^1,\ldots,W^m$ are Wiener processes. We assume that all the processes are defined on a probability space $(\Omega,\mathcal{A},P)$, and progressively measurable with respect to an increasing family of σ-algebras $\mathcal{A}_t$. Let V be another Wiener process on $(\Omega,\mathcal{A},P)$, and independent from the W^i and x_o. Consider the observation process y given by

$$dy_t = h(x_t)dt + dV , \quad y(0) = 0 .$$

For $0 \leq t$, let $\mathcal{Y}_t$ denote the σ-subalgebra of $\mathcal{A}_t$ generated by the y_s, $0 \leq s \leq t$. The DMZ equation is supposed to give the conditional probability of x_t given $\mathcal{Y}_t$, except for a normalization constant. That is, if a particular path $y(\cdot)$ of the observation process has been observed, and if we plug this path into (2.a) and solve (with the initial condition being the probability distribution of x_0), then the solution will be a measure-valued function $t \to \mu_t$, such that $\mu_t(\mathbb{R}^n)^{-1}\mu_t$ is the conditional probability distribution of x_t given that the particular path $y(\cdot)$ has been observed.

The y process is not a Wiener process. However, it is well known that, for each T, there exists a probability measure π_T on $(\Omega,\mathcal{A}_T)$, such that π_T and P are mutually absolutely continuous, and that $\{y_t\}$ is a Wiener process with respect to this new measure. Therefore, any property of Wiener paths which holds almost surely, is also true almost surely for the y paths.

REFERENCES

[1] E. Pardoux, Stochastic R.D.E.'s and filtering of diffusion processes, Stochastics 3 (1979), pp. 147-182.

[2] M. Hazewinkel, S. Marcus, H.J. Sussmann, Nonexistence of finite dimensional filters for the cubic sensor problem, to appear.

[3] H.J. Sussmann, Rigorous results on robust nonlinear filtering, to appear.

H.J. SUSSMANN

Mathematics Department
Rutgers University
New Brunswick
New Jersey 08903

U.S.A.

Abstracts—Résumés

I.J. BAKELMAN. VARIATIONAL PROBLEM CONNECTED WITH THE MONGE-AMPERE EQUATION

Abstract : Let G be an n-dimensional convex bounded domain in the Euclidian space R^n with Cartesian coordinates $x_1, x_2, \ldots, x_n$ and H be any convex subdomain of G such that $\mathrm{dist}(\bar{H}, \partial G) = h_H > 0$. In this paper it is proved that the convex generalized solution $u(x_1, x_2, \ldots, x_n)$ of the Dirichlet problem

$$\det \| u_{x_i x_j} \| = f(x_1, .., x_n) ,$$

$$u|_{\partial G} = 0$$

(where f is a positive summable function in G), can be obtained as $\inf_H u_H(x_1, x_2, \ldots, x_n)$, where $u_H(x_1, x_2, \ldots, x_n)$ is the solution of the variational problem

$$I_H(u) = - \int_G u(w, de) + (n+1) \int_G f_H u dx$$

$$u|_{\partial G} = 0$$

in the class of all continuous non-positive functions $u(x_1, x_2, \ldots, x_n)$, satisfying the condition $u|_{\partial G} = 0$. Here $w(x_1, x_2, \ldots, x_n)$ is the convex function, spanned of the graph of the fonction

$$v(x_1, \ldots, x_n) = \begin{cases} u(x_1, \ldots, x_n) & \text{if} \quad (x_1, \ldots, x_n) \in H \\ 0 & \text{if} \quad (x_1, \ldots, x_n) \in G\text{-}H \end{cases}$$

from below, $\omega(w, e)$ is the area of the generalized tangential mapping of the function $w(x_1, \ldots, x_n)$, constructed by the supporting hyperplanes to the graph of $w(x_1, \ldots, x_n)$ and

$$f_H(x_1, \ldots, x_n) = \begin{cases} f(x_1, \ldots, x_n) & \text{if} \quad (x_1, \ldots, x_n) \in H \\ 0 & \text{if} \quad (x_1, \ldots, x_n) \in G\text{-}H \end{cases}$$

Résumé : Soit G un domaine convexe borné de l'espace $\mathbb{R}^n$ de coordonnées cartésiennes $x_1,\ldots,x_n$. Soit H un sous-domaine convexe de G tel que $\mathrm{dist}(\overline{H},\partial H) = h_H > 0$. On prouve que la solution généralisée convexe $u(x_1,\ldots,x_n)$ du problème de Dirichlet

$$\det \|u_{x_i x_j}\| = f(x_1,\ldots,x_n)$$

$$u|_{\partial G} = 0$$

(où f est une fonction positive sommable dans G), peut être obtenue comme $\inf_H u_H(x_1,\ldots,x_n)$ où $u_H(x_1,\ldots,x_n)$ est la solution du problème variationnel :

$$I_H(u) = -\int_G u(w,de) + (n+1)\int_G f_H u dX$$

$$u|_{\partial G} = 0$$

dans la classe des fonctions non-positives continues $u(x_1,\ldots,x_n)$, satisfaisant $u|_{\partial G} = 0$. Ici $w(x_1,\ldots,x_n)$ est l'enveloppe convexe inférieure du graphe de la fonction

$$v(x_1,\ldots,x_n) = \begin{cases} u(x_1,\ldots,x_n) & \text{si } (x_1,\ldots,x_n) \in H \\ 0 & \text{si } (x_1,\ldots,x_n) \in G-H \end{cases}$$

$\omega(w,e)$ est l'aire de l'application tangente généralisée de la fonction $w(x_1,\ldots,x_n)$, construite par les hyperplans supportant le graphe de $w(x_1,\ldots,x_n)$ et

$$f_H(x_1,\ldots,x_n) = \begin{cases} f(x_1,\ldots,x_n) & \text{si } (x_1,\ldots,x_n) \in H \\ 0 & \text{si } (x_1,\ldots,x_n) \in G-H \end{cases}$$

BEN-ARTZI. SOME NEW RESULTS IN THE SPECTRAL AND SCATTERING THEORY OF STARK-LIKE HAMILTONIANS

Abstract : Let $H = H_o + V(x)$ be a Schrödinger operator in $\mathbb{R}^n$, where $H_o = -\Delta + V_o(x_1)$ is the Laplacian perturbed by a one-dimensional potential, whereas $V(x)$ depends on all coordinates, and satisfies $(1+|x|)^{-N}V(x) \in L^2(\mathbb{R}^n)$ for some integer N. Assume that :

(i) $V_o(t) \in C^\infty(\mathbb{R})$ and for some $\delta > 0$ and all $j > 0$,

$$V_o^{(j)}(t).\ (1+|V_o(t)|)^{-1} = O(|t|^{-j\delta}) \text{ as } |t| \to \infty .$$

(ii) $V_o(t) = O(t^2)$ as $t \to \infty$ and $V_o(t) \to \mp\infty$ as $t \to \pm\infty$.

(these assumptions can be much relaxed). If, in addition,

(iii) $|V_o(t)| > |t|^\varepsilon$, some $\varepsilon > 0$, near $\pm\infty$ and $V \in L^2(R^n)$,

then H, H_o are essentially self-adjoint (when restricted to $C_o^\infty(\mathbb{R}^n)$) and the wave-operators

$$W_\pm = \underset{t \to \pm\infty}{s - \lim} \exp(itH) \exp(-itH_o)$$

exist (H_o is totally spectrally absolutely continuous). More general conditions are given for the existence of $W_\pm(H,H_o)$, allowing V to grow in directions orthogonal to x_1.

If, in addition to (i) - (iii), V_o is monotone non-increasing, the resolvent $R_o(z) = (H_o-z)^{-1}$ can be extended continuously (from the upper or lower half-plane) to $\operatorname{Im} z = 0$ in some suitable weighted - L^2 topology (limiting absorption principle). If

$$|V(x_1,x')| \le C(1+|x_1|)^{-1-\varepsilon}(1+|x'|)^{-\varepsilon}(1+|V_o(x_1)|)^{1/2} \qquad \varepsilon > 0$$

then the same is true for $R(z) = (H-z)^{-1}$, except possibly for a discrete sequence of eigenvalues corresponding to eigenfunctions of rapid decay. The wave-operators are then complete.

All these results apply to the Stark case $H = -\Delta - Ex_1 - \frac{1}{|x|}$.

Résumé : Soit $H = H_o + V(x)$ un opérateur de Schrödinger dans $\mathbb{R}^n$; $H_o = -\Delta + V_o(x_1)$ est le laplacien perturbé par un potentiel unidimensionnel, $V(x)$ dépend de toutes les coordonnées et satisfait :
$(1+|x|)^{-N} V(x) \in L^2(\mathbb{R}^n)$ pour un entier N. On suppose :

(i) $V_o(t) \in C^\infty(\mathbb{R})$ et pour $\delta > 0$ et $\forall j > 0$:

$$V_o^{(j)}(t). (1+|V_o(t)|)^{-1} = O(|t|^{-j\delta}) \text{ pour } |t| \to \infty.$$

(ii) $V_o(t) = O(t^2)$ pour $t \to \infty$ et $V_o(t) \to \mp \infty$ si $t \to \pm \infty$

(Ces hypothèses peuvent être affaiblies).

Si de plus :

(iii) $|V_o(t)| > |t|^\varepsilon$ pour un $\varepsilon > 0$ au voisinage de $\pm\infty$ et $V \in L^2(\mathbb{R}^n)$

alors : H, H_o sont essentiellement auto-adjoints (restreints à $C_o^\infty(\mathbb{R}^n)$)
et les opérateurs d'ondes

$$W_\pm = s - \lim_{t \to \pm\infty} \exp(it\, H) \exp(-it\, H_o)$$

existent (H_o est spectralement totalement absolumment continu).

Des conditions plus générales sont données pour assurer l'existence de $W_\pm(H,H_o)$, permettant à V de croître dans des directions orthogonales à x_1.

Si, de plus de (i) - (iii) , V_o est monotone non-croissant la résolvante $R_o(z) = (H_o-z)^{-1}$ peut être prolongée continûement (du semi plan inférieur ou supérieur) à $\text{Im } z = 0$ dans une topologie appropriée à poids dans L^2 (le principe d'absorbtion limitée). Si

$$|V(x_1,x')| \leq C(1+|x_1|)^{-1-\varepsilon} (1+|x'|)^{-\varepsilon}(1+|V_o(x_1)|)^{1/2} , \quad \varepsilon > 0$$

alors la même chose est vraie pour $R(z) = (H-z)^{-1}$ à l'exception possible d'une suite discrète de valeurs propres correspondant à des fonctions propres à décroissance rapide. Les opérateurs des ondes sont alors complets.

Tous les résultats s'appliquent au cas de Stark : $H = -\Delta - Ex_1 - \frac{1}{|x|}$.

P.M. FITZPATRICK. GLOBAL MULTIDIMENSIONAL EXISTENCE RESULTS FOR m-PARAMETER COMPACT VECTOR FIELDS.

Abstract : Let X be a Banach space, m be a positive integer, and $\mathcal{O} \subseteq \mathbb{R}^m \times X$ be open. Suppose $f : \mathcal{O} \to X$ is an m-parameter compact vector field : i.e. $f(\lambda,x) = x-F(\lambda,x)$, for $(\lambda,x) \in \mathcal{O}$, where $F : \mathcal{O} \to X$ is compact. A mapping $g : \mathcal{O} \to \mathbb{R}^m$ is called a complement for $f : \mathcal{O} \to X$ if the Leray-Schauder degree of (g,f) on $\mathcal{O}$ with respect to 0 is non-zero : here, $(g,f)((\lambda,x)) = (g(\lambda,x),f(\lambda,x))$, for $(\lambda,x) \in \mathcal{O}$. When f has a complement, g, it is possible to prove the existence of global m-dimensional components of $f^{-1}(0)$ which emanate from $g^{-1}(0)$. Such a result yields sharp m-dimensional extensions of the continuation theorem of Leray and Schauder and of the global bifurcation theorem of Rabinowitz. In addition, one can also deduce multiplicity results for the set of zeros of compact perturbations of Fredholm mappings of positive index.

Résumé : Soit X un espace de Banach, m un entier positif, et $\mathcal{O}$ un ouvert de $\mathbb{R}^n \times X$. Supposons que $f : \mathcal{O} \to X$ est un champ de vecteurs compact à m-paramètres : i.e. $f(\lambda,x) = x-F(\lambda,x)$, pour $(\lambda,x) \in \mathcal{O}$, où $F : \mathcal{O} \to X$ est compact. Une application $g : \mathcal{O} \to \mathbb{R}^m$ est dite un complement pour $f : \mathcal{O} \to X$ si le degré de Leray et Schauder de (g,f) sur $\mathcal{O}$ par rapport à 0 est non nul : ici nous avons $(g,f)((\lambda,x)) = (g(\lambda,x),f(\lambda,x))$, pour $(\lambda,x) \in \mathcal{O}$. Quand f a un complement, g, il est possible de démontrer l'existence de composantes globales m-dimensionnelles de $f^{-1}(0)$ qui émanent de $g^{-1}(0)$. Ce résultat donne une forte extension m-dimensionnelle du théorème de continuation de Leray et Schauder et du théorème de bifurcation globale de Rabinowitz. De plus il est aussi possible de déduire des résultats sur la multiplicité de l'ensemble des zéros des perturbations compactes des applications de Freholm d'indice positif.

S. KAMIN. ELLIPTIC SINGULAR PERTURBATION PROBLEMS WITH TURNING POINTS.

Abstract : The boundary value problem for the elliptic equation

$$\varepsilon(\sum a_{ij} u_{x_i x_j} + \sum a_i u_{x_i}) + \sum b_i u_{x_i} + cu = f$$

is considered. We study the asymptotic behavior of the solution $u_\varepsilon(x)$ as $\varepsilon \to 0$. It is known that the behavior of the characteristics of the reduced equations $\sum b_i U_{x_i} = 0$ is of decisive importance in this connection. Different cases are discussed.

Résumé : On considère le problème aux limites pour l'équation elliptique

$$\varepsilon(\sum a_{ij} u_{x_i x_j} + \sum a_i u_{x_i}) + \sum b_i u_{x_i} + cu = f$$

On étudie le comportement asymptotique des solutions $u_\varepsilon(x)$ quand $\varepsilon \to 0$. Celui-ci est déterminé par le comportement des caractéristiques de l'équation réduite $\sum b_i U_{x_i} = 0$. Différents cas sont discutés.

P.L. LIONS. OPTIMAL CONTROL OF DIFFUSION PROCESSES AND HJB EQUATION

Abstract : We consider general optimal stochastic control problems and we study the regularity of the optimal cost function. Under optimal assumptions we prove regularity results which, combined with the notion of viscosity solutions, yield immediately that the optimal cost function satisfies the Hamilton-Jacobi-Bellman equations.

Résumé : Nous considérons des problèmes généraux de contrôle optimal stochastique et nous étudions la régularité de la fonction coût optimum. Sous des hypothèses optimales nous prouvons des résultats de régularité qui, combinés avec la notion de solution de viscosité, impliquent immédiatement que la fonction coût optimum satisfait l'équation de Hamilton-Jacobi-Bellman associée au problème de contrôle. Nous démontrons également des résultats d'unicité généraux pour les solutions des équations de Hamilton-Jacobi-Bellman.

F. MIGNOT - J.P. PUEL. FLAMBAGE D'UNE TIGE VISCOELASTIQUE

Abstract : We consider here a model describing the quasistatic equilibrium of a viscoelastic rod under the action of an axial thurst P(.) depending on time. In particular we study the buckling of such a rod and we point out some "bifurcation" phenomena which are qualitatively very different from the ones observed in the purely elastic model. We show the existence of a critical value $\lambda_1 (\lambda_1 > 0)$ and of a critical slope $\alpha_1 (\alpha_1 < 0)$ such that :

(i) If $P(t) < \lambda_1$ on $(0,T)$ the only solution is the trivial one.

(ii) If $P(t) < \lambda_1$ on $(0,t_o)$ and

$$P(t) = \lambda_1 + p_1(t-t_o) + (t-t_o)^2 Q(t) \text{ on } (t_o,T), \text{ then}$$

- if $p_1 < \alpha_1$ there is no "bifurcation" at t_o.
- if $\alpha_1 < p_1 < 0$ there exist infinitely many bifurcated branches at t_o.
- if $p_1 \geq 0$ there exists a unique bifurcated branch at t_o.

Résumé : On considère ici un modèle décrivant l'équilibre quasistatique d'une tige viscoelastique sous l'action d'une charge axiale P(.) dépendant du temps. En particulier, on étudie le flambage d'une telle tige et on met en évidence des phénomènes de "bifurcation" qualitativement très différents de ceux obtenus pour le modèle purement élastique. On montre l'existence d'une valeur critique λ_1 ($\lambda_1 > 0$) et d'une pente critique α_1 ($\alpha_1 < 0$) telles que :

(i) Si $P(t) < \lambda_1$ sur $(0,T)$ la seule solution est la solution triviale.

(ii) Si $P(t) < \lambda_1$ sur $(0,t_o)$ et

$$P(t) = \lambda_1 + p_1(t-t_o) + (t-t_o)^2 Q(t) \quad \text{sur} \quad (t_o,T) \text{ , alors :}$$

- Si $p_1 < \alpha_1$ il n'y a pas de "bifurcation" en t_o.
- Si $\alpha_1 < p_1 < 0$ il y a une infinité de branches bifurquées en t_o.
- Si $p_1 \geq 0$ il y a une branche bifurquée unique en t_o.

J.C. SAUT, B. SCHEURER. UNIQUE CONTINUATION AND UNIQUENESS OF THE CAUCHY PROBLEM FOR ELLIPTIC OPERATORS WITH UNBOUNDED COEFFICIENTS

Abstract : Let Ω be a connected open set in $\mathbb{R}^n$, $n \geq 2$ and P a uniformly elliptic operator of order $m \geq 2$. P is said to have the (weak) unique continuation property if every solution of $Pu = 0$ which vanishes on a non empty open set in Ω, vanishes identically on Ω. The aim of this lecture is to present some recent results on this property for operators having irregular coefficients.

Résumé : Soit Ω un ouvert connexe de $\mathbb{R}^n$, $n \geq 2$ et P un opérateur uniformément elliptique d'ordre $m \geq 2$. On dit que P a la propriété de prolongement unique (faible) si toute solution de $Pu = 0$ qui s'annule sur un ouvert non vide de Ω est identiquement nulle dans Ω. Le but de cet exposé et de présenter des résultats récents sur cette propriété pour des opérateurs à coefficients le moins réguliers possible.

M. SCHATZMAN. SPATIAL STRUCTURING IN A MODEL IN NEUROPHYSIOLOGY

Abstract : Several systems modeling some spatial structuration phenomena in neurophysiology are considered. The typical equation is written as

$$u_t = (w * u)(1-u^2), \quad u(x,0) = u_o(x).$$

Swindale introduced this system [11], and observed the formation of a pattern of stripes for large time, in numerical simulations. We show existence and uniqueness for the evolution problem, and state several results on the asymptotic behavior. In particular, a stable non zero solution of the above system